WORKSHOPS IN COMPUTING
Series edited by C. J. van Rijsbergen

Also in this series

ALPUK91, Proceedings of the 3rd UK Annual Conference on Logic Programming, Edinburgh, 10–12 April 1991
Geraint A.Wiggins, Chris Mellish and Tim Duncan (Eds.)

Specifications of Database Systems
International Workshop on Specifications of Database Systems, Glasgow, 3–5 July 1991
David J. Harper and Moira C. Norrie (Eds.)

7th UK Computer and Telecommunications Performance Engineering Workshop
Edinburgh, 22–23 July 1991
J. Hillston, P.J.B. King and R.J. Pooley (Eds.)

Logic Program Synthesis and Transformation
Proceedings of LOPSTR 91, International Workshop on Logic Program Synthesis and Transformation, University of Manchester, 4–5 July 1991
T.P. Clement and K.-K. Lau (Eds.)

Declarative Programming, Sasbachwalden 1991
PHOENIX Seminar and Workshop on Declarative Programming, Sasbachwalden, Black Forest, Germany, 18–22 November 1991
John Darlington and Roland Dietrich (Eds.)

Building Interactive Systems: Architectures and Tools
Philip Gray and Roger Took (Eds.)

Functional Programming, Glasgow 1991
Proceedings of the 1991 Glasgow Workshop on Functional Programming, Portree, Isle of Skye, 12–14 August 1991
Rogardt Heldal, Carsten Kehler Holst and Philip Wadler (Eds.)

Object Orientation in Z
Susan Stepney, Rosalind Barden and David Cooper (Eds.)

Code Generation – Concepts, Tools, Techniques
Proceedings of the International Workshop on Code Generation, Dagstuhl, Germany, 20–24 May 1991
Robert Giegerich and Susan L. Graham (Eds.)

Z User Workshop, York 1991, Proceedings of the Sixth Annual Z User Meeting, York, 16–17 December 1991
J.E. Nicholls (Ed.)

Formal Aspects of Measurement
Proceedings of the BCS-FACS Workshop on Formal Aspects of Measurement, South Bank University, London, 5 May 1991
Tim Denvir, Ros Herman and R.W. Whitty (Eds.)

AI and Cognitive Science '91
University College, Cork, 19–20 September 1991
Humphrey Sorensen (Ed.)

5th Refinement Workshop, Proceedings of the 5th Refinement Workshop, organised by BCS-FACS, London, 8–10 January 1992
Cliff B. Jones, Roger C. Shaw and Tim Denvir (Eds.)

Algebraic Methodology and Software Technology (AMAST'91)
Proceedings of the Second International Conference on Algebraic Methodology and Software Technology, Iowa City, USA, 22–25 May 1991
M. Nivat, C. Rattray, T. Rus and G. Scollo (Eds.)

ALPUK92, Proceedings of the 4th UK Conference on Logic Programming, London, 30 March – 1 April 1992
Krysia Broda (Ed.)

Logic Program Synthesis and Transformation
Proceedings of LOPSTR 92, International Workshop on Logic Program Synthesis and Transformation, University of Manchester, 2–3 July 1992
Kung-Kiu Lau and Tim Clement (Eds.)

Formal Methods in Databases and Software Engineering, Proceedings of the Workshop on Formal Methods in Databases and Software Engineering, Montreal, Canada, 15–16 May 1992
V.S. Alagar, Laks V.S. Lakshmanan and F. Sadri (Eds.)

continued on back page...

S. Purushothaman and Amy Zwarico (Eds.)

NAPAW 92

Proceedings of the First North American Process Algebra Workshop, Stony Brook, New York, USA, 28 August 1992

Published in collaboration with the British Computer Society

BCS

Springer-Verlag
London Berlin Heidelberg New York
Paris Tokyo Hong Kong
Barcelona Budapest

S. Purushothaman, PhD
Department of Computer Science, 333 Whitmore Laboratory
The Pennsylvania State University, University Park, PA 16802, USA

Amy Zwarico
Department of Computer Science, The Johns Hopkins University
Baltimore, MD 21218, USA

ISBN 978-3-540-19822-2 ISBN 978-1-4471-3217-2 (eBook)
DOI 10.1007/978-1-4471-3217-2

British Library Cataloguing in Publication Data
NAPAW '92:Proceedings of the First North American Process Algebra Workshop, Stony Brook, New York, USA, 28 August 1992. – (Workshops in Computing Series)
I. Purushothaman, Sahasranaman II. Zwarico, Amy III. Series
512.00285
ISBN-13:978-3-540-19822-2

Library of Congress Cataloging-in-Publication Data
North American Process Algebra Workshop (1st: 1992 : Stony Brook, N.Y.)
NAPAW 92 : proceedings of the First North American Process Algebra Workshop, Stony Brook, New York, USA, 28 August 1992 / S. Purushothaman and Amy Zwarico, eds.
p. cm. – (Workshops in computing)
"Published in collaboration with the British Computer Society."
Includes bibliographical references and index.
ISBN-13:978-3-540-19822-2

1. Real-time data processing–Congresses. 2. Electronic data processing–Distributed processing–Congress. 3. Algebra–Congresses.
I. Purushothaman, S. (Sahasranaman), *1957–* . II. Zwarico, Amy. *1960–* .
III. British Computer Society. IV. Title. V. Series.
QA76.54.N67 1992 92-44291
004'.33'01512–dc20 CIP

Typesetting: Camera ready by contributors
34/3830-543210 Printed on acid-free paper

Preface

This proceedings contains fourteen papers on process algebras presented at the First North American Process Algebra Workshop, held on 28 August 1992 in Stony Brook, New York. NAPAW was held in conjunction with CONCUR 92. It is hoped that NAPAW will be held yearly to bring together researchers from the North American continent interested in process algebras.

We would like to thank the program committee members for reading through the abstracts, and Scott Smolka whose encouragement and assistance was critical to the success of the workshop.

November 1992

S. Purushothaman
University Park, PA

Amy Zwarico
Baltimore, MD

Program Committee

Rance Cleaveland, North Carolina State University

Richard Gerber, University of Maryland at College Park

Faron Moller, University of Edinburgh

S. Purushothaman, Pennsylvania State University

Amy Zwarico, Johns Hopkins University

Contents

Session 4

Session 1

Real–Time Calculi and Expansion Theorems
Extended Abstract

Jens Chr. Godskesen* and Kim G. Larsen
Aalborg University, Denmark †

1 Introduction

During the last few years various process calculi have been extended to include real–time in order to handle quantitative aspects of real–time systems, for instance that some critical event must not or should happen within a certain time period. A number of extensions of classical process calculi with explicit timing constraints has emerged in the last few years. We mention the calculi defined in [Wan90] and in [Che91], all being extensions of CCS [Mil89], and the ones defined in [NSY91] and [BB89]. Common to these real–time calculi is that time is represented by some dense time domain, e.g. as in this paper the non–negative reals. Other timed calculi based on discrete time include [NRSV90, DS89, HR91].

Axiomatization and decidability of various equivalences (in particular bisimulation equivalence) between real–time processes based on dense time domains have proven notoriously hard problems [1]. Normally, when introducing a new process calculus, the problems of axiomatization and decidability are solved in two stages: the problems are first solved for the class of regular processes, i.e. processes with no parallel composition, after which it is shown how to remove parallel composition through the use of a so–called *Expansion Theorem.*

In this paper we introduce a real–time calculus, TCCS, being a slightly simplified version of Wang's calculus [Wan90]. The calculus uses a dense time domain. In [HLW91] it has been shown that bisimulation equivalence *is* decidable for the regular part of TCCS. Thus, compared with the traditional strategy above, the only thing remaining is to establish an Expansion Theorem. However, as a main result of this paper we show that no such Expansion Theorem can exist! That is, parallel composition can not in general be removed from terms. In fact, we show a more general result, which we call the *Gap Theorem*: not only

*On leave from TFL, Denmark. TFL is the Danish telecommunication research laboratory.

†Address:Dep. of Math. and Comp. Sc., Aalborg University, Fredrik Bajers Vej 7, 9220 Aalborg, Denmark. Tlf. +45 98158522. Email:{jcg,kgl}@iesd.auc.dk. This work has been supported by the Danish Research Council through the DART project. The first author is supported by a scholarship from Aalborg University.

[1] Recent work by Čerāns [Č92] and Cheng [Che91] offers the first examples of decidability and axiomatization for real–time calculi based on dense time.

can parallel composition not be *removed* from terms; in general we can not even hope to *reduce* the number of parallel compositions in terms. The Gap Theorem is achieved by the definition of a translation from TCCS to another model for real–time systems called *timed graphs*.
We claim that this negative result of TCCS having no Expansion Theorem carries over to most other similar timed process calculi. At least when the calculus is restricted to standard operators found in the literature. An Expansion Theorem for TCCS can be established however by adding to the language a (very powerful) mechanism to record the explicit time occurrence of events. In [Wan91] for instance the calculus is extended with a notion called time variables on which an Expansion Theorem is based.

2 Timed processes

By a *timed process* we shall understand a process as defined in [Wan90]. In [Wan90] time is incorporated into an untimed calculus in accordance with the two–phase functioning principle mentioned above. In this paper however, only a minimal set of finite timed processes will be considered.
We presuppose a set, Δ, of *action names* and define the set $\mathcal{A}$ to be $\Delta \cup \{\epsilon(d) \mid d \in \mathbf{R} \setminus \{0\}\}$ where $\mathbf{R}$ is the set of non–negative reals. An element α of $\mathcal{A}$ is referred to as a *delay* if $\alpha = \epsilon(d)$ for some d, otherwise α will be called an *action*. We let μ range over Δ and take α to range over $\mathcal{A}$.

Definition 1 *The set of (finite)* timed processes, *TCCS, is defined inductively by*

$$P ::= Nil \mid \alpha.P \mid P + P \mid P|P$$

We often write α for the process $\alpha.Nil$.
The semantics of timed processes is given by a labeled transition system $\mathbf{P} = (\text{TCCS}, \mathcal{A}, \{\xrightarrow{\alpha} \subseteq \text{TCCS} \times \text{TCCS} \mid \alpha \in \mathcal{A}\})$, where the relations $\xrightarrow{\alpha}$ are the least sets satisfying

$$\frac{}{Nil \xrightarrow{\epsilon(d)} Nil} \qquad \frac{}{\mu.P \xrightarrow{\mu} P} \qquad \frac{}{\mu.P \xrightarrow{\epsilon(d)} \mu.P}$$

$$\frac{}{\epsilon(d).P \xrightarrow{\epsilon(d)} P} \qquad \frac{}{\epsilon(d+e).P \xrightarrow{\epsilon(d)} \epsilon(e).P} \qquad \frac{P \xrightarrow{\epsilon(e)} P'}{\epsilon(d).P \xrightarrow{\epsilon(d+e)} P'}$$

$$\frac{P \xrightarrow{\mu} P'}{P + Q \xrightarrow{\mu} P'} \qquad \frac{Q \xrightarrow{\mu} Q'}{P + Q \xrightarrow{\mu} Q'} \qquad \frac{P \xrightarrow{\epsilon(d)} P' Q \xrightarrow{\epsilon(d)} Q'}{P + Q \xrightarrow{\epsilon(d)} P' + Q'}$$

$$\frac{P \xrightarrow{\mu} P'}{P \mid Q \xrightarrow{\mu} P' \mid Q} \qquad \frac{Q \xrightarrow{\mu} Q'}{P \mid Q \xrightarrow{\mu} P \mid Q'} \qquad \frac{P \xrightarrow{\epsilon(d)} P' Q \xrightarrow{\epsilon(d)} Q'}{P \mid Q \xrightarrow{\epsilon(d)} P' \mid Q'}$$

Intuitively, one may think of a transition $P \xrightarrow{\epsilon(d)} P'$ as P may idle d time units and in so doing become P'. From the semantics it follows that a process P

always can make time progress by idling some arbitrary time unit. Moreover, we see that for the constructs + and | processes are forced to synchronize on delays. In this note we consider a simplified parallel operator compared with [Wan90] because synchronization of actions is not considered.
Strong bisimulation equivalence between two timed processes is defined as usual.

Definition 2 *A binary relation $\mathcal{R} \subseteq TCCS \times TCCS$ is a strong bisimulation if $(P, Q) \in \mathcal{R}$ implies for all α*

$$\begin{array}{l} \forall P'.\, P \xrightarrow{\alpha} P' \;\; \textit{implies} \;\; \exists Q'.\, Q \xrightarrow{\alpha} Q' \;\; \textit{and} \;\; (P', Q') \in \mathcal{R} \\ \forall Q'.\, Q \xrightarrow{\alpha} Q' \;\; \textit{implies} \;\; \exists P'.\, P \xrightarrow{\alpha} P' \;\; \textit{and} \;\; (P', Q') \in \mathcal{R} \end{array}$$

We say that two timed processes P and Q are *strong equivalent* if there exists a strong bisimulation $\mathcal{R}$ containing (P, Q), in this case we write $P \sim Q$. As is often the case, $\sim$ is the largest strong bisimulation and a congruence (i.e. preserved by all operators of TCCS).
It is obvious that $P_1 = a \,|\, b$ is strong equivalent to $a.b + b.a$. This may suggest an *Expansion Theorem*, like the one for CCS, eliminating parallel composition. Now, consider the processes $P_2 = \epsilon(1).a \mid b$ and $P_3 = \epsilon(1).a \mid b.\epsilon(1).c$. The naive expansions of P_2 and P_3 would be $\epsilon(1).(a \mid b) + b.\epsilon(1).a$ and $\epsilon(1).(a \mid b.\epsilon(1).c) + b.(\epsilon(1).a \,|\, \epsilon(1).c)$ respectively. However, it is easy to see that P_2 and P_3 are not strong equivalent to their naive expansions: P_2 and P_3 can perform the computations below which their naive expansions clearly cannot match.

$$\begin{array}{l} P_2 \xrightarrow{\epsilon(.5)} \epsilon(.5).a \mid b \xrightarrow{b} \epsilon(.5).a \mid Nil \xrightarrow{\epsilon(.5)} a \mid Nil \xrightarrow{a} Nil \mid Nil \\ P_3 \xrightarrow{\epsilon(.5)} \epsilon(.5).a \mid b.\epsilon(1).c \xrightarrow{b} \epsilon(.5).a \mid \epsilon(1).c \xrightarrow{\epsilon(.5)} a \mid \epsilon(.5).c \xrightarrow{a} Nil \mid \epsilon(.5).c \end{array}$$

This leaves open the question as to whether there do exist expansions of P_2 and P_3 and more general if any TCCS process can be expanded. As we shall see later, the answer to the latter question is negative.

3 Timed Graphs

Timed graphs [Dil89] provide an alternative method for describing finite state real–time systems that may change state by performing action transitions taking no time and delay transitions modelling the progress of time. The system has a finite set of real valued clocks. The transitions a system may perform depends on the value of the clocks. After a transition some of the clocks may be reset.
Before defining the notion of a timed graph we need to introduce time vectors, enabling conditions and reset functions. By a *time vector* $\mathbf{t}$ of length i we understand an element of $\mathbf{R}^i$. $\mathbf{t}_j$ denotes the j'th element of $\mathbf{t}$. We write $\mathbf{t} :: \mathbf{u}$ for the concatenation of time vectors and we take $\mathbf{t} + d$, $d \in \mathbf{R}$, to mean $(\mathbf{t}_1 + d, \ldots, \mathbf{t}_i + d)$. $\mathbf{0}$ is the time vector where all elements are zero. An *enabling condition* for time vectors of length i is a subset of $\mathbf{R}^i$ of the form $b^j_c = \{\mathbf{t} \in \mathbf{R}^i \mid \mathbf{t}_j \geq c\}$ where $c \in \mathbf{R}$ and $j \leq i$. We let $\mathcal{B}_i$ denote the set of all enabling conditions for time vectors of length i. Moreover, we use the notation $b^j_c(\mathbf{t})$ to indicate that $\mathbf{t} \in b^j_c$. We also write *tt* for the trivial conditions b^j_0 (j arbitrary). Often, when no confusion will arise, we will write $\mathbf{t}_j \geq c$ for the

enabling condition b_c^j. A *reset function*, r, for time vectors of length i is an element of $\mathcal{R}_i = 2^{\{1,\dots,i\}}$. We define $r(\mathbf{t})$ to mean the time vector $\mathbf{t}'$ where

$$\mathbf{t}'_j = \begin{cases} 0 & \text{if } j \in r \\ \mathbf{t}_j & \text{otherwise} \end{cases}$$

Definition 3 *A timed graph is a structure $G = (N, i, n_0, \rightarrow)$ where N is a finite set of nodes, i is the number of clocks, n_0 is the initial node and $\rightarrow \subseteq N \times \Delta \times \mathcal{B}_i \times \mathcal{R}_i \times N$ is the transition relation.*

Whenever $(n, \mu, b, r, n') \in \rightarrow$ we write $n \xrightarrow{(\mu,b,r)} n'$ instead and say that we have a transition from n to n' with label μ, enabling condition b and reset function r. The class of timed graphs is denoted $\mathcal{G}$.
Intuitively, a timed graph is interpreted as follows. The timed graph is started in its initial node with its clocks set to 0. Then the clocks begin to run, all with the same speed. At any moment a transition from one node to another can be performed if the enabling condition of the transition is satisfied by the current values of the clocks. The execution of the transition itself takes no time. Immediately after the performance of a transition the clocks indexed by the resetting function are reset to 0. Formally, the interpretation of a timed graph, $G = (N, i, n_0, \rightarrow)$, is a labelled transition system

$$I_G = (S, \mathcal{A}, \{\xrightarrow{\alpha} \subseteq S \times S \mid \alpha \in \mathcal{A}\})$$

where $S = N \times \mathbf{R}^i$ is the set of *states* and the transition relations $\xrightarrow{\alpha}$ are the least sets satisfying

$$\frac{n \xrightarrow{(\mu,b,r)} n'}{(n, \mathbf{t}) \xrightarrow{\mu} (n', r(\mathbf{t}))}\ b(\mathbf{t}) \qquad\qquad \frac{}{(n, \mathbf{t}) \xrightarrow{\epsilon(d)} (n, \mathbf{t} + d)}$$

The first rule states, that a timed graph in state $(n, \mathbf{t})$ may perform an action μ, if the enabling condition b holds at $\mathbf{t}$, after which a new state $(n', \mathbf{t}')$ with the clocks indexed by r reset to 0 is entered. The second rule says that time may always progress.
Strong bisimulation equivalence between two graphs G and H can be defined as usual over the obvious disjoint union of I_G and I_H such that G and H are strong equivalent, denoted $G \sim H$, whenever $((n_G, \mathbf{0}), (n_H, \mathbf{0}))$ belongs to a strong bisimulation, where n_G and n_H are the initial nodes in G and H respectively. Similarly, strong bisimulation equivalence between a process P and a graph G can be defined over the obvious disjoint unioun of $\mathbf{P}$ and I_G. We write $P \sim G$ whenever there exists a strong bisimulation containing $(P, (n_G, \mathbf{0}))$.
Recall the processes P_1, P_2 and P_3 from the previous section. There exists timed graphs describing the same real–time behaviour (up to strong bisimulation equivalence) as those processes. The timed graphs for P_1, P_2 and P_3 could be G_1, G_2 and G_3 respectively in figure 1.
The graph G_1 contains a single clock whereas the graphs G_2 and G_3 each contain two clocks. This may suggest a direct relation between the minimal number of parallel components needed to express a real–time system as a timed process and the number of clocks needed to express the same system as a timed

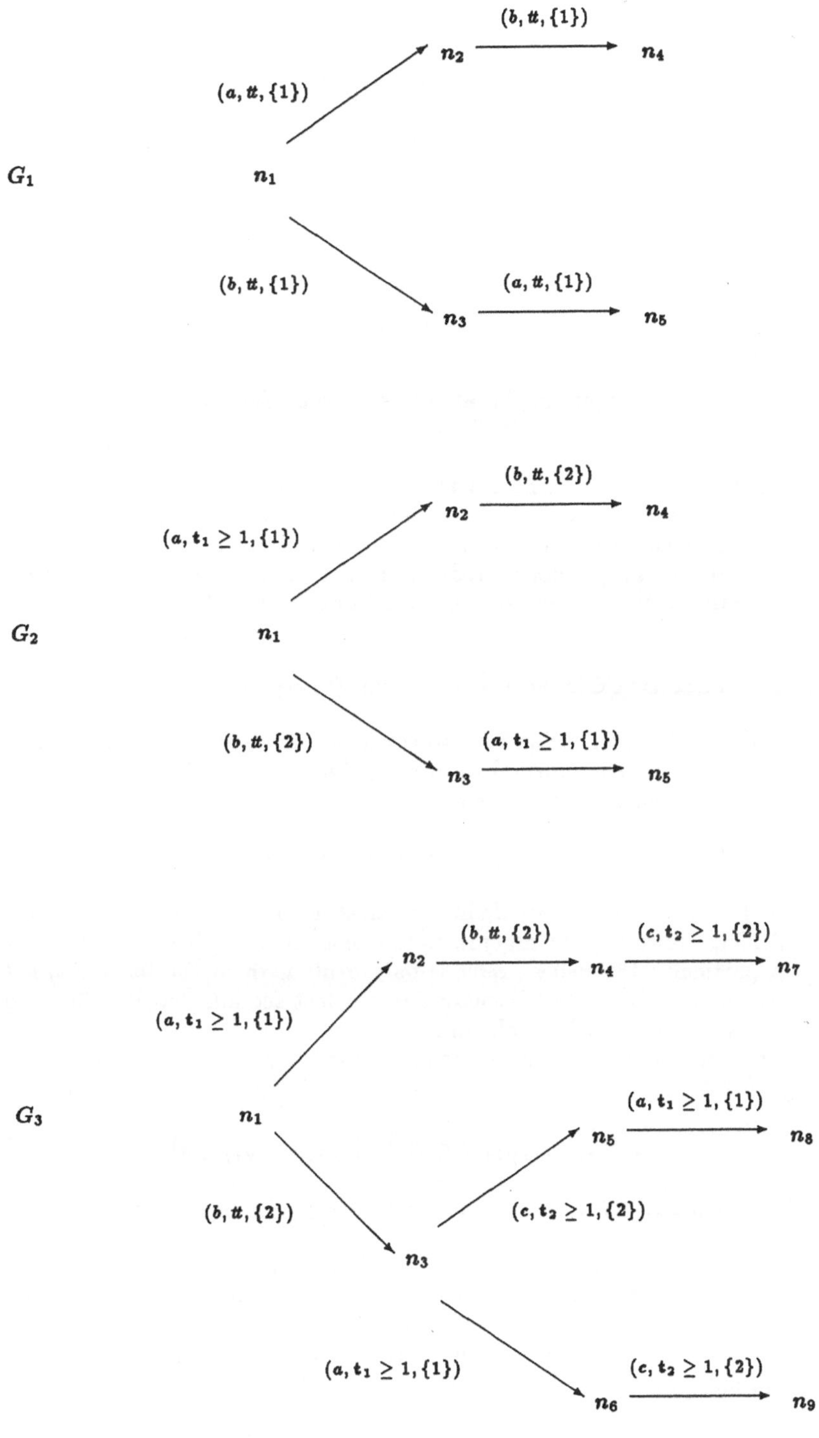

Figure 1: Graphs for the processes P_1, P_2 and P_3.

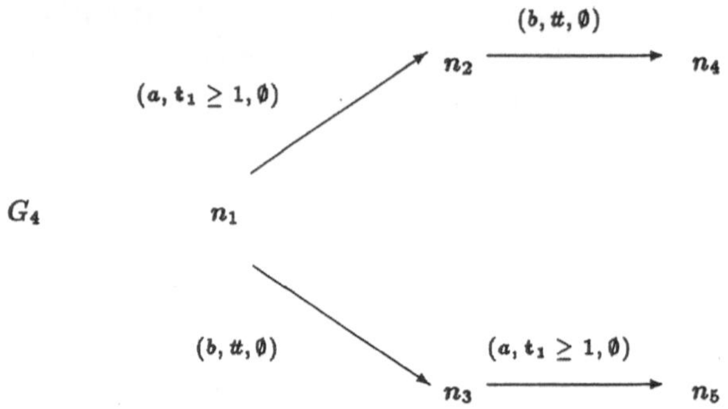

Figure 2: An alternative graph for the process P_2.

graph. However, it turns out that the graph G_4 in figure 2 with only one clock is strong equivalent to P_2. An immediate question is now if it is possible to perform a similar reduction in the number of clocks of G_3 and more generally if any timed graph can be reduced in the number of clocks. As we shall show in section 5 the answer to these questions are negative.

4 An algebra of timed graphs

In this section we define how to construct a timed graph $G[\![P]\!]$ for any TCCS process P in delay normal form such that $P \sim G[\![P]\!]$. A process is said to be in *delay normal form* if it is generated by

$$P ::= Nil \mid \mu.P \mid \epsilon(d).\mu.P \mid P+P \mid P|P$$

To be more precise, we define an algebra of timed graphs over operators in TCCS. However, the algebra is not defined over all the operators in TCCS; in particular the delay prefix is only dealt with in the form $\epsilon(d).\mu.P$. It will become clear from the discussion in the last section that it suffices to consider only this partly defined algebra.
First some auxiliary definitions are needed. Let $b \in \mathcal{B}_i$. Define $b \triangleright j \in \mathcal{B}_{i+j}$ and $j \triangleleft b \in \mathcal{B}_{i+j}$ by

$$b \triangleright j =_{def} \{(d_1, \ldots, d_{i+j}) \in \mathbf{R}^{i+j} \mid (d_1, \ldots, d_i) \in b\}$$

$$j \triangleleft b =_{def} \{(d_1, \ldots, d_{i+j}) \in \mathbf{R}^{i+j} \mid (d_{j+1}, \ldots, d_{i+j}) \in b\}$$

Thus the enabling condition $b \triangleright j$ holds for any $i+j$ time vector if and only if b is true for the first i clock values and $j \triangleleft b$ holds for any $i+j$ time vector if and only if b holds for the last i clock values. Let $r \in \mathcal{R}_i$ and let $j \in \mathbf{N}$. Define $j + r \in \mathcal{R}_{i+j}$ to be $\{j + k \mid k \in r\}$, that is $j + r$ resets the $(j+k)$'th clock if and only if r resets the k'th clock.

The algebra of timed graphs is defined below.

$$G[\![Nil]\!] = (\{n\}, 1, n, \emptyset)$$

Let $G[\![P]\!] = (N, i, n_P, \rightarrow_P)$ be the timed graph for P, then

$$G[\![\mu.P]\!] = (N \cup \{n\}, i, n, \rightarrow)$$

where $n \notin N$ and $\rightarrow\ =\ \rightarrow_P \cup \{n \xrightarrow{(\mu,b,r)} n_P\}$, in which b is $\mathbf{t}_1 \geq 0$ and $r = \{1, \ldots, i\}$.

$$G[\![\epsilon(d).\mu.P]\!] = (N \cup \{n\}, i, n, \rightarrow)$$

where $n \notin N$ and $\rightarrow\ =\ \rightarrow_P \cup \{n \xrightarrow{(\mu,b,r)} n_P\}$, in which b is $\mathbf{t}_1 \geq d$ and $r = \{1, \ldots, i\}$.
Let $G[\![P]\!] = (N_P, i_P, n_P, \rightarrow_P)$ and $G[\![Q]\!] = (N_Q, i_Q, n_Q, \rightarrow_Q)$ be such that $N_P \cap N_Q = \emptyset$. Assume without loss of generality that $i_P \geq i_Q$ and let $j = i_P - i_Q$. Then

$$G[\![P + Q]\!] = (N, i_P, n, \rightarrow)$$

where $N = N_P \cup N_Q \cup \{n\}$, $n \notin N_P \cup N_Q$ and $\rightarrow$ is the least set satisfying

$$\frac{n_P \xrightarrow{(\mu,b,r)}_P n'}{n \xrightarrow{(\mu,b,r)} n'} \qquad \frac{n_Q \xrightarrow{(\mu,b,r)}_Q n'}{n \xrightarrow{(\mu,b \triangleright j,r)} n'} \qquad \frac{n' \xrightarrow{(\mu,b,r)}_P n''}{n' \xrightarrow{(\mu,b,r)} n''} \qquad \frac{n' \xrightarrow{(\mu,b,r)}_Q n''}{n' \xrightarrow{(\mu,b \triangleright j,r)} n''}$$

Let $G[\![P]\!] = (N_P, i, n_P, \rightarrow_P)$ and $G[\![Q]\!] = (N_Q, j, n_Q, \rightarrow_Q)$, then

$$G[\![P \mid Q]\!] = (N_P \times N_Q, i + j, (n_P, n_Q), \rightarrow)$$

where $\rightarrow$ is the least set satisfying

$$\frac{n \xrightarrow{(\mu,b,r)}_P n'}{(n,m) \xrightarrow{(\mu,b \triangleright j,r)} (n',m)} \qquad \frac{m \xrightarrow{(\mu,b,r)}_Q m'}{(n,m) \xrightarrow{(\mu,i \triangleleft b,i+r)} (n,m')}$$

Note, that only when graphs are put together in parallel the number of clocks in the resulting graph is increased.
Throughout this paper we let $\mathcal{G}_{TCCS}$ denote the set all graphs generated by the above constructions and we let $\mathcal{G}^i_{TCCS} \subseteq \mathcal{G}_{TCCS}$ be those graphs with i clocks.
Timed graphs can be viewed as alternative representations for timed processes which is correct in the sense that the interpretation of the timed graph representation of a TCCS process is strongly equivalent to the standard operational interpretation of the process.

Theorem 1 *For any P in delay normal form, $P \sim G[\![P]\!]$.*

The graphs $G[\![P_2]\!]$ and $G[\![P_3]\!]$ for the processes P_2 and P_3 defined in section 2 are G_2 and G_3 in figure 1.

5 The Gap Theorem

In this section we prove that an increase in the number of clocks used in timed graphs does lead to a genuine increase in expressibility. More precisely, we show that for any given number of clocks there exists a timed graph which cannot be expressed (up to strong equivalence) by any timed graph with fewer clocks.

Definition 4 *Let $\mathcal{G}_1, \mathcal{G}_2 \subseteq \mathcal{G}$. Then $\mathcal{G}_1$ is at least as expressive as $\mathcal{G}_2$ if for all $G \in \mathcal{G}_2$ there exists $G' \in \mathcal{G}_1$ such that $G \sim G'$ and in that case we write $\mathcal{G}_2 \preceq \mathcal{G}_1$. If $\mathcal{G}_2 \preceq \mathcal{G}_1$ and $\mathcal{G}_1 \not\preceq \mathcal{G}_2$ we write $\mathcal{G}_2 \prec \mathcal{G}_1$ and say that $\mathcal{G}_1$ is more expressive than $\mathcal{G}_2$.*

The main theorem is

Theorem 2 $\forall i.\, \mathcal{G}^i_{TCCS} \prec \mathcal{G}^{i+1}_{TCCS}$

Proof: We do not give a full proof in this version of the paper. The complete proof can be found in [GL92].
Obviously, $\mathcal{G}^i_{TCCS} \preceq \mathcal{G}^{i+1}_{TCCS}$ so we just have to establish the existence of some graph $G \in \mathcal{G}^{i+1}_{TCCS}$ that cannot be matched up to bisimulation equivalence by any graph in $\mathcal{G}^i_{TCCS}$. It turns out that $G[\![P_{i+1}]\!]$ is such a graph, with

$$P_{i+1} = \epsilon(1).\nu_1 \mid \mu_2.\epsilon(1).\nu_2 \mid \ldots \mid \mu_{i+1}.\epsilon(1).\nu_{i+1}$$

in which all action names are distinct. □

Staying in the subcalculus of TCCS processes in delay normal form it follows from the proof of the above theorem that an increase in the number of parallel components in a process does lead to a genuine increase in expressibility. To be more precise, it follows that for any given number of parallel components there exists a TCCS process which is strongly inequivalent to any TCCS process with fewer parallel components. The result however carries over to general TCCS processes since we for any TCCS process can find an equivalent TCCS process in delay normal form with exactly the same number of parallel components. In particular it follows that TCCS does not have an expansion theorem.
The process calculus TCCS considered in this paper is fairly minimal in terms of the number of constructs, in particular the parallel composition considered is a pure interleaving operator allowing no synchronization between components. In [Wan90] the parallel composition is a conservative extension of the parallel composition in CCS, allowing synchronization between complementary actions with an internal action τ being the result of the synchronization. For the process P_{i+1} above it is clear that the addition of synchronization does not enable a reduction of timers (or parallel components): as P_{i+1} cannot ever perform a τ action, P_{i+1} will be inequivalent to any synchronizing process. We also conjecture that the addition of recursion to TCCS does not invalidate our Gap Theorem. Any (truly) recursive process will possess some infinite computation with infinitely many action–transitions; as P_{i+1} possess no such infinite computations P_{i+1} will be inequivalent to any (truly) recursive process.

References

[ACD90] Rajeev Alur, Costas Courcoubetis, and David Dill. Model–checking for real–time systems. In *Proceedings of the Fifth IEEE Symposium on Logic in Computer Science*, 1990.

[AD90] Rajeev Alur and David Dill. Automata for modelling real–time systems. In *Automata, Languages and Programming: Proceedings of the 17th ICALP*, LNCS 443. Springer-Verlag, 1990.

[BB89] J.C.M. Baeten and J.A. Bergstra. Real time process algebra. Technical Report P8916, University of Amsterdam, 1989.

[Che91] Liang Chen. An interleaving model for real–time systems. Technical report, LFCS, University of Edinburgh, Scotland, 1991. Preliminary version.

[Dil89] D.L. Dill. Timing assumptions and verification of finite–state concurrent systems. *Lecture Notes in Computer Science*, 407, 1989.

[DS89] Jim Davis and Steve Schneider. An introduction to timed CSP. Technical Report PRG–75, Oxford University Computing Laboratory, 1989.

[GL92] Jens Chr. Godskesen and Kim G. Larsen. Real–time calculi and expansion theorems. In *Twelfth Conference on the FST and TCS*, Lecture Notes in Computer Science. Springer-Verlag, December 1992. To appear.

[HLW91] Uno Holmer, Kim Larsen, and Yi Wang. Deciding properties of regular timed processes. In Kim G. Larsen and Arne Skou, editors, *Proceedings of the Third Workshop on Computer Aided Verification,*, volume 575 of *Lecture Notes in Computer Science*. Springer-Verlag, 1991.

[HR91] Matthew Hennessy and Tim Regan. A process algebra for timed systems. Technical Report 5/91, University of Sussex, 1991.

[Mil89] Robin Milner. *Communication and Concurrency*. Series in Computer Science. Prentice–Hall International, 1989.

[MT90] Faron Moller and Chris Tofts. A temporal calculus of communicating systems. In *CONCUR'90*, volume 458 of *Lecture Notes in Computer Science*. Springer-Verlag, 1990.

[NRSV90] X. Nicollin, J.-L. Richier, Joseph Sifakis, and J. Voiron. ATP: an algebra for timed processes. In *Proceedings of the IFIP TC 2 Working Conference on Programming Concepts and Methods*, Sea of Gallilee, Israel, April 1990.

[NSY91] Xavier Nicollin, Joseph Sifakis, and Sergio Yovine. From ATP to timed graphs and hybrid systems. In *Real–Time: Theory in Practice*, volume 600 of *Lecture Notes in Computer Science*. Springer-Verlag, 1991.

[Č92] Kārlis Čerāns. Decidability of bisimulation equivalences for processes with parallel timers. *To appear in Proceedings of CAV'92*, 1992.

[Wan90] Yi Wang. Real–time behaviour of asynchronous agents. In *CONCUR '90*, volume 458 of *Lecture Notes in Computer Science*. Springer-Verlag, 1990.

[Wan91] Yi Wang. *A Calculus of Real Time Systems*. PhD thesis, Chalmers University of Technology, Göteborg, Sweden, 1991.

Modal Logics in Timed Process Algebras

Luboš Brim
Department of Computing Science, Masaryk University
Brno, Czechoslovakia

Abstract

Processes of timed process algebras may be modelled as timed transition systems. A timed transition system is a labelled transition system whose sort is timed, that is labels are either action labels or elements from a time domain.

Modal logics can be used to express properties of transition systems. Timed modal logics are modal logics whose sort is timed. In this paper we consider timed propositional modal logic and timed modal μ-calculus. Definability of important timed properties is the main body of the paper. The properties under consideration include time determinism, time additivity, maximal progress and finite variability.

Finally, minimal timed logics for the classes of time deterministic and time additive transition systems are chracterised axiomatically.

1 Introduction

Processes of timed process algebras may be modelled as timed transition systems. A *timed transition system* is a labelled transition system whose labels are either elements from a set of action names A or elements of a *time domain* D. A timed transition system with label set $\mathcal{L}$ is said to have *timed sort* $\mathcal{L}$.

Timed process algebras differ in their time domains (discrete or continuous) as well as in the underlying model properties with respect to time. For an overview of timed process algebras see [12].

Modal logics can be used to express properties of transition systems (see e.g [2, 4, 9, 13]). *Timed modal logics* are classical modal logics whose language has timed sort. We shall consider timed propositional modal logic and timed modal μ-calculus and show how to define model properties of timed process algebras with these logics.

When speaking about properties of timed transition systems, two types of properties arise. The first group consists of fundamental properties with respect to time. These are properties the particular process algebras are built around (time additivity, time determinism, maximal progress, finite variability). The second group includes properties that a particular process is very often supposed to have (such as persistency, freedom from time deadlock). Some of these properties are not expressible in timed propositional modal logic and a richer logic is needed.

For each timed sort $\mathcal{L}$ and class Λ of timed transition systems of sort $\mathcal{L}$ there is the *minimal* timed modal logic. In some cases the minimal logic can be characterized axiomatically and we present axiom systems for time deterministic and time additive transition systems.

In section 2, we recall basic facts about timed transition systems. For detailed explanation we refer to the work by Nicollin and Sifakis [11, 12]. Section 3 defines the properties of timed transition systems under consideration. Then, section 4 introduces timed propositional modal logic and timed modal μ-calculus with intention to capture quantitative properties of timed systems. In section 5, we give the corresponding definability results for the properties from section 3. Finally, in the last section we shall shortly examine the possibility of axiomatizing minimal timed modal logics.

2 Timed transition systems

Operational semantics of process algebras are based on transition systems. In the case of timed process algebras the corresponding semantic structures are called *timed transition systems*. Timed transition systems differ from (untimed) transition systems in that they carry additional structure on the set of labels, reflecting the idea that some of the actions are considered to be time delays, the nature of which is different from actions. We shall therefore consider transition systems whose labels are either elements a of a vocabulary of actions A or elements d of a *time domain* $\mathcal{D}$. A may contain non-visible (internal) action denoted by τ. $s \xrightarrow{a} s'$ means that state s may become s' by performing the atomic action a and $s \xrightarrow{d} s'$ means that s may become s' by *idling* for d time units.

What conditions should be put on a time domain? The most important aim of introducing time into process algebras is to have the possibility of dealing with quantitative aspects of the behaviour of processes. To that end we need a structure with addition and zero element, satisfying some simple conditions. Furthermore timed transition systems are usually considered to be systems with a global conception of time. Therefore we assume that the set of time elements can be ordered in a global way - the relation $\leq$ will be assumed to be total (i.e. transitive, irreflexive and comparable).

Definition 2.1 *A time domain $\mathcal{D}$ is a commutative monoid $\mathcal{D} = (D, +, 0)$ satisfying the following requirements:*

1. $d + d' = d \Leftrightarrow d' = 0$
2. *the preorder $\leq$ defined by $d \leq d' \Leftrightarrow \exists d'' : d + d'' = d'$ is a total preorder*

Lemma 2.2 *The following properties hold:*

1. 0 *is the least element*
2. *for any d, d', if $d \leq d'$, then the element d'' such that $d + d'' = d'$ is unique. It is denoted by $d' - d$.*

We denote $D - \{0\}$ by D_*. We write $d < d'$ instead of $d \leq d' \wedge d \neq d'$. Time domain $\mathcal{D}$ is called *dense* if $\forall d, d' \in D.(d < d' \Rightarrow \exists d'' \in D.d < d'' < d')$. $\mathcal{D}$ is called *discrete* if $\forall d \in D.\exists d' \in D.(d < d' \wedge \forall d'' \in D.d < d'' \Rightarrow d' \leq d'')$.

Definition 2.3 A timed transition system *is a quadruple* $\mathcal{T} = (S, A, \mathcal{D}, R)$ *where*

1. S *is a non-emtpy set (of* states*)*
2. A *is a non-empty set (of* action labels*)*
3. $\mathcal{D} = (D, +, 0)$ *is a time domain*
4. $\forall l \in A + D,\ R_l$ *is a binary relation on* S

The set $\mathcal{L} = A + D$ *is called the* sort *of the timed transition system.*

Notation. We shall denote elements from A by $a, b, \ldots$, elements from D by d, elements from S by s, t, all possibly indexed.

Operational semantics for a timed process algebra model processes as timed transition systems. The technique used to present corresponding classes of timed transition systems is a language of expressions for states and a language for expressions for delays, together with rules for generating transitions. The languages used for timed process algebras can be often viewed as extensions to the languages for untimed process algebras by adding some specific constructs or by assuming that in some cases prefixing by an action may imply delay.

As an example consider the timed extension to CCS - TCCS by Moller and Tofts [10]. The language of states includes for example the following prefixing constructors:

$$P ::= \ldots \mid (t).P \mid \delta.P \mid \ldots$$

The meaning to the these constructors is given by the following rules generating corresponding transition:

$$\frac{}{a.P \xrightarrow{a} P} \qquad \frac{}{\delta.P \xrightarrow{d} \delta.P}$$

$$\frac{P \xrightarrow{a} P'}{\delta.P \xrightarrow{a} P'} \qquad \frac{}{(s+t).P \xrightarrow{s} (t).P}$$

$$\frac{}{(t).P \xrightarrow{t} P} \qquad \frac{P \xrightarrow{s} P'}{(t).P \xrightarrow{s+t} P'}$$

Further examples of classes of timed transition systems are TCSP [5, 13, 14], TiCCS [16], TPL [6, 7], U-LOTOS [1], ATP [9].

Definition 2.4 *Let* $\mathcal{T}$ *be a timed transition system. For any* $L \subseteq \mathcal{L}$, $s, s' \in S$, $w = a_1 \ldots a_n \in A^*$ *we say that*

1. $s \xrightarrow{L} s'$ *iff* $\exists l \in L. s \xrightarrow{l} s'$
2. $s \xrightarrow{(a,d)} s'$ *iff* $\exists s_1, s_2 \in S, d_1, d_2 \in D. (s \xrightarrow{d_1} s_1 \wedge s_1 \xrightarrow{a} s_2 \wedge$
   $s_2 \xrightarrow{d_2} s'' \wedge d = d_1 + d_2)$
3. $s \stackrel{(w,d)}{\Longrightarrow} s'$ *iff* $s \xrightarrow{(a_1,d_1)} \ldots \xrightarrow{(a_n,d_n)} s'$ *where* $d = d_1 + \ldots + d_n$

Definition 2.5 *A* timed path σ *through a timed transition system is a finite or ω length sequence of the form*

$$s \stackrel{(a_1,d_1)}{\longrightarrow} s_1 \stackrel{(a_2,d_2)}{\longrightarrow} s_2 \ldots$$

We denote by $MTP(\mathcal{T})$ the set of maximal timed paths through the timed transition system $\mathcal{T}$.

3 Properties of timed transition systems

Transition systems we meet in timed process algebras satisfy several important conditions. These conditions are to be considered as the basic properties of timed processes, because they ensure that processes behave consistently with respect to time. In that sense they represent the basic assumptions a timed process algebra is built around.

The most important conditions, which are satisfied by most timed process algebras, are time determinism and time additivity. Further properties we shall examine include action urgency, persistency, finite variability or bounded control. For each property we shall give an intuitive explanation and then we describe the property formally as a property of transition relations. In most cases there are more possible formalizations.

Time determinism

When a system is in a state s and is willing to idle (not to perform any actions) for some duration d, then the resulting behaviour is completely determined from s and d.

Definition 3.1 *Let $\mathcal{T}$ be a timed transition system and $l \in \mathcal{L}$. We say that* $\stackrel{l}{\longrightarrow}$ *is* deterministic *if for all $s, s' \in S$*

$$s \stackrel{l}{\longrightarrow} s' \wedge s \stackrel{l}{\longrightarrow} s'' \Rightarrow s' = s''$$

We say that $\mathcal{T}$ is time deterministic *if* $\stackrel{d}{\longrightarrow}$ *is deterministic for every $d \in D$.*

Time additivity

In order to ensure soundness of the notion of time, it is usually required that a system which can idle for $d + d'$ time units can idle for d time units and then for d' time units and vice-versa. In both cases the resulting behaviour is the same.

Definition 3.2 *We say that a timed transition system $\mathcal{T}$ is* time additive *if for all $s, s' \in S, d, d' \in D$*

$$(\exists s'' \in S. s \stackrel{d}{\longrightarrow} s'' \wedge s'' \stackrel{d'}{\longrightarrow} s') \Leftrightarrow s \stackrel{d+d'}{\longrightarrow} s'$$

Time additivity is also called *time continuity.*

Tau-urgency

Tau-urgency requires internal communications to be performed as soon as possible. That is whenever τ can be performed it *must* be performed without letting time pass.

Definition 3.3 *Timed transition system $\mathcal{T}$ is said to be* tau-urgent *if for every $s, s', s'' \in S, d \in D$*

$$s \xrightarrow{\tau} s' \Rightarrow s \not\xrightarrow{d} s''$$

Tau-urgency is also called *minimal delay* or *maximal progress*. Sometimes a more general notion of *action urgency* is considered.

Persistency

Persistency requires that the progress of time cannot suppress the ability to perform an action. It is also called *non-retractiveness.*

Definition 3.4 *Timed transition system $\mathcal{T}$ is said to be* persistent *if for all $s, s', s'' \in S, d \in D, a \in A$:*

$$s \xrightarrow{a} s' \wedge s \xrightarrow{d} s'' \Rightarrow \exists s''' \in S.s'' \xrightarrow{a} s'''$$

Definition 3.5 *Timed transition system $\mathcal{T}$ is said to be* weakly persistent, *if for all $s_1, \ldots, s_n, s' \in S, a \in A, d_1, \ldots, d_n \in D$:*

$$s_1 \xrightarrow{a} s' \wedge s_1 \xrightarrow{d_1} \ldots \xrightarrow{d_n} s_n \Rightarrow \exists s'' \in S.s_n \xrightarrow{a} s''$$

Sometimes even a weaker notion of persistency is considered. The weaker requirement is that any action that can be performed remains possible during some time interval.

Definition 3.6 *The relation $\xrightarrow{a}$ is said to be* d-interval persistent *if for all $d' \in (0, d), s', s'' \in S, w \in A^*$:*

$$s \xrightarrow{a} s' \wedge s \stackrel{(w,d')}{\Longrightarrow} s'' \Rightarrow \exists s''' \in S.s'' \xrightarrow{a} s'''$$

Time transition system is said to be d-interval persistent *if for each $a \in A$, $\xrightarrow{a}$ is d-interval persistent. It is called* interval persistent *there is a $d \in D$ such that it is d-interval persistent.*

Finite variability

Timed transition system is *finite variable* (*non-Zenon, well-timed*) if it can perfom only finitely many actions in a finite time interval. There are several ways of formalizing this notion. One of the possibilities is to require a system delay to be performed between any two visible actions.

Definition 3.7 *$\mathcal{T}$ satisfies the* strong finite variablity *property if and only if for all d and for all $\sigma \in MTP(\mathcal{T})$*

$$(i < j \leq length(\sigma) \wedge \sum_{k=k+1}^{j} d_k \leq d) \Rightarrow j - i < \infty$$

A weaker notion of finite variability might require a system delay to appear between any two finite sequences of actions.

Bounded control

Bounded control requires that for any d there is a bounded number of changes in initial actions in any time interval of duration d. Let

$$init(s) = \{a \in A \mid \exists s' \in S. s \xrightarrow{a} s' \}$$

Definition 3.8 *Timed transition system $\mathcal{T}$ has the* bounded control *property if there is $d \in D$ such that*

$$s \xrightarrow{d_1} s' \wedge init(s) \neq init(s') \Rightarrow d_1 \geq d_2$$

According to the properties investigated we can categorize timed transition systems into classes. In section 6 we shall consider the following classes.

Definition 3.9

$$\Lambda_{TD} \stackrel{\text{df}}{=} \{\mathcal{T} \mid \mathcal{T} \textit{ is time deterministic} \}$$
$$\Lambda_{TA} \stackrel{\text{df}}{=} \{\mathcal{T} \mid \mathcal{T} \textit{ is time additive} \}$$

Very often we meet properties that describe some specific pattern of behaviour of the system with respect to time. These are not properties every process in the class under consideration should have, however they represent good behaviour of processes. Typically, *responsiveness* and *promptness*, belong to this category. We shall not deal with them in this paper.

All properties of timed transition systems fall into three categories:

- *action properties* - properties of relations $\xrightarrow{a}$ for $a \in A$
- *time properties* - properties of relations $\xrightarrow{d}$ for $d \in D$
- *action-time properties* - properties which involve both action and delay transitions

One of the reasons for introducing time into process algebras is to capture *quantitative* properties of systems. From this point of view are time and time-action properties fundamental. However, their "quantitative" expressibility differ. Time determinism is in fact qualitative property, whilst d-persistency, interval persistency or finite variability are really quantitative properties.

4 Timed modal logics

One question raised when designing a real-time extension to a temporal or modal logic is how to incorporate timing requirements in a formula. There are several solutions to this problem. Here we first introduce a simple variant of propositional modal μ-calculus called *Timed modal μ-calculus* (TMμC). Timed modal μ-calculus is propositional modal μ-calculus whose sort is *timed*, that is the labels are taken either from a set of actions or from a time domain. For example, the timed modal operator [3.5] is interpreted as "after every delay of 3.5 time units". The introduction of time structure on a part of labels seems to be necessary to capture quantitave properties.

The syntax for formulae Φ of *timed modal μ-calculus* TMμC is as follows:

$$\Phi ::= Z \mid \neg\Phi \mid \Phi_1 \wedge \Phi_2 \mid [K]\Phi \mid \nu Z.\Phi$$

where Z ranges over a denumerable family of propositional variables and K over *subsets* of the set $\mathcal{L} = A + D$, where $\mathcal{D} = (D, +, 0)$ is a time domain. The set $\mathcal{L}$ is called the *(timed) sort.* We say that the language has time domain $\mathcal{D}$. A formula is of sort $\mathcal{L}$ if it belongs to a language of sort $\mathcal{L}$.

A restriction on $\nu Z.\Phi$ is that each free occurence of Z in Φ should lie within the scope of an even number of negations.

Derived operators are defined as follows:

$$\begin{array}{ll} tt = \nu Z.Z & \Phi \vee \Psi = \neg(\neg\Phi \wedge \neg\Psi) \\ \langle K\rangle\Phi = \neg[K]\neg\Phi & \mu Z.\Phi = \neg\nu Z.\neg\Phi[Z := \neg Z] \end{array}$$

where $\neg\Phi[Z := \neg Z]$ is the result of substituing $\neg Z$ for each free occurence of Z. Further useful abbreviations (which also apply to the $\langle K\rangle$ modalities) are:

$$\begin{array}{rclrcl} [-K]\Phi & = & [\mathcal{L} - K]\Phi & [-]\Phi & = & [\mathcal{L}]\Phi \\ [l_1, \ldots, l_n]\Phi & = & [\{l_1, \ldots, l_n\}]\Phi & [d \leq t]\Phi & = & [\{d \mid d \leq t\}]\Phi \end{array}$$

A sublanguage of timed modal μ-calculus which does not contain fix-point operators is called *timed propositional modal logic* (TPML).

Formulas of TMμC are interpreted on timed transition systems. Let $\mathcal{T} = (S, A, \mathcal{D}, R)$ be a timed transition system. A *timed model* $\mathcal{M}$ on the timed transition system $\mathcal{T}$ is a pair $(\mathcal{T}, \mathcal{V})$ where $\mathcal{V}$ is a valuation assigning sets of states to propositional variables: $\mathcal{V}(Z) \subseteq S$. We use the notation $\mathcal{V}[Z := \mathcal{E}]$ for the valuation $\mathcal{V}'$ which agrees with $\mathcal{V}$ except that $\mathcal{V}'(Z) = \mathcal{E}$.

Definition 4.1 *The set of states of $\mathcal{T}$ having the property Φ in the timed model $(\mathcal{T}, \mathcal{V})$ is inductively defined as $||\Phi||_{\mathcal{V}}$:*

$$\begin{array}{rcl} ||Z||_{\mathcal{V}} & = & \mathcal{V}(Z) \qquad (Z \in Var) \\ ||\neg\Phi||_{\mathcal{V}} & = & \mathcal{P} - ||\Phi||_{\mathcal{V}} \\ ||\Phi_1 \wedge \Phi_2||_{\mathcal{V}} & = & ||\Phi_1||_{\mathcal{V}} \cap ||\Phi_2||_{\mathcal{V}} \\ ||[K]\Phi||_{\mathcal{V}} & = & \{s \in S \mid \forall s' \in S. \forall l \in K.\ s \xrightarrow{l} s' \Rightarrow s' \in ||\Phi||_{\mathcal{V}}\} \\ ||\nu Z.\Phi||_{\mathcal{V}} & = & \bigcup\{\mathcal{E} \subseteq \mathcal{P} \mid ||\Phi||_{\mathcal{V}[Z:=\mathcal{E}]} \supseteq \mathcal{E}\} \end{array}$$

From this definition we can easily get clauses for derived operators. For the operator $\mu Z.\Phi$ we have:

$$||\mu Z.\Phi||_{\mathcal{V}} = \bigcap\{\mathcal{E} \subseteq \mathcal{P} \mid ||\Phi||_{\mathcal{V}[Z:=\mathcal{E}]} \subseteq \mathcal{E}\}$$

Properties, we assume, are only expressed by *closed* formulas, i.e. those without free variables. This means that the set of processes having a property Φ is independent of valuations.

From the three types of properties of timed transition systems mentiond in section 3 we shall deal just with time and action-time properties. Here are some very simple examples of what properties can be expressed in timed modal logics.

$\langle d\rangle \mathrm{tt}$	can delay d
$[d > t]\mathrm{ff}$	cannot delay more then t
$\langle D\rangle \mathrm{tt}$	can delay
$\langle d\rangle \mathrm{tt} \wedge [L - \{d\}]\mathrm{ff}$	delays exactly d
$[d]\mathrm{ff}$	cannot delay d
$\langle a\rangle \mathrm{tt} \wedge [-a]\mathrm{ff}$	must perform a immediately

Now we present examples of composite schemas involving ν and μ operators. The formula $\nu Z.\langle K\rangle Z$ expresses the *capacity* to perform K transitions endlessly. Therefore the special case $\nu Z.\langle D\rangle Z$ represents the capacity to perform delays endlessly. The formula $\nu Z.A \vee \langle K\rangle Z$ expresses the capacity to perform K transitions continuously until A holds. Hence $\nu Z.A \vee \langle D\rangle Z$ expresses that after some delay A can become true.

Definition 4.2 *Let Φ be a formula and $\mathcal{M} = (\mathcal{T}, \mathcal{V})$ be a timed model. We say that*

1. Φ *is* true at s under $\mathcal{M}$, *written* $s \models_{\mathcal{M}} \Phi$, *if* $s \in ||\Phi||^{\mathcal{T}}_{\mathcal{V}}$
2. Φ *is* $\mathcal{M}$-valid, *written* $\models_{\mathcal{M}} \Phi$, *if for all states* $s, s \models_{\mathcal{M}} \Phi$
3. Φ *is* $\mathcal{T}$-valid, *written* $\models_{\mathcal{T}} \Phi$, *if for all models* $\mathcal{M}$ *on* $\mathcal{T}, \models_{\mathcal{M}} \Phi$
4. Φ *is* $\mathcal{T}$-satisfiable *if for some* $\mathcal{M}$ *on* $\mathcal{T}$ *and state* $s, s \models_{\mathcal{M}} \Phi$

The set of all formulas that are true at s under the timed model $\mathcal{M}$ is denoted by $||s||^{\mathcal{M}}$.

Definition 4.3 *Let Λ be a class of timed transition systems (of sort $\mathcal{L}$). We say that Φ is* Λ-valid, *written $\Lambda \models \Phi$, if Φ is $\mathcal{T}$-valid for all $\mathcal{T} \in \Lambda$. If Λ contains all timed transition systems of an appropriate sort, we call Φ* valid, *written $\models \Phi$.*

For timed models we can look for truth-preserving operations, i.e. operations on timed models such that $\mathcal{M}, s \models \Phi$ is preserved. There are variants of the three classical operations of *generation, zig-zag connection* and *filtration.* These properties are the main framework in proving non-definability of properties. In the sequel we shall use timed zig-zag connections and relations.

Definition 4.4 *A* timed zig-zag morphism *from $\mathcal{T}_1$ to $\mathcal{T}_2$ is a function $f : S_1 \to S_2$ satisfying*

1. $s \xrightarrow{a}_1 s'$ *implies* $f(s) \xrightarrow{a}_2 f(s')$
2. $f(s) \xrightarrow{a}_2 s'$ *implies* $\exists s''.s \xrightarrow{a}_1 s''$ *and* $f(s'') = s'$
3. $s \xrightarrow{d}_1 s'$ *implies* $f(s) \xrightarrow{d}_2 f(s')$
4. $f(s) \xrightarrow{d}_2 s'$ *implies* $\exists s''.s \xrightarrow{d}_1 s''$ *and* $f(s'') = s'$

Definition 4.5 *A* timed zig-zag *on a timed model* $\mathcal{M} = (\mathcal{T}, \mathcal{V})$ *is a binary relation* Z *on* S *such that if* sZs' *then for all* $a \in A, d \in D$ *and atomic* Q

1. $s \in \mathcal{V}(Q)$ *iff* $s' \in \mathcal{V}(Q)$
2. $s \xrightarrow{a} s_1$ *implies* $\exists s''.s' \xrightarrow{a}_1 s''$ *and* $s_1 Z s_1'$
3. $s' \xrightarrow{a} s_1'$ *implies* $\exists s''.s \xrightarrow{a} s''$ *and* $s_1 Z s_1'$
4. $s \xrightarrow{d} s_1$ *implies* $\exists s''.s' \xrightarrow{d}_1 s''$ *and* $s_1 Z s_1'$
5. $s' \xrightarrow{d} s_1'$ *implies* $\exists s''.s \xrightarrow{d} s''$ *and* $s_1 Z s_1'$

When there is a timed zig-zag relation on $\mathcal{M}$ relating s and s' we write $s \equiv_z s'$.

Theorem 4.6 *If* f *is a zig-zag morphism from* $\mathcal{T}_1$ *onto* $\mathcal{T}_2$ *then for all timed modal (μ-)formulas* Φ,

$\mathcal{T}_1 \models \Phi$ *implies* $\mathcal{T}_2 \models \Phi$

Theorem 4.7 *If* $s \equiv_z s'$ *then* $||s||^{\mathcal{M}} = ||s'||^{\mathcal{M}}$. *If* $\mathcal{M}$ *is image finite and* $||s||^{\mathcal{M}} = ||s'||^{\mathcal{M}}$ *then* $s \equiv_z s'$.

5 Definability of properties

One of the central theme in modal and temporal logics is to examine which model properties are definable in the particural logic. In this section we examine which of the properties defined in section 3 has corresponding formulas in timed modal μ-calculus, or more precisely which of them are definable.

Definition 5.1 *A class* $\mathcal{U}$ *of timed transition systems is* definable relative *to the class* $\mathcal{U}_1$, *if there is a set* Γ *of formulas such that*

$\mathcal{U} = \{\mathcal{T} \in \mathcal{U}_1 \mid \forall \Phi \in \Gamma . \mathcal{T} \models \Phi \}$

If $\mathcal{U}_1$ *is the class of all timed transition systems,* $\mathcal{U}$ *is called* definable.

The table in Figure 1 lists properties of timed transition systems from section 3 which are definable with TMμC.

Property	Defined by formula
time additivity	$\langle d + d' \rangle \Phi \leftrightarrow \langle d \rangle \langle d' \rangle \Phi$
time determinism	$\langle d \rangle \Phi \rightarrow [d] \Phi$
tau-urgency	$\langle \tau \rangle \mathrm{tt} \rightarrow [d] \mathrm{ff}$
persistency	$\langle a \rangle \mathrm{tt} \rightarrow [d] \langle a \rangle \mathrm{tt}$
weak-persistency	$\nu Y_1.(\mu Y_2.(\langle D \rangle \mathrm{tt} \wedge [-a] \mathrm{ff}) \wedge [-] Y_2) \wedge [-] Y_1$

Figure 1: *Definability of timed properties in TMμC*

As an example how one proves such correspondences we prove the definability for time determinism and time additivity.

Lemma 5.2 $\mathcal{T} \models \langle d\rangle\Phi \rightarrow [d]\Phi$ *iff* $\mathcal{T}$ *is time deterministic.*

Proof. Suppose $\langle d\rangle\Phi \rightarrow [d]\Phi$ for every Φ is $\mathcal{T}$-valid. Assume $s \xrightarrow{d} s'$ and $s \xrightarrow{d} s''$. We need to show $s = s''$. Let $\mathcal{M}$ be a modal timed model on $\mathcal{T}$ with valuation $\mathcal{V}(Q) = \{s'\}$. We know $s \models_{\mathcal{M}} \langle d\rangle Q \rightarrow [d]Q$. By definition of $\mathcal{V}$, $s' \models_{\mathcal{M}} Q$. Since $s \xrightarrow{d} s'$, $s \models_{\mathcal{M}} \langle d\rangle Q$. Therefore s also has the property $[d]Q$ under $\mathcal{M}$. But $s \xrightarrow{d} s''$ and so $s'' \in Q$. That is $s = s''$ by definition of $\mathcal{V}$. Suppose now $\mathcal{T}$ is time deterministic. To show $\mathcal{T} \models \langle d\rangle\Phi \rightarrow [d]\Phi$ consider any model $\mathcal{M}$ on $\mathcal{T}$ and any state s. Suppose $s \models_{\mathcal{M}} \langle d\rangle\Phi$, then we need to show $s \models_{\mathcal{M}} [d]\Phi$. Because $s \models_{\mathcal{M}} \langle d\rangle\Phi$, for some s', $s \xrightarrow{d} s'$ and $s' \models_{\mathcal{M}} \Phi$. By the time determinism of $\mathcal{T}$ we know that for every s'' such that $s \xrightarrow{d} s''$ is $s = s''$ and therefore for every s'' such that $s \xrightarrow{d} s''$, $s'' \models_{\mathcal{M}} \Phi$, hence $s \models_{\mathcal{M}} [d]\Phi$. ∎

Lemma 5.3 $\mathcal{T} \models \langle d + d'\rangle\Phi \leftrightarrow \langle d\rangle\langle d'\rangle\Phi$ *iff* $\mathcal{T}$ *is time additive.*

Proof. Suppose $\langle d+d'\rangle\Phi \leftrightarrow \langle d\rangle\langle d'\rangle\Phi$ for every Φ is $\mathcal{T}$-valid. Assume $s \xrightarrow{d} s''$ and $s'' \xrightarrow{d} s'$ for some s''. We need to show $s \xrightarrow{d+d'} s'$. Let $\mathcal{M}$ be a modal timed model on $\mathcal{T}$ with valuation $\mathcal{V}$ such that $\mathcal{V}(Q) = \{s'\}$. We have $s' \models_{\mathcal{M}} Q$ and from $s'' \xrightarrow{d'} s'$, $s'' \models_{\mathcal{M}} \langle d'\rangle Q$. Since $s \xrightarrow{d} s''$, $s \models_{\mathcal{M}} \langle d\rangle\langle d'\rangle Q$. By definition of $\mathcal{V}(Q)$ it follows that $s \xrightarrow{d+d'} s'$.

Now suppose $\mathcal{T}$ is time additive. We need to show $\mathcal{T} \models \langle d + d'\rangle\Phi \leftrightarrow \langle d\rangle\langle d'\rangle\Phi$. Consider any model $\mathcal{M}$ on $\mathcal{T}$ and any state s. There are two cases. First assume $s \models_{\mathcal{M}} \langle d + d'\rangle\Phi$ and we have to show $s \models_{\mathcal{M}} \langle d\rangle\langle d'\rangle\Phi$. Because $s \models_{\mathcal{M}} \langle d + d'\rangle\Phi$ for some s', $s \xrightarrow{d+d'} s'$ and $s' \models_{\mathcal{M}} \Phi$. By the time additivity there is an s'' with the features $s \xrightarrow{d} s''$ and $s'' \xrightarrow{d'} s'$. Then $s'' \models_{\mathcal{M}} \langle d'\rangle\Phi$ and $s \models_{\mathcal{M}} \langle d + d'\rangle\Phi$. For the second case assume $s \models_{\mathcal{M}} \langle d\rangle\langle d'\rangle\Phi$ and we need to show $s \models_{\mathcal{M}} \langle d + d'\rangle\Phi$. Since $s \models_{\mathcal{M}} \langle d\rangle\langle d'\rangle\Phi$, $s \xrightarrow{d} s''$ and $s'' \models_{\mathcal{M}} \langle d'\rangle\Phi$ for some s''. Because $s'' \models_{\mathcal{M}} \langle d'\rangle\Phi$ for some s', $s'' \xrightarrow{d'} s'$ and $s' \models_{\mathcal{M}} \Phi$. By time additivity $s \xrightarrow{d+d'} s'$. Now from $s' \models_{\mathcal{M}} \Phi$ it follows that $s \models_{\mathcal{M}} \langle d + d'\rangle\Phi$. ∎

That the others properties are not definable by a TMμC-formula can be proved using the preservation results. The proof proceeds through counter-example to the preservation behaviour, particularly to zig-zag morphism. To see how such a negative conclusion is reached we prove as an example the case of finite variability.

Theorem 5.4 *Finite variability is not definable with TMμC.*

Proof. Consider the following TTS $\mathcal{T}_1$ with time domain R (real numbers) and where

$$
\begin{array}{lcl}
S_1 & = & \{s_0, s_1', s_2', s_4', \ldots, s_1'', s_2'', s_4'', \ldots\} \\
\mathcal{L} & = & \{a, 1, 1/2, 1/4, 1/8, \ldots\} \\
\xrightarrow{a} & = & \{(s_0, s_1'), (s_1'', s_1'), \\
 & & (s_0, s_2'), (s_2'', s_2'), \\
 & & \ldots \\
\xrightarrow{1} & = & \{(s_1', s_1'')\} \\
\xrightarrow{1/2} & = & \{(s_2', s_2'')\} \\
\xrightarrow{1/4} & = & \{(s_4', s_4'')\} \\
\ldots & &
\end{array}
$$

$\mathcal{T}_1$ is finite variable. Now, let $\mathcal{T}_2$ be

$$
\begin{array}{lcl}
S_2 & = & \{t_0, t_1\} \\
\mathcal{L} & = & \{a, 1, 1/2, 1/4, 1/8, \ldots\} \\
\xrightarrow{a} & = & \{(t_0, t_1)\} \\
\xrightarrow{1} & = & \{(t_1, t_0)\} \\
\xrightarrow{1/2} & = & \{(t_1, t_0)\} \\
\xrightarrow{1/4} & = & \{(t_1, t_0)\} \\
 & & \ldots
\end{array}
$$

$\mathcal{T}_2$ is a zig-zag morphic image of $\mathcal{T}_1$:

$$
\begin{array}{lcl}
f(s_0) & = & t_0 \\
f(s_i') & = & t_1 \\
f(s_i'') & = & t_0
\end{array}
$$

However, $\mathcal{T}_2$ is *not* finite variable.

Sometimes we are able to stay relative results. For example, on time additive transition systems weak persistency is equivalent to persistency. We have thus the following relative definability result.

Theorem 5.5 *Weak persistency is TPML definable relative to the class of time additive transition systems by the formula*

$$\langle a\rangle \mathrm{tt} \rightarrow [d]\langle a\rangle \mathrm{tt}$$

6 Minimal timed modal logics

For each timed sort $\mathcal{L}$ there is a *minimal* timed modal logic. In the case of TPML we shall denote it by $\mathrm{TK}_{\mathcal{L}}$, in the case of TM$\mu$C by $\mu\mathrm{TK}_{\mathcal{L}}$. Minimal logics are the smallest sets of valid formulas.

Definition 6.1 .

$$
\begin{array}{lcl}
\mu TK_{\mathcal{L}} & \stackrel{\mathrm{df}}{=} & \{\Phi \mid \Phi \textit{ is TM}\mu\textit{C-formula of sort } \mathcal{L} \textit{ and } \models \Phi\} \\
TK_{\mathcal{L}} & \stackrel{\mathrm{df}}{=} & \{\Phi \mid \Phi \textit{ is TPML-formula of sort } \mathcal{L} \textit{ and } \models \Phi\}
\end{array}
$$

Sets of valid formulas could be characterised by axiom systems. An *axiom system* J consists of a finite set of axiom schemas, and a finite set of inference rules. A *proof* of Φ in J (written $J \vdash \Phi$ or $\vdash_J \Phi$) is a finite sequence of formulas $\Phi_1, \ldots, \Phi_n$ such that $\Phi = \Phi_n$ and each $\Phi_i (1 \leq i \leq n)$ is either an axiom instance, or the result of an application of an inference rule to formulas in the set $\{\Phi_1, \ldots, \Phi_n\}$. Φ is a *theorem*, written $J \vdash \Phi$, if there is a proof of Φ in J. We call the axiom system *sound* (with respect to a class of models Λ) if $J \vdash \Phi \Rightarrow \Lambda \models \Phi$. We call it *complete* if $\Lambda \models \Phi \Rightarrow J \vdash \Phi$.

A complete and sound axiomatization for minimal timed modal μ-calculus $\mu TK_{\mathcal{L}}$ is not known. Axiomatic characterisation of the logic $\mathrm{TK}_{\mathcal{L}}$ is the set of *theorems* of the following axiom system TK, where l ranges over members of $\mathcal{L}$ and Φ, Ψ over TPML-formulas:

Axioms	1.	Any tautology instance
	2.	$[l](\Phi \rightarrow \Psi) \rightarrow ([l]\Phi \rightarrow [l]\Psi)$
Rules	MP	if Φ and $\Phi \rightarrow \Psi$ then Ψ
	NEC	if Φ then $[l]\Phi$

Axiom system TK is sound and complete with respect to the class Λ_T of timed transition systems (of sort $\mathcal{L}$).

Theorem 6.2 *For all TPML-formulas* Φ

$$TK \vdash \Phi \text{ iff } \Lambda_T \models \Phi$$

Now, we shall briefly deal with the question of the existence of sound and complete axiom systems for the following two classes of TPML-formulas (of sort $\mathcal{L}$):

Definition 6.3

$$\Gamma_{TD} \stackrel{\mathrm{df}}{=} \{\Phi \mid \Lambda_{TD} \models \Phi\} \quad \textit{(time deterministic)}$$
$$\Gamma_{TA} \stackrel{\mathrm{df}}{=} \{\Phi \mid \Lambda_{TA} \models \Phi\} \quad \textit{(time additive)}$$

Axiom system TK_{TD} is the axiom system TK extended by the schema $\langle d\rangle\Phi \rightarrow [d]\Phi$. Axiom system TK_{TA} is the axiom systems TK extended by the schema $\langle d+d'\rangle\Phi \leftrightarrow \langle d\rangle\langle d'\rangle\Phi$.

Theorem 6.4 *For all TPML-formulas* Φ

1. $TK_{TD} \vdash \Phi$ *iff* $\Lambda_{TD} \models \Phi$
2. $TK_{TA} \vdash \Phi$ *iff* $\Lambda_{TA} \models \Phi$

Proof of soundness is straightforward. To show completeness of the appropriate axiomatic characterisation we use the Henkin method.

Definition 6.5 *Let J be an axiomatic system. A formula Φ is said to be J-*consistent *if $J \not\vdash \neg\Phi$. A set Γ of formulas is said to be J-consistent if* $\Gamma \not\vdash_J$ff.

We can alternatively state the completeness property: if Φ is J-consistent then Φ is Λ-satifiable. The Henkin method for showing an axiom system is complete involves building a model from sets of J-consistent formulas.

A set Γ of formulas is a *maximal* T-consistent set of TPML-formulas if Γ is J-consistent and $\Gamma \vdash_J \neg\Phi$ for any $\Phi \in \overline{\Gamma}$. Let Jmc be the set of all maximal J-consistent sets of TPML-formulas.

Definition 6.6 *The* canonical *modal timed model with respect and* J *is* $\mathcal{M}_J = (\mathcal{T}, \mathcal{V})$ *where:*

1. $\mathcal{S}_\mathcal{T} = \ Jmc$
2. $\mathcal{L}_\mathcal{T} = \{l \mid l$ *occurs in some formula*
3. $\mathcal{D}_\mathcal{T} = \mathcal{D}$
4. $s \xrightarrow{l} s'$ *iff for all finite* $\beta \subseteq s'.s \cup \{\langle l\rangle\beta\}$ *is* J*-consistent*
5. $\mathcal{V}(Q) = \{s \mid Q \in s\}$

Theorem 6.7 *For all TPML-formulas* Φ

$s \models_{\mathcal{M}_J} \Phi$ *iff* $\Phi \in s$

To show the completeness we need to show that the resulting timed model $\mathcal{M}_J$ is of the appropriate kind.

Lemma 6.8 *If* J *is* KT_{TD} *then* $\mathcal{M}_J$ *is time deterministic.*

Proof. Suppose $s \xrightarrow{d} s', s \xrightarrow{d} s''$ within $\mathcal{M}_J$. We have to show $s' = s''$. Let $\Phi \in s'$. Then $\langle d\rangle\Phi \in s$. By the schema $\langle d\rangle\Phi \to [d]\Phi$, $[d]\Phi \in s$. But then $\Phi \in s''$ as $s \xrightarrow{d} s''$. The other case is similar. ∎

Lemma 6.9 *If* J *is* KT_{TA} *then* $\mathcal{M}_J$ *is time additive.*

7 Conlusion

We have presented the timed variant of propositional modal μ-calculus. With this logic we can define many properties of timed transition systems. Especially, the most crucial properties, e.g. time determinism and time additivity, have been shown to be definable and even more, there are complete and sound axiomatisations of time deterministic and time additive systems.

On the other side, some very important quantitative properties of timed transition systems are not definable with this logic and more powerfull framework is needed.

References

[1] Bolognesi, T. and Lucidi, F.: *LOTOS-like process algebra with urgent or timed interactions.* In Proc. REX Workshop "Real-Time: Theory in Practice". Mook, the Netherlands, 1991.

[2] Bradfield, J.: *Verifying Temporal Properties of Systems.* Birkhäuser, 1992.

[3] Bradfield, J. and Stirling, C.: *Verifying temporal properties of processes.* Springer-Verlag, LNCS **485**, 1990.

[4] Brim, L.: *Analyzing Time Aspects of Dynamic Systems.* EMCSR'92, Vienna, Austria, April 1992.

[5] Davis, J. and Schneider, S.: *An Introduction to Timed CSP.* Technical Monograph PRG-75, Oxford University, UK, 1989.

[6] Hennessy, M. and Regan, T.: *A Temporal Process Algebra.* Technical Report 2/90, University if Sussex, UK, 1990.

[7] Hennessy, M. and Regan, T.: *A Process Algebra for Timed Systems.* Technical Report 5/91, University if Sussex, UK, 1991.

[8] Kozen, D.: *Results on the propositional mu-calculus.* TCS **27**, 1983.

[9] Manna, Z. and Pnueli, A.: *The anchored version of the temporal framework.*, Springer-Verlag, LNCS **354**, 1989.

[10] Moller, F. and Tofts, C.: *A Temporal Calculus of Communicating Systems.* University of Edinburgh, Report No. LFCS-89-104, 1989.

[11] Nicollin, X. and Sifakis, J.: *The algebra of timed processes ATP: Theory and Application.* Technical Report RT-C26, Laboratoire de Génie Informatique de Grenoble, 1990.

[12] Nicollin, X. and Sifakis, J.: *An Overview and Synthesis on Timed Process Algebras.* CAV'91, Alborg, Denmark, July 1991. Springer-Verlag, LNCS **391**, 1989.

[13] Reed, G.M. and Roscoe, A.W.: *A timed model for Communicating Sequential Processes.* TCS, 58, 1988.

[14] Schneider, S.: *An Operational Semantics for Timed CSP.* Technical Report PRG, Oxford University, UK, 1991.

[15] Stirling, C.: *An Introduction to Modal and Temporal Logics for CCS.*

[16] Wang Yi: *Real-time behaviour of asynchronous agents.* In proc. CONCUR'90, LNCS 458, 1990.

Process Communication Environment*

Damas P. Gruska
Institute of Informatics, Comenius University
842 15 Bratislava, Czecho-Slovakia

Andrea Maggiolo-Schettini
Dipartimento di Informatica, Università di Pisa
56 125 Pisa, Italy

Abstract

A real time extension of CCS which takes into account properties of an interconnection communication network is presented. With the help of this calculus several semantics for CCS are defined. Limitations of CCS are discussed.

1 Introduction

A common feature of different types of parallel machines and distributed systems is the presence of many processing elements and interconnections among them. It does not matter whether the systems are of one chip size or of a worldwide telecommunication network size.

Process Algebras belong to basic tools for specification and verification of these systems viewing them as "communicating systems". In general, standard Process Algebras as *Communicating Sequential Processes* [8], *Algebra of Communicating Processes* [1], *Calculus of Communicating Systems* [10] abstract many real properties of actual systems as duration and structure of actions, location of processes, properties of an interconnection network and so on.

Since for many applications these properties are crucial there is a research going on extending the existing abstract calculi, defining more sofisticated semantics for them and so on.

But, while abstract calculi express system properties which are more or less independent on level of system description, the less abstract calculi depend much more on those levels. For example, on one level we can incorporate reasoning on durations of actions (expressed in some time units) into a calculus and we can still abstract real properties of an interconnection network. For another level, speaking about durations of actions (using finer time units) we cannot neglect time needed for start of communications since (re)configuration of a network might take some time units.

The aim of this paper is to introduce a real time extension of Milner's CCS describing communicating systems on the level where we have to take into account also the properties of the network.

*Work partially supported by CNR - Progetto Finalizzato Informatica and by the grant MSMS SR 01/72

One of the main areas of successful application of Process Algebras is specification of communication protocols. We start with motivating the presented calculi by a typical communication network [12] (See Fig. 1).

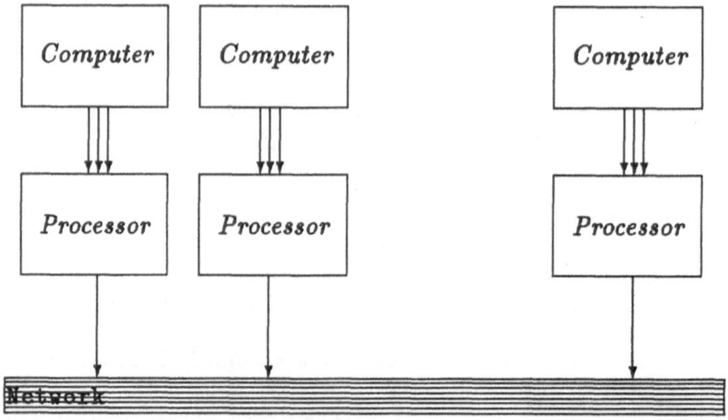

Figure 1: Computer network using communications processors

The system consists of computers, their communication processors and a communication network. Requests for data transmision are sent from a computer to its communication processor. For these transmissions there are fast local physical links connecting both these components (Physical Communications) [2]. On the other side, communication processors are usually connected via a network of logical links (Logical Communications). While for communications between the computers and their communication processors we can accept the handshaking paradigm (as a complete inteconnection network) for communications among processors this paradigm seems to be unrealistic. In the presented calculus we assume that the communication processors are connected via a network with a limited capacity. By such a capacity we mean a number of independent communication paths. In the presented description level we do not distinguish between active interconnection networks, such as crossbars, n-dimensional cubes, omega networks, and passive ones, such as multiple global buses, rings, trees and so on. We only require that n links can transmit n pieces of information, i.e. that they can mediate n communications at the same time. ¿From now on speaking of a communication network we will mean a network of a limited capacity. We consider two kinds of atomic actions. Actions representing communications via a complete network (*eager actions*) and actions representing communications via a limited capacity network (*lazy actions*). We assume that performing actions of the second kind decreases the number of free communication links. It means that a communication consumes a communication link and therefore the communication is possible only if there is free link in a net.

For our purpose we need a way how to model also a duration of actions (communications). For this we use a simple timed model of a process algebra capable of expressing duration of actions. Among many timed models [7, 11, 4, 5, 13] we choose TiCCS, a slight modification of the one in [5], which seems to be fully suitable.

The main idea of TiCCS is to introduce a t action into a standard process algebra, such as CCS. The execution of a t action by a process indicates that the process is idling until the next clock cycle. We continue to use the CCS theory underlying assumption that all processes may idle indefinitely. Using the CCS syntax the process $a.P$ can idle, i.e. it can perform a t action. In our calculus we can simulate duration of actions. For example, the process $P = a.t.b.t.Nil$ can perform the action a anytime but the second atomic action b can be performed not earlier then after one tick of a clock. In this way we will model that a has a duration of one time unit.

Moreover we assume that a lazy action has not to occupy a link for the whole time of its duration. For example, proper communication can represent only a part of the time needed for execution of the action. To express finish of communication we use a time action t_f. After performing this action a link is again free even though the corresponding action does not finish. Since all lazy actions represent internal communications it is enough to express finish of communications in, for example, output actions. For instance the fact that action $\bar{a}$ takes five time units but the communication part takes only two time units will be expressed by writing $\bar{a}.t.t_f.t.t.t$. Moreover we require that for each output action the duration of the communication part is not shorter than the duration of the whole complementary input action. It means that the duration of the action a (co-action to the action from the previous example) must be at least two time units.

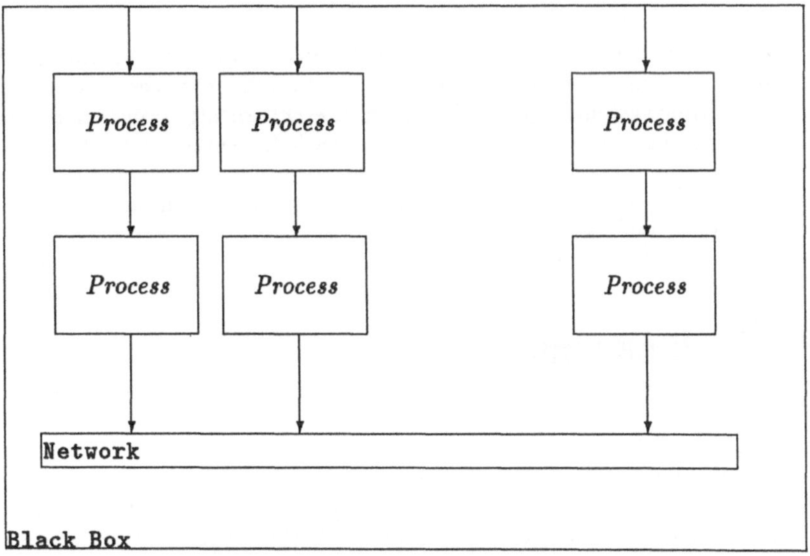

Figure 2: Communicating Processes

We consider timed processes running in a net offering at the beginning n communication links (*netted processes*, for short). Every lazy action beginning decreases the number of free links by one. After the action has been completed the link is again free. Then if there are no free links a netted process cannot perform any lazy action except idling. It means that communication by means of lazy actions can be performed only in the case that there is a free communication link. For eager actions there are no restrictions.

In Fig. 2 we show our abstraction of the system from Fig. 1. Lazy actions represents communications via the net. All these communications are internal. Eager actions represent communications of computers with an environment or with their communication processors. They are either internal or external.

An action τ has usually two meanings. It may be the result of an internal communication or it expresses a lack of information. To avoid problems with these two different meanings from now on we assume that τ represents only communications.

A concept of netted process has been firstly indroduced in [6] where a net architecture simpler with respect to the one we present here has been considered. In [6] the attention was mainly concentrated on algebraic semantics and a proof system for netted processes, and for simplicity only uniform timing was considered. In the present paper we start from a calculus for netted processes considering different types and durations of actions, and with the help of this calculus we define more and more abstract semantics for CCS. More precisely, we show that semantics which abstract net properties are finer than CCS strong bisimulation and a semantics abstracting also duration of actions coincides with the strong bisimulation.

The paper is organized as follows. In Section 2 we describe the language TiCCS and give an operational semantics for netted processes in terms of labeled transition systems. In Section 3 we present an algebraic net semantics and its properties. In Sections 4 we define various semantics for CCS by abstracting properties as capacity of a net and durations of actions. Due to lack of space we have omitted all the proofs, as almost all of them are standard but rather technical and long.

2 The language

To define the language TiCSS, we first presuppose two sets of atomic action symbols E (for eager actions) and L (for lazy actions) not containing symbols τ_E, τ_L, t, t_f such that for every $a \in E$ $(a \in L)$, there exists an $\overline{a} \in E$ $(\overline{a} \in L)$ and $\overline{\overline{a}} = a$. We define $A = E \cup L$, $Act = A \cup \{\tau_E\} \cup \{\tau_L\}$, $Actt = Act \cup \{t\} \cup \{t_f\}$. We let $a, b, \ldots$ range over A; τ ranges over $\{\tau_E, \tau_L\}$; $u, v, \ldots$ range over Act; and $x, y \ldots$ range over $Actt$.

The set TiCCS of timed terms is defined by the following BNF expression where $x \in Actt, X \in Var$, with Var a set of process variables, S ranging over relabelling functions, with $S : Actt \to Actt$ such that $\overline{S(a)} = S(\bar{a})$ for $a \in A, S(\tau) = \tau, S(t) = t$, and $M \subseteq A$:

$$P ::= Nil \mid X \mid x.P \mid P + P \mid P \mid P \mid \mu X P \mid P \setminus M \mid P[S]$$

We use the usual definitions for free and bound variables, open and closed terms and guarded recursion. The set **TiCSS** of *processes* consists of closed and guarded TiCSS terms.

Now we are ready to define a set of TiCCS processes communicating via a net with capacity n. ¿From now on we will call them *n-netted processes* or just *netted processes*. It means that the starting capacity is n links, and, depending on communications, it may vary between 0 and n.

Definition 1.1 $\|\mathbf{TiCCS}\|^n = \{\|P\|_k^n \mid P \in \mathbf{TiCCS}, 0 \leq k \leq n\}$ is the set of (timed) n-netted processes.

We interpret $\|P\|_k^n$ as a timed process P in a n-net which has at the moment still k links free, i.e. $n - k$ links are in use.

A structural operational semantics of $\| \mathbf{TiCCS} \|^n$ is defined in terms of Labeled Transition Systems.

Definition 1.2 *A Labeled Transition System* is a triple $(S, T, \{\xrightarrow{x}, x \in T\})$ where S is a set of states, T is a set of labels and $\{\xrightarrow{x}, x \in T\}$ is the transition relation such that each $\xrightarrow{x}$ is a binary transition relation on S. We write $P \xrightarrow{x} P'$ instead of $(P, P') \in \xrightarrow{x}$ and $P \not\xrightarrow{x}$ if there is no P' such that $P \xrightarrow{x} P'$.

Definition 1.3 The net transition relation $\{\xrightarrow{x}_{\|TiCCS\|^n}, x \in Actt\}$ is defined as the least relation satisfying the following inference rules:

Rules for handling actions from the set *Act*:

$$\frac{u \in E}{\|u.P\|^n_k \xrightarrow{u} \|P\|^n_k} \quad Pr1$$

$$\frac{k > 0, u \text{ is an ouput action from } L}{\|u.P\|^n_k \xrightarrow{u} \|P\|^n_{k-1}} \quad Pr2$$

$$\frac{u \text{ is an input action from } L}{\|u.P\|^n_k \xrightarrow{u} \|P\|^n_k} \quad Pr3$$

$$\frac{\|P\|^n_k \xrightarrow{u} \|P'\|^n_{k'}}{\|P \mid Q\|^n_k \xrightarrow{u} \|P' \mid Q\|^n_{k'}} \quad Pa1$$

$$\frac{\|P\|^n_k \xrightarrow{u} \|P'\|^n_{k'}}{\|Q \mid P\|^n_k \xrightarrow{u} \|Q \mid P'\|^n_{k'}} \quad Pa2$$

$$\frac{a \in E, \|P\|^n_k \xrightarrow{a} \|P'\|^n_k, \|Q\|^n_k \xrightarrow{\bar{a}} \|Q'\|^n_k}{\|P \mid Q\|^n_k \xrightarrow{\tau_E} \|P' \mid Q'\|^n_k} \quad Pa3$$

$$\frac{a \in L, \|P\|^n_k \xrightarrow{a} \|P'\|^n_{k'}, \|Q\|^n_k \xrightarrow{\bar{a}} \|Q'\|^n_{k'}}{\|P \mid Q\|^n_k \xrightarrow{\tau_L} \|P' \mid Q'\|^n_{k'}} \quad Pa4$$

$$\frac{\|P\|^n_k \xrightarrow{u} \|P'\|^n_{k'}}{\|P + Q\|^n_k \xrightarrow{u} \|P'\|^n_{k'}} \quad S1$$

$$\frac{\|P\|^n_k \xrightarrow{u} \|P'\|^n_{k'}}{\|Q + P\|^n_k \xrightarrow{u} \|P'\|^n_{k'}} \quad S2$$

Rules for handling the action t:

$$\frac{}{||t.P||^n_k \xrightarrow{t} ||P||^n_k} \qquad Pr4$$

$$\frac{}{||t_f.P||^n_k \xrightarrow{t} ||P||^n_{k+1}} \qquad Pr5$$

$$\frac{}{||\tau_L.P||^n_0 \xrightarrow{t} ||\tau_L.P||^n_0} \qquad Pr6$$

$$\frac{}{||a.P||^n_k \xrightarrow{t} ||a.P||^n_k} \qquad Pr7$$

$$\frac{||P||^n_k \xrightarrow{t} ||P'||^n_{k_1},\ ||Q||^n_k \xrightarrow{t} ||Q'||^n_{k_2},\ ||P \mid Q||^n_k \overset{\tau}{\not\rightarrow}}{||P \mid Q||^n_k \xrightarrow{t} ||P' \mid Q'||^n_{k_1+k_2-k}} \qquad Pa5$$

$$\frac{||P||^n_k \xrightarrow{t} ||P'||^n_{k'},\ ||Q||^n_k \xrightarrow{t} ||Q'||^n_{k'}}{||P + Q||^n_k \xrightarrow{t} ||P' + Q'||^n_{k'}} \qquad S3$$

$$\frac{||P||^n_k \xrightarrow{t} ||P'||^n_{k'},\ ||Q||^n_k \xrightarrow{t} ||Q'||^n_{k'}}{||Q + P||^n_k \xrightarrow{t} ||Q' + P'||^n_{k'}} \qquad S4$$

Common rules:

$$\frac{||P||^n_k \xrightarrow{x} ||P'||^n_{k'}}{||P \setminus M||^n_k \xrightarrow{x} ||P' \setminus M||^n_{k'}},\ (x, \overline{x} \notin M) \qquad Res$$

$$\frac{||P[\mu X P/X]||^n_k \xrightarrow{x} ||P'||^n_{k'}}{||\mu X P||^n_k \xrightarrow{x} ||P'||^n_{k'}} \qquad Rec$$

$$\frac{||P||^n_k \xrightarrow{x} ||P'||^n_{k'}}{||P[S]||^n_k \xrightarrow{S(x)} ||P'[S]||^n_k} \qquad Rl$$

As it was mentioned earlier, we assume that eager actions do not consume communication links (see rule $Pr1$). The fact that lazy actions consume them is expressed only with the help of output lazy actions (see rule $Pr2$ and rule $Pr3$). The result of a communication depends on what kind of actions communicate (see rule $Pa3$ and rule $Pa5$). The action t_f increases the number of free communication links (see rule $Pr5$). An internal action τ_L may idle only if there is not a free link at a moment (see rule $Pr6$).

In the definition of the labeled transition system we used negative premises (see $Pa5$). In general this may lead to problems, for example with consistency of the defined system. We avoid these dangers by making derivations of τ independent on derivations of t. For an explanation and details see [3].

The set of CCS processes consists of closed and guarded CSS terms which are terms without t actions. An operational semantics for CCS processes is given by means of a labeled transition system similar to the one we presented for netted processes except that it does not handle a t action and a net capacity (see [10]).

In general, not all TiCCS processes are meaningful. We assume that processes do not share processors and so durations are static properties, namely durations of the same actions are the same. Moreover, as it was mentioned earlier, we require that the communication part of any output lazy action is not shorter then the duration of the corresponding input action. To ensure all these conditions we will consider only processes obtained from CCS ones by expressing durations of actions. As it was said earlier, we express durations of actions by postfixing them by sequences of t actions. (Sometimes it might be more appropriate for expressing durations of output actions to prefix the actions by sequence of t. The presented technique is easily extendable to this case.) Moreover every otput lazy action is postfixed also with one t_f action. This can be done by means of a time refinement as a syntactical substitution.

Definition 1.4 Let us consider a mapping $r : u \mapsto u.t^{i_u}, i_u \geq 0$ for every $u \in Act$ except when u is a lazy output action. In that case $r : \bar{u} \mapsto \bar{u}.t^{i_{a_1}}.t_f.t^{i_{a_2}}, i_{\bar{u}_1}, i_{\bar{u}_2} \geq 0$. Moreover for lazy actions we assume that if $r : u \mapsto u.t^{i_u}$ and $r : \bar{u} \mapsto \bar{u}.t^{i_{a_1}}.t_f.t^{i_{a_2}}$ then $i_{\bar{u}_1} \leq i_u$.

We define an extension of r on the set of TiCCS processes:

$$\begin{aligned}
r(Nil) &= Nil \\
r(X) &= X \\
r(a.P) &= r(a).r(P) \\
r(P+Q) &= r(P)+r(Q) \\
r(P \mid Q) &= r(P) \mid r(Q) \\
r(P[S]) &= P[r \circ S] \\
r(\mu XP) &= \mu X r(P) \\
r(P \setminus M) &= r(P) \setminus M
\end{aligned}$$

where $r \circ S$ is a composition of mappings. We shall call the mapping r a *time refinement.*

¿From now on we will consider only a TiCCS processes of the form $P \setminus L$ where P is given from CCS process by a time refinement.

3 Algebraic net semantics

In this section we define an algebraic semantics by means of a bisimulation, i.e. as an equivalence relation on the set of netted processes. Following the usual observational scenario we do not distinguish netted processes which cannot be distinguished by an observer. E.

Definition 2.1 Let $(\|\mathbf{TiCCS}\|^n, Actt, \rightarrow)$ be a labeled transition system. A relation $\Re \subseteq\| \mathbf{TiCCS} \|^n \times \| \mathbf{TiCCS} \|^n$ is called a *net bisimulation* if it is symmetric and it satisfies the following condition: if $\| P \|_k^n \ \Re \ \| Q \|_k^n$ and $\|$

$P\|_k^n \xrightarrow{x} \|P'\|_{k'}^n$, $x \in Actt$, then there exists a process Q' such that $\|Q\|_k^n \xrightarrow{x} \|Q'\|_{k'}^n$ and $\|P'\|_{k'}^n \; \Re \; \|Q'\|_{k'}^n$.

Two netted processes $\| P \|_k^n, \| Q \|_k^n$ are *net bisimilar*, abbreviated $\|P\|_k^n \simeq \|Q\|_k^n$, if there exists a net bisimulation relating $\|P\|_k^n$ and $\|Q\|_k^n$.

It is easy to prove that the arbitrary union of net bisimulations is again a net bisimulation; $\simeq$ is the maximal net bisimulation on $\|\mathbf{TiCCS}\|^n$, namely $\simeq = \bigcup\{\Re, \Re \text{ is a net bisimulation }\}$.

We can extend the definition of net bisimulation to the set of netted terms, i.e. timed terms within a n-net. Let X be a free variable occurring in $\|T_1\|_n^n$ or $\|T_2\|_n^n$ or both. Then $\|T_1\|_n^n \simeq \|T_2\|_n^n$ iff for every TiCCS process P $\|T_1[P/X]\|_n^n \simeq \| T_2[P/X]\|_n^n$.

The next theorem states that the relation $\simeq$ is a congruence. It means that the corresponding semantics is suitable for bottom-up specification techniques.

Theorem 2.1 Let $\| T_1 \|_k^n, \| T_2 \|_k^n, \| T_3 \|_k^n$ be guarded netted terms and $\|T_1\|_k^n \simeq \|T_2\|_k^n$. Then

$$
\begin{aligned}
\|x.T_1\|_k^n &\simeq \|x.T_2\|_k^n \\
\|T_1 + T_3\|_k^n &\simeq \|T_2 + T_3\|_k^n \\
\|T_1 \mid T_3\|_k^n &\simeq \|T_2 \mid T_3\|_k^n \\
\|T_1[S]\|_k^n &\simeq \|T_2[S]\|_k^n \\
\|T_1 \setminus L\|_k^n &\simeq \|T_2 \setminus L\|_k^n \\
\|\mu X T_1\|_k^n &\simeq \|\mu X T_2\|_k^n \;.
\end{aligned}
$$

4 Net semantics for CCS

In this section we define semantics for CCS with the help of net bisimulation. Behaviour of any process depends on properies of a net, durations and kinds of actions. So each relation on the set of CCS processes will be parametrized by $p = (n, r, d)$ where n is a capacity of the net, r is a time refinement and d is a mapping dividing CCS actions into sets of eager and lazy ones. By $d(P)$ we denote the process with such division.

We define a p-net semantics for CCS processes. We consider two CCS processes equivalent if the corresponding timed processes, i.e. the processes with an additional information on durations of actions, are net bisimilar.

Definition 3.1 For CCS processes P_1, P_2 we define *p-net bisimulation* (denoted by $\overset{p}{\sim}$) in the following way:

$$P_1 \overset{p}{\sim} P_2 \text{ iff } \|r(d(P_1))\|_n^n \simeq \|r(d(P_2))\|_n^n$$

where $p = (n, r, d)$.

An interpretation of $P_1 \overset{p}{\sim} P_2$ is such that processes P_1, P_2 behave in the same manner under the condition p i.e. that properties of actions and of a net are p.

Example 3.1 CCS processes $P_1 = (a.\bar{b}.c.Nil \mid \bar{b}.Nil \mid b.Nil) \setminus L$ and $P_2 = (a.Nil \mid \bar{b}.Nil \mid b.Nil) \setminus L$ are p-net bisimilar if $p = (n, r, d)$, where $r(a) = a.t, r(b) = b.t, r(\bar{b}) = \bar{b}.t_f$, $d(a) \in E, d(b) \in L$ and n is arbitrary. Roughly speaking these processes behave in the same manner since if a (which has the nonzero duration) starts there is no reason why to wait with communication by means $b, \bar{b}$ and so there is no chance of communicating by means of $\bar{b}$ from $a.\bar{b}.c.Nil$. Note that the processes are not bisimilar.

Also p-net semantics support bottom-up specification for any p.

Theorem 3.1 $\overset{p}{\sim}$ is a congruence relation on the set **CCS** of CCS terms for every p.

We recall the definition of strong bisimulation for CCS processes (see [10]).

Definition 3.2 Let $(\mathbf{CCS}, Act, \rightarrow)$ be a labelled transition system. A relation $\Re \subseteq \mathbf{CCS} \times \mathbf{CCS}$ is called a *(strong) bisimulation* if it is symmetric and it satisfies the following condition: if $P \Re Q$ and $P \overset{u}{\rightarrow} P'$, then there exists a process Q' such that $Q \overset{u}{\rightarrow} Q'$ and $P' \Re Q'$.

Two processes P, Q are *(strongly) bisimilar*, abbreviated $P \sim Q$, if there exists a strong bisimulation relating P and Q.

It is well known that the arbitrary union of strong bisimulations is again a strong bisimulation; $\sim$ is the maximal strong bisimulation on **CCS**, namely $\sim = \bigcup\{\Re, \Re$ is a strong bisimulation $\}$.

For an intuition for the following theorem see Example 3.1.

Theorem 3.2 $\sim \not\subseteq \overset{p}{\sim}, \overset{p}{\sim} \not\subseteq \sim$ if $p = (n, r, d)$ where r is different from identity and $n > 1$.

It means that if we consider nontrivial properties of actions and of the net then behaviour of CCS process (under these properties) is different from the behaviour of abstract CCS process, i.e. of process with semantics (observer) abstracting or which does not know the properties.

Now we try to formalize this abstracting. ¿From an operational point of view a division of actions into two groups and the net capacity have influence only on performing τ and t actions since some communications are possible only in the case that there is a free link for them at a moment, otherwise communications can idle, i.e. t can be performed. When we abstract this in the sense that we do not know the division of the actions and momentary net capacity we must allow idling to τ in general.

Let us consider set of netted processes $||\mathbf{TiCCS}||^{?} = \{||P||^{?}_{?} \mid P \in \mathbf{TiCCS}\}$ with the operational semantics given by labelled transition system $(\rightarrow_?)$ similar to one from Definition 1.2 except that momentary net capacity is ignored as well kinds of actions. It means that whatever is possible for some momentary capacity and for some kind of actions is allowed in general.

Definition 3.3 Let $(||\mathbf{TiCCS}||^{?}, Actt, \rightarrow_?)$ be a labeled transition system. A relation $\Re \subseteq ||\mathbf{TiCCS}||^{?} \times ||\mathbf{TiCCS}||^{?}$ is called a (n, d)*-abstracting bisimulation* if it is symmetric and it satisfies the following condition: if $||P||^{?}_{?} \; \Re \; ||Q||^{n}_{k}$ and

$||P||^?_? \xrightarrow{x}_? ||P'||^?_?$, $x \in Actt$, then there exists a process Q' such that $||Q||^?_? \xrightarrow{x}_? ||Q'||^?_?$ and $||P'||^?_? \,\Re\, ||Q'||^?_?$.

Two netted processes $||P||^?_?, ||Q||^?_?$ are *(n, d)-abstracting bisimilar*, abbreviated $||P||^?_? \simeq_{(n,d)} ||Q||^?_?$, if there exists a (n, d)-abstracting bisimulation relating $||P||^?_?$ and $||Q||^?_?$.

It is easy to prove that the arbitrary union of (n, d)-abstracting bisimulations is again a (n, d)-abstracting bisimulation; $\simeq_{(n,d)}$ is the maximal (n, d)-abstracting bisimulation on $||\,\mathbf{TiCCS}\,||^?$, namely $\simeq_{(n,d)} = \bigcup\{\Re, \Re$ is a (n, d)-abstracting bisimulation $\}$.

Now we can define a semantics for CCS which abstracts n and d but does not abstract durations of actions (so it is parametrized by r).

Definition 3.4 For CCS processes P_1, P_2 we define *r-bisimulation* (denoted by $\overset{r}{\sim}$) in the following way

$$P_1 \overset{r}{\sim} P_2 \text{ iff } ||r(P_1)||^?_? \simeq_{n,d} ||r(P_2)||^?_?$$

where r is a time refinement.

An interpretation of $P_1 \overset{r}{\sim} P_2$ is that processes P_1, P_2 behave in the same manner under the condition that net properties and an action division are abstracted.

Theorem 3.3 $\overset{r}{\sim}$ is a congruence relation on the set **CCS** of CCS terms for every r.

For any durations of actions $\overset{r}{\sim}$ is finer or equal to bisimulation. Moreover if r is different from identity the semantics is not interleaving, i.e. it distinguishes between $a.Nil \mid b.Nil$ and $a.b.Nil + b.a.Nil$.

Theorem 3.4 $\overset{r}{\sim} \subseteq \sim$ for every r.

We formalize abstraction of time properties similarly as a weak bisimulation abstracts internal actions.

For a set S, let S^* to denote the set of finite sequences of elements of S. We denote the concatenation of sequences by juxtaposition. By ϵ we will denote the empty sequence.

If $s = x_1 x_2 \ldots x_n$, $x_i \in Act$; then we write

$$P \overset{s}{\Rightarrow}_? Q \text{ instead of } P(\xrightarrow{t}_?)^* \xrightarrow{x_1}_? (\xrightarrow{t}_?)^* \ldots (\xrightarrow{t}_?)^* \xrightarrow{x_n}_? (\xrightarrow{t})^* Q.$$

Definition 3.5 Let $(||\mathbf{TiCCS}||^?, Actt, \rightarrow_?)$ be a labeled transition system. A relation $\Re \subseteq ||\,\mathbf{TiCCS}\,||^? \times ||\,\mathbf{TiCCS}\,||^?$ is called a *p-abstracting bisimulation* if it is symmetric and it satisfies the following condition: if $||\,P\,||^?_? \,\Re\, ||\,Q\,||^n_k$ and $||\,P\,||^?_? \overset{s}{\Rightarrow}_? ||\,P'\,||^?_?$, $s \in Act \cup \{\epsilon\}$, then there exists a process Q' such that $||Q||^?_? \overset{s}{\Rightarrow}_? ||Q'||^?_?$ and $||P'||^?_? \,\Re\, ||Q'||^?_?$.

Two netted processes $||P||^?_?, ||Q||^?_?$ are *p-abstracting bisimilar*, abbreviated $||\,P\,||^?_? \simeq_p ||\,Q\,||^?_?$, if there exists a p abstracting bisimulation relating $||\,P\,||^?_?$ and $||Q||^?_?$.

Again it is easy to prove that the arbitrary union of p-abstracting bisimulations is again a p-abstracting bisimulation; $\simeq_p$ is the maximal p-abstracting bisimulation on $||\mathbf{TiCCS}||^?$, namely $\simeq_p = \bigcup\{\Re, \Re$ is a p-abstracting bisimulation $\}$.

Theorem 3.5 Let r_1, r_2 are time refinements and P be a CCS process. Then $||r_1(P)||^?_? \simeq_p ||r_2(P)||^?_?$.

The previous Theorem allows as define a semantics for CCS which abstracts also durations of actions.

Definition 3.6 For CCS processes P_1, P_2 we define *abstract-bisimulation* (denoted by $\overset{a}{\sim}$) in the following way

$$P_1 \overset{a}{\sim} P_2 \text{ iff } ||r(P_1)||^?_? \simeq_p ||r(P_2)||^?_?$$

where r is a time refinement.

Theorem 3.6 $\overset{a}{\sim} = \sim$.

It means that if we abstract all properties we get strong bisimulation semantics for CCS, what exactly corresponds to the CCS paradigm.

5 Conclusions

In this paper we have defined a real time extension of CCS which takes into account properties of an interconnection communication network. Every timed calculus has to deal with communications. If it allows idling to elementary actions then there is problem whether also an internal action may idle. The usual solution is that of restricting idling for τ (see [7, 13]). It is based on an idea that an environment has no influence on performing any internal actions and so there is no reason for idling. A possible background of this assumption is that if two processes can communicate than they will communicate immediately, i.e. there is always a free link for them. We can model this situation by $||\mathbf{TiCCS}||^\infty_k$ or considering all actions to be eager and the resulting calculus is equal to ACP [7] or time part of [9].

In our calculus we have assumed that some actions represent communications via a net of limited capacity. We developed operational semantics in Plotkin style and bisimulation semantics for the calculus.

Moreover we have defined several semantics for CCS. Starting from the assumption that CCS semantics abstract properties of a net and durations of actions we defined semantics which take into account also these properties. Then we have defined CCS semantics which are based on the idea that we abstract or do not know net properties and later also durations of actions. The last semantics (the most abstracting) coincides with strong bisimulation. Semantics which abstract only net properties are finer or equal to strong bisimulation. But semantics which take into account all properties are different from any one of CCS. The only exception is the case that all actions are lazy and net is of capacity 1 (single global bus). Here we come to limitations of using CCS like calculi. In the case that an interconnection network is known and it can be viewed as a network of a limited capacity it seems to be more convenient to use for time calculi semantics tailored for this network. On the other side, standard semantics can be used if we abstract or do not know interconnection network properties. This might be the case of a multiple bus shared with other communicating system which an observer does not observe, and so the number of free links is left to chance. Or, alternatively, the net might be a kind of a crossbar interconnection network in which a communication path can be created only in the case that it does not cross another one and only if all switches are in order. This means that a momentary existence or nonexistence of a communication path strongly depends on location of processes with respect to a given net topology. Also by abstracting this we get a random capacity network. So, as a result, we have that CCS communication environment is either the simple global bus or the parallel non determined network.

References

[1] J.A. Bergstra and J.W. Klop, Algebra of communicating processes with abstraction, *Theoretical Computer Science*, 37, pp. 77- 21, 1985.

[2] U. Black, *Computer Networks*, Prentice-Hall International, New York, 1987.

[3] J.F. Groote, Transition systems specification with negative premises, In *CONCUR'90*, LNCS 458, pp. 332-341, 1990.

[4] J.F. Groote, Specification and verification of real time systems in ACP, Technical report CS-R9015, CWI, Amsterdam, 1990.

[5] D.P. Gruska and A. Maggiolo-Schettini, A timed process description language based on CCS, Technical Report TR-9/91, Dipartimento di Informatica, Università di Pisa, 1991.

[6] D.P. Gruska and A. Maggiolo-Schettini, Net semantics for communicating processes, Technical Report 8/92, Institute of Informatics, Comenius University, 1992.

[7] M. Hennessy and T. Regan, A temporal process algebra, In *FORTE '90*, Third International Conference on Formal Description Techniques, Madrid, November 5-8, pp. 25-40, 1990.

[8] C.A.R. Hoare, *Communicating Sequential Processes*, Prentice-Hall International, New York,1985.

[9] H. Hansson and B. Jonsson, A calculus for communicating systems with time and probabilities, In *Proceedings of 11th IEEE Real - Time Systems Symposium*, Orlando, 1990.

[10] R. Milner, *Communication and Concurrency*, Prentice-Hall International, New York,1989.

[11] F. Moller and C. Tofts, A temporal calculus of communicating programs, In *CONCUR'90*, LNCS 458, pp. 401-415, 1990.

[12] C. Weitzman, *Distributed Mocro/Minicomputer Systems*, Prentice-Hall International, New York,1980.

[13] W. Yi, CCS+time = an interleaving model for real time systems, In *ICALP'91*, LNCS 510, pp. 217-228, 1991.

Session 2

A Process Calculus with Incomparable Priorities*

Hans Hansson and Fredrik Orava

Swedish Institute of Computer Science
Box 1263, S-164 28 Kista, SWEDEN, E-mail: {hansh,fredrik}@sics.se
and
Department of Computer Systems, Uppsala University

Abstract

We define an asynchronous process calculus in which priorities are associated to transitions. The calculus distinguishes between transitions originating from different processes by regarding their priorities to be incomparable. We define a prioritized bisimulation equivalence and discuss its relation to "the natural" congruence relation for our calculus. We illustrate the calculus with two small examples.

1 Introduction

Priorities are used in most computer systems to achieve efficient sharing of limited resources (e.g. processors). Most formal models lack an explicit notion of priorities, thus making it difficult to model for instance interrupts and priority based scheduling. In this paper we present a process calculus where priorities are associated to transitions.

The motivation behind our calculus is that the scheduling of events at different physical locations should be independent, e.g. priority information in one processing element should not influence executions in other processing elements. As an example consider the system in Figure 1, where A and B are processing elements. A choice between events α and β in A is resolved by the priority

Figure 1: Two independently executing processing elements.

*This is a substantially revised version of a paper that appeared under the same title and by the same authors in the NAPAW participant's proceedings.

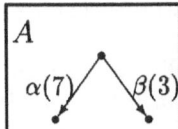

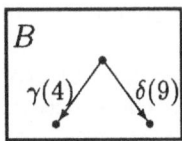

information provided i.e., in element A, event α will be scheduled for execution (since the priority value associated to α, 7, is higher than that associated to β, 3). Similarly, event δ will be scheduled for execution in B. Since A and B are independent, no global scheduling of the two events should be enforced, even though the priority associated to δ is higher than that associated to α i.e., priorities associated to events in different processing elements should be incomparable.

The priority mechanism in our calculus has the following characteristics:

1. It is *globally dynamic* [SS90], i.e. the relation allows an a-action to have priority over a b-action in some system states, and the converse to hold in some other system states.

2. It is *location dependent*, i.e. the priorities in two processes executing at different locations are incomparable. This is a natural requirement for distributed systems where processes are executing in different nodes (processing elements), the processes in each node being scheduled by a separate scheduler.

3. Priorities are *adjusted* by communications. We use a synchronous communication mechanism. Thus, communications introduce dependencies among components executing in parallel since a communication event is shared by all communicating components. It is therefore important that the components have a coherent view of communication events i.e., they should agree on the priority of communication events. We have chosen to achieve this by adjusting the priority of the communication event to be the sum of the (local) priorities of the communication events in the communicating components. The adjustment has the effect that priorities of communication events may increase when composing components in parallel. Other adjustments are of course possible, e.g., taking the arithmetic mean of the priorities.

4. Priorities have effect on the behaviour of a system by *excluding* events for which there are comparable identical events with higher priority. As an illustration, consider the process

$$P \;=\; a(7).0 + a(4).0 + \tau(5).0 + \tau(3).0 + b(1).0 \;\Big|\; \tau(7).0$$

   where a prefix $\alpha(n)$ expresses a possibility to perform an α event with priority n. The process P can perform any of its events except $a(4)$ and $\tau(3)$. The event $a(4)$ is excluded by the comparable event $a(7)$ and $\tau(3)$ is excluded by the comparable event $\tau(5)$. Note however that event $\tau(7)$ will not exclude event $\tau(5)$ since these events are performed by components executing in parallel and are thus incomparable.

In our calculus we use a CSP-like communication paradigm, in which processes synchronize (cooperate) by simultaneously performing transitions labeled with identical actions. Such synchronizations can in a loosely coupled distributed environment not be directly implemented. Some handshaking procedure (protocol) is needed. We will assume the existence of such a procedure involving the schedulers of both communicating nodes. As an effect of the handshaking the local priorities for the involved communication events will be adjusted. As an illustration, consider the system $T|U$:

$$T|U \;=\; a(2).0 + b(1).0 \;\big|\; b(2).0 + c(1).0$$

Locally (not assuming any particular environment), T is more willing to perform an a-event than a b-event (the priority associated to the a-action is higher than that associated to the b-action), and U is locally more willing to perform a b-event than a c-event. In $T|U$ however, the priority of a b-communication will be 3 (with respect to both T's and U's scheduler). Hence, in this context both T and U are more willing to perform a b-event than an a-event or a c-event respectively. In the communication the b-actions will be joined forming a single b-event with priority 3. The priority of this event will be comparable with the priorities of both the a and c-events. The priority of the a and c-events will however still be incomparable. In fact, $T|U$ behaves as

$$\Big(a(2).0 \;\big|\; c(1).0\Big) + b(3).0$$

We use hiding to convert communication events to internal (τ) events. This operation will not change the associated priority. As seen above, the priority associated with external (non-τ) events can increase via parallel compositions. This is not the case for internal (τ) events since they cannot participate in further communications. As a consequence, for comparable τ-events the priorities will give us *a-priori* information about events that can be excluded, e.g.

$$\tau(7).P + \tau(6).Q \quad \text{behaves exactly as} \quad \tau(7).P$$

since the $\tau(7)$-event will regardless of context exclude the $\tau(6)$-event. This also holds for comparable transitions labeled with identical external events, e.g.

$$a(7).P + a(6).Q \quad \text{behaves exactly as} \quad a(7).P$$

since, even though the priority of the a-events can be increased, the priorities of the $a(7)$ and $a(6)$ events will in all contexts increase with the same amount, hence the "$a(6)$" event will regardless of context have lower priority than the "$a(7)$" event.

Related work

Several process calculi with priorities have been defined, e.g. [BBK86, Cam91, CH88, CW91, GL90, Jef92, SS90, Tof90]. One can distinguish between calculi in which:

- priorities are associated with *choices*. In such a calculus, an expression of type $P +_l Q$ (where $+_l$ is a choice operator giving the left-hand-side process priority) means that if ambiguities occur between P and Q in synchronizing with an environment then P will always be chosen. This is the approach taken in [Cam91, CW91, SS90, Tof90].

- priorities are associated with *events*. The main approaches are:
  - To have a priority function from events to priorities. This yields a *static* priority structure with the same priority ordering in all system states. This is the approach taken in [BBK86, GL90, Jef92].
  - To have a priority function from events and system states to priorities. This yields a *dynamic* priority structure with different priority orderings in different system states. This is the approach taken by us and in [CH88].

In [BBK86] an arbitrary partial priority order between events can be defined, thus events can be *defined* to be incomparable. Partial orderings are also claimed to be definable in [CW91] and [SS90]. In our approach comparability/incomparability is induced by the structure of terms.

The approach closest to ours is Gerber and Lee's synchronous calculus CCSR [GL90]. In CCSR, priorities associated to actions are used to arbitrate between processes competing for the same resource (e.g. processing element). Resources are executing independently. Synchronization between processes executing on different resources introduces dependencies. Non-synchronizing events on different resources are independent (incomparable in our terminology).

Outline

In Section 2 we define the syntax and operational semantics for our priority calculus (PC). In Section 3 we define a priority bisimulation equivalence and relate it to the "standard" strong bisimulation congruence. In Section 4 we give two examples to illustrate the calculus. In Section 5 we discuss alternative formulations of priority bisimulation equivalence. Finally in Section 6 we make some concluding remarks.

2 The Calculus

In this section we define our calculus, the Priority Calculus (PC).

2.1 Syntax

Let Λ be a set of symbols denoting actions ($\tau \notin \Lambda$), ranged over by a, b, etc. Let Λ_τ denote $\Lambda \cup \{\tau\}$, ranged over by α, β, etc. Let $\mathcal{P}ri$ be a totally ordered set with a binary operation, $+$, of priority values ranged over by π_1, π_2, etc. Let $\mathcal{P}ri_\perp$ denote $\mathcal{P}ri \cup \{\perp\}$, where $\perp$ is the least element in $\mathcal{P}ri_\perp$. We also assume that $+$ is monotonic, commutative, associative, that there exists an identity with respect to +, and that for each element π there exists an inverse element, denoted $-\pi$. In this paper we assume $\mathcal{P}ri$ to be the integers. Let L range over subsets of Λ.

Let $\mathcal{P}roc$ be the set of processes, ranged over by P, Q, P_1 etc. The syntax for PC is given by:

$$P \quad ::= \quad \mathbf{0} \;\Big|\; \alpha(\pi).P \;\Big|\; P+P \;\Big|\; P\backslash L \;\Big|\; P \,|_L\, P$$

Intuitively, $\mathbf{0}$ denotes inaction (corresponding to NIL in CCS), "." denotes action prefixing, "+" non-deterministic choice, "$\backslash L$" hiding (renaming the actions in L to the internal action τ), and " $|_L$ " denotes the parallel composition of two processes synchronizing on the set of actions in L.

2.2 Semantics

We use *Structural Operational Semantics* (e.g. Plotkin [Plo81]) to define the semantics for PC. For technical reasons we define the semantics in two steps. First we define the semantics of an *unconstrained* transition relation, in which comparable α-transitions with different priorities may occur.. Then we define a *prioritized* transition relation, where among comparable α-transitions only the ones with highest priority can occur.

Definition 1 (Priority structure)
An occurrence *is an element in the language generated by* $\{v, h, l, r\}^*$, *ranged over by* σ, σ_1, *etc.*[i]. *We use* $\mathcal{O}cc$ *to denote the set of finite sets of occurrences, ranged over by* $\mathcal{O}$, $\mathcal{O}'$, *etc. A* priority structure *is a pair* $\langle \mathcal{O}, \pi \rangle$ *where*

- $\mathcal{O}$ *is a set of occurrences i.e.,* $\mathcal{O} \in \mathcal{O}cc$, *and*
- π *is a priority value i.e.,* $\pi \in \mathcal{P}ri$.

For $\mathcal{O} \in \mathcal{O}cc$ *and* $x \in \{v, h, l, r\}$ *we use* $\mathcal{O} \bullet x$ *to denote the concatenation of* x *to each element of* $\mathcal{O}$. *Let* σ_1 *and* σ_2 *be occurrences. The relation*

$$\mathbf{Comp} \subseteq \{v, h, l, r\}^* \times \{v, h, l, r\}^*$$

is the least symmetric and reflexive binary relation satisfying:

- $(\sigma_1 \bullet v, \sigma_2 \bullet h) \in \mathbf{Comp}$
- $(\sigma_1 \bullet a_1, \sigma_2 \bullet a_2) \in \mathbf{Comp}$ *if* $a_1 = a_2$ *and* $(\sigma_1, \sigma_2) \in \mathbf{Comp}$

[i]The notation v and h originates from the Swedish words for left (vänster) and right (höger), and l and r originates from the corresponding German words.

We write **Comp** (σ_1, σ_2) *to denote* $(\sigma_1, \sigma_2) \in$ **Comp**. *We extend* **Comp** (σ_1, σ_2) *to sets of occurrences as follows:*

Comp $(\mathcal{O}_1, \mathcal{O}_2)$ *if there exists* $\sigma_1 \in \mathcal{O}_1$ *and* $\sigma_2 \in \mathcal{O}_2$ *such that* **Comp** (σ_1, σ_2).

We say that $\mathcal{O}_1$ *and* $\mathcal{O}_2$ *are* comparable *if* **Comp** $(\mathcal{O}_1, \mathcal{O}_2)$ *otherwise* $\mathcal{O}_1$ *and* $\mathcal{O}_2$ *are* incomparable.

In the definition of the operational semantics for PC we use occurrences to record information about "locations" of transitions. For example, if $P(\in \mathcal{P}roc)$ can perform a transition t with occurrence σ then $P + Q$ can perform t with occurrence $\sigma \bullet v$ and $P\,|_L\, Q$ can perform t with occurrence $\sigma \bullet l$. Intuitively, two transitions are comparable if they occur on different sides of the same +-operator. If they occur on different sides of the same parallel composition operator they are executing in parallel and are incomparable.

Definition 2 (The unconstrained transition relation)
Let $\longmapsto \subseteq \mathcal{P}roc \times \mathcal{O}cc \times \mathcal{P}ri \times \Lambda_\tau \times \mathcal{P}roc$ *be the* unconstrained transition relation. *A tuple in* $\longmapsto$ *is written* $P \overset{\langle \mathcal{O}, \pi \rangle \alpha}{\longmapsto} Q$. *The unconstrained transition relation is defined as the least relation satisfying the laws in Table 1, where for a set of priority values* S

$$max\ S = \begin{cases} \pi\,.\,(\pi \in S \ \wedge\ \forall \pi' \in S\ .\ \pi \geq \pi') & \textit{if } S \neq \emptyset \\ \bot & \textit{if } S = \emptyset \end{cases}$$

$$max_\alpha(P, \mathcal{O}) = max\{\pi_i \mid P \overset{\langle \mathcal{O}_i, \pi_i \rangle \alpha}{\longmapsto} P_i \ \wedge\ \mathbf{Comp}\,(\mathcal{O}, \mathcal{O}_i)\}$$

$$max_\alpha(P) = max\{\pi_i \mid P \overset{\langle \mathcal{O}_i, \pi_i \rangle \alpha}{\longmapsto} P_i\}$$

Intuitively, $max_\alpha(P)$ denotes the maximal priority with which P can perform a α-transition, and $max_\alpha(P, \mathcal{O})$ denotes the maximal priority with which P can perform a α-transition comparable to $\mathcal{O}$.

The rules in Table 1, except **hide**$_\tau$, are standard with the exception that they record location and priority information. Note that, in the rule **par**$_{\mathbf{com}}$ the resulting transition is considered to occur *both* to the left and to the right of the parallel composition operator and that the priority is "the sum" of the priorities of the component transitions. The rule **hide**$_\tau$ in Table 1 states that the hide operation makes transitions internal (unobservable) and the priority information in the transitions applies. Thus, $P\backslash L$ can only perform a τ-transition if the associated priority is higher (or equal) than that of all comparable τ-transitions from P. To determine this we have to consider all "old" τ-transitions as well as all "new" τ-transitions (i.e., τ-transitions that are consequences of the hiding operation).

prefix :

$$\frac{-}{\alpha(\pi).P \xrightarrow{\langle\{\epsilon\},\pi\rangle\alpha} P}$$

choice$_\alpha$**l** :

$$\frac{P_1 \xrightarrow{\langle\mathcal{O},\pi\rangle\alpha} P_1'}{P_1 + P_2 \xrightarrow{\langle\mathcal{O}\bullet v,\pi\rangle\alpha} P_1'}$$

choice$_\alpha$**r** :

$$\frac{P_1 \xrightarrow{\langle\mathcal{O},\pi\rangle\alpha} P_1'}{P_2 + P_1 \xrightarrow{\langle\mathcal{O}\bullet h,\pi\rangle\alpha} P_1'}$$

par$_{\Lambda_\tau}$**l**:

$$\frac{P_1 \xrightarrow{\langle\mathcal{O},\pi\rangle\alpha} P_1'}{P_1 \,|_L\, P_2 \xrightarrow{\langle\mathcal{O}\bullet l,\pi\rangle\alpha} P_1' \,|_L\, P_2} \quad \alpha \notin L$$

par$_{\Lambda_\tau}$**r**:

$$\frac{P_1 \xrightarrow{\langle\mathcal{O},\pi\rangle\alpha} P_1'}{P_2 \,|_L\, P_1 \xrightarrow{\langle\mathcal{O}\bullet r,\pi\rangle\alpha} P_2 \,|_L\, P_1'} \quad \alpha \notin L$$

par$_{\mathbf{com}}$:

$$\frac{P_1 \xrightarrow{\langle\mathcal{O}_1,\pi_1\rangle a} P_1' \;,\; P_2 \xrightarrow{\langle\mathcal{O}_2,\pi_2\rangle a} P_2'}{P_1 \,|_L\, P_2 \xrightarrow{\langle\mathcal{O}_1\bullet l\,\cup\,\mathcal{O}_2\bullet r,\;\pi_1+\pi_2\rangle a} P_1' \,|_L\, P_2'} \quad \alpha \in L$$

hide$_\Lambda$:

$$\frac{P \xrightarrow{\langle\mathcal{O},\pi\rangle a} P'}{P\backslash L \xrightarrow{\langle\mathcal{O},\pi\rangle a} P'\backslash L} \quad a \notin L$$

hide$_\tau$:

$$\frac{P \xrightarrow{\langle\mathcal{O},\pi\rangle\gamma} P'}{P\backslash L \xrightarrow{\langle\mathcal{O},\pi\rangle\tau} P'\backslash L} \quad \begin{cases} \gamma \in L \cup \{\tau\} \\ \wedge \\ \pi = max\left\{\pi' \;\middle|\; \begin{array}{l} \pi' = max_\beta(P,\mathcal{O}) \,\wedge \\ \beta \in L \cup \{\tau\} \end{array}\right\} \end{cases}$$

Table 1: The unconstrained transition relation.

Definition 3 (Semantics for PC)
Let $\longrightarrow \subseteq \mathcal{P}roc \times \mathcal{O}cc \times \mathcal{P}ri \times \Lambda_\tau \times \mathcal{P}roc$ *be the* prioritized transition relation. *A tuple in* $\longrightarrow$ *is written* $P \xrightarrow{\langle \mathcal{O}, \pi \rangle \alpha} Q$. *The operational semantics for PC is defined as the least relation satisfying the law in Table 2.*

$$\textbf{prio}: \qquad \frac{P \stackrel{\langle \mathcal{O}, \pi \rangle \alpha}{\longmapsto} P'}{P \xrightarrow{\langle \mathcal{O}, \pi \rangle \alpha} P'} \; \pi = max_\alpha(P, \mathcal{O})$$

Table 2: The prioritized transition relation.

Intuitively, the rule **prio** states that among comparable α-transitions only transitions with the highest priority can occur.

Proposition 4
Let $P \in \mathcal{P}roc$ *such that* $P \xrightarrow{\langle \mathcal{O}, \pi \rangle \alpha} P'$ *and* $P \xrightarrow{\langle \mathcal{O}', \pi' \rangle \alpha} P''$ *then either*

$$\neg \mathbf{Comp}(\mathcal{O}, \mathcal{O}') \qquad or \qquad \pi = \pi'$$

Intuitively, Proposition 4 states that if there are more than one α-transition from P then either these transitions are incomparable i.e., they arise from components executing in parallel, or they have the same priority. This formalizes our intuition that transitions arising from components executing in parallel are independent i.e., the priority information associated with these transitions are incomparable.

3 Priority Equivalence

In this section we define an equivalence relation between processes in PC. We start with the usual (strong) bisimulation equivalence (e.g. [Mil89]) and demonstrate that it is not a congruence for PC. Then we define a strong congruence relation in the usual way (e.g. [Mil80]) by taking as congruence the relation included in the strong bisimulation equivalence which is preserved by all operators in PC. Since this is not a constructive definition we define an equivalence, called the priority equivalence, and show that it is a congruence with respect to action prefix, summation and parallel composition. We also show that our equivalence is contained in the strong congruence, which is the "natural" congruence for our calculus. The proofs are omitted here, but given in the full paper [HO].

First we define what we can observe of transitions. Intuitively, the priority information associated with transitions is not observable. An observer can only perceive the action associated with a transition.

Definition 5 (Observable Transitions)
The observable transition relation $\hookrightarrow \subseteq \mathcal{P}roc \times \Lambda_\tau \times \mathcal{P}roc$ *is defined by*

$$P \stackrel{\alpha}{\hookrightarrow} P' \quad \textit{if} \quad P \xrightarrow{\langle \mathcal{O}, \pi \rangle \alpha} P' \ \textit{for some} \ \mathcal{O} \ \textit{and} \ \pi,$$

where $P \stackrel{\alpha}{\hookrightarrow} P'$ *denotes* $(P, \alpha, P') \in \hookrightarrow$.

Definition 6 (Bisimulation)
The relation $\mathcal{R} \subseteq \mathcal{P}roc \times \mathcal{P}roc$ *is a* bisimulation *if* $(P, R) \in \mathcal{R}$ *implies:*

$$\textit{if } P \stackrel{\alpha}{\hookrightarrow} P' \textit{ then } \exists Q' \,.\, Q \stackrel{\alpha}{\hookrightarrow} Q' \ \wedge \ (P', Q') \in \mathcal{R}$$

and

$$\textit{if } Q \stackrel{\alpha}{\hookrightarrow} Q' \textit{ then } \exists P' \,.\, P \stackrel{\alpha}{\hookrightarrow} P' \ \wedge \ (Q', P') \in \mathcal{R}$$

Two processes P *and* Q *are* strongly equivalent, *written* $P \sim Q$, *if there exists a bisimulation* $\mathcal{R}$ *such that* $(P, R) \in \mathcal{R}$.

Proposition 7
$\sim$ *is an equivalence relation.*

Observation 8
Strong equivalence, $\sim$, *is not a congruence. For example, suppose* $P \not\sim \mathbf{0}$ *then*

$$a(2).\mathbf{0} + b(1).P \sim a(2).\mathbf{0} + b(3).P$$

but in the context of the hiding operator, $\backslash\{a, b\}$, *the processes are not strongly equivalent*

$$(a(2).\mathbf{0} + b(1).P)\backslash\{a, b\} \not\sim (a(2).\mathbf{0} + b(3).P)\backslash\{a, b\}$$

because the first process has only one observable transition

$$(a(2).\mathbf{0} + b(1).P)\backslash\{a, b\} \stackrel{\tau}{\hookrightarrow} \mathbf{0}$$

which cannot be matched by the only transition of the second process

$$(a(2).\mathbf{0} + b(3).P)\backslash\{a, b\} \stackrel{\tau}{\hookrightarrow} P.$$

Moreover, it is not sufficient to require strong equivalence in all hiding-contexts to get a congruence. For example:

$$a(2).\mathbf{0} + b(3).P \sim a(2).\mathbf{0} + b(5).P$$

are strongly equivalent in all hiding-contexts, but

$$(a(2).0 \mid_a (a(2).0 + b(3).P))\backslash\{a,b\} \not\sim (a(2).0 \mid_a (a(2).0 + b(5).P))\backslash\{a,b\}$$

because the first process has only one observable transition:

$$(a(2).0 \mid_a (a(2).0 + b(3).P))\backslash\{a,b\} \overset{\tau}{\hookrightarrow} 0 \mid_a 0$$

which cannot be matched by the only transition of the second process:

$$(a(2).0 \mid_a (a(2).0 + b(5).P))\backslash\{a,b\} \overset{\tau}{\hookrightarrow} a(2).0 \mid_a P.$$

Definition 9 (Context)
A context $\mathcal{C}(\cdot)$ is a process with a "hole" where a process P can be plugged in. The context $\mathcal{C}(\cdot)$ is a distinguishing context *for P and Q if $P \sim Q$ but $\mathcal{C}(P) \not\sim \mathcal{C}(Q)$.*

Definition 10 (Strong Congruence)
P and Q are strongly congruent, *written $\sim^c$, if $\mathcal{C}(P) \sim \mathcal{C}(Q)$ for all contexts $\mathcal{C}(\cdot)$.*

From Observation 8 we see that strong equivalence is not a congruence for PC due to the priority information associated to transitions. Furthermore, it is not sufficient to consider priorities of individual transitions only, we must also consider priorities of comparable alternative transitions. Below, we define a priority sensitive equivalence taking the above into account. In the definition of the equivalence we will use the following:

Definition 11 (Minimal distance)
The minimal priority distance from a transition $P \xrightarrow{\langle \mathcal{O}, \pi \rangle \alpha} P'$ to a comparable β-transition is given by $\mathcal{D}_{min} : (\longrightarrow \times \Lambda_\tau) \rightarrow (\mathcal{P}ri \cup \{\infty\})$, where

$$\mathcal{D}_{min}(P \xrightarrow{\langle \mathcal{O}, \pi \rangle \alpha} P', \beta) = \begin{array}{l} \textit{if } P \xrightarrow{\langle \mathcal{O}', \pi' \rangle \beta} P'' \wedge \mathbf{Comp}(\mathcal{O}, \mathcal{O}') \\ \textit{then } \pi - max_\beta(P, \mathcal{O}) \\ \textit{else } \infty \end{array}$$

where ∞ is the top element in $\mathcal{P}ri \cup \{\infty\}$, i.e., $\forall \pi \in \mathcal{P}ri \,.\, \pi < \infty$.

$\mathcal{D}_{min}(P \xrightarrow{\langle \mathcal{O}, \pi \rangle \alpha} P', \beta)$ is a measure of how much the priorities of β-transitions has to be increased to make the priority of the comparable β-transition with highest priority equal to π. Intuitively, for a transition $tr = P \xrightarrow{\langle \mathcal{O}, \pi \rangle \alpha} P'$ $\mathcal{D}_{min}(tr, \beta)$ is a measure of the robustness of tr with respect to β-transitions, i.e., it gives information about the contexts in which tr will be excluded by β-transitions. For instance, if $\mathcal{D}_{min}(tr, \beta) = 1$ then tr will be excluded by a β-transition in the context $\left(\cdot \mid_\beta \beta(k).Q \right)\backslash\{\alpha, \beta\}$ if $k > 1$.

Definition 12 (Priority Bisimulation)
The relation $\mathcal{R} \subseteq \mathcal{P}roc \times \mathcal{P}roc$ is a priority bisimulation *if $(P, Q) \in \mathcal{R}$ implies:*

$\forall \alpha \in \Lambda_\tau$

$$\textit{if } P \xrightarrow{\langle \mathcal{O}, \pi \rangle \alpha} P'$$

$$\textit{then} \begin{cases} \exists Q' \,.\, Q \xrightarrow{\langle \mathcal{O}', \pi' \rangle \alpha} Q' \;\wedge\; (P', Q') \in \mathcal{R} \;\wedge \\ \quad \pi \leq \pi' \;\wedge \\ \quad \forall\, \beta \in \Lambda_\tau . \mathcal{D}_{min}(P \xrightarrow{\langle \mathcal{O}, \pi \rangle \alpha} P', \beta) \leq \mathcal{D}_{min}(Q \xrightarrow{\langle \mathcal{O}', \pi' \rangle \alpha} Q', \beta) \end{cases}$$

and vice versa.

Two processes P and Q are priority equivalent *(written $P \sim_\pi Q$) if there exists a priority bisimulation $\mathcal{R}$ such that $(P, Q) \in \mathcal{R}$.*

We illustrate $\sim_\pi$ by a few examples. In the examples we will describe processes as transition diagrams in which transitions connected by a • denote comparable transitions and transitions touching the same oval denote incomparable transitions. Transitions originating from the center of an oval denote transitions comparable with all transitions touching the same oval. Also, we will sometimes annotate transitions with $\mathcal{D}_{min}$, e.g. if transition tr is annotated with $[2]_b$ then $\mathcal{D}_{min}(tr, b) = 2$.

Figure 2 (left) illustrates that for incomparable transitions labeled with the same action (a) leading to equivalent states, only the transition with highest priority is significant. Figure 2 (right) illustrates that putting the processes to the left in a (choice) context with a comparable b-transition will not invalidate the equivalence.

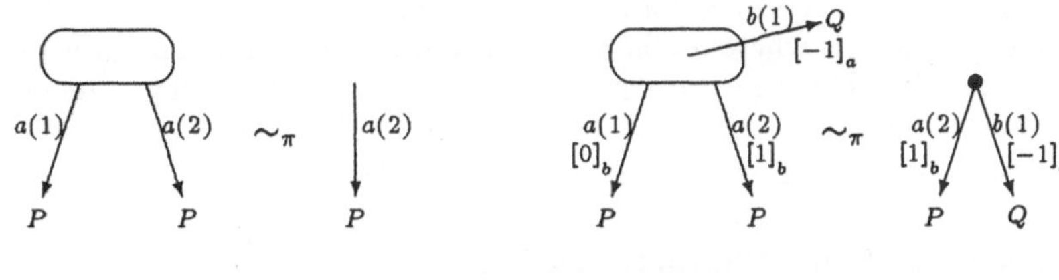

Figure 2: Equivalent processes.

Figure 3 illustrates that the distances of comparable transitions must match, and that the highest priority values must match. The $a(1)$ transition to the left is simulated by the $a(3)$ transition to the right, since $3 \geq 1$ and the distance to the comparable b-transitions are the same (-1). Figure 4 show that if we replace the $a(1)$-transition to the left in Figure 3 by an $a(2)$-transition the two processes will no longer be equivalent, since the distance to comparable b-transitions to the left will be greater than the corresponding distance to the right. Hence, the $a(2)$-transition cannot be simulated by any transition to the right. Figure 4 also shows the existence of a distinguishing context.

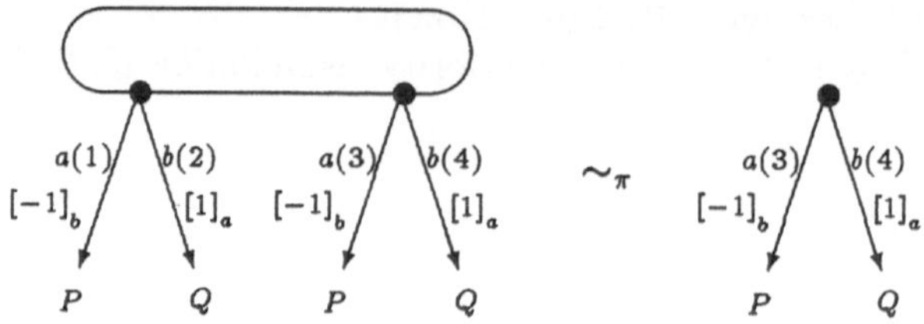

Figure 3: Equivalent processes.

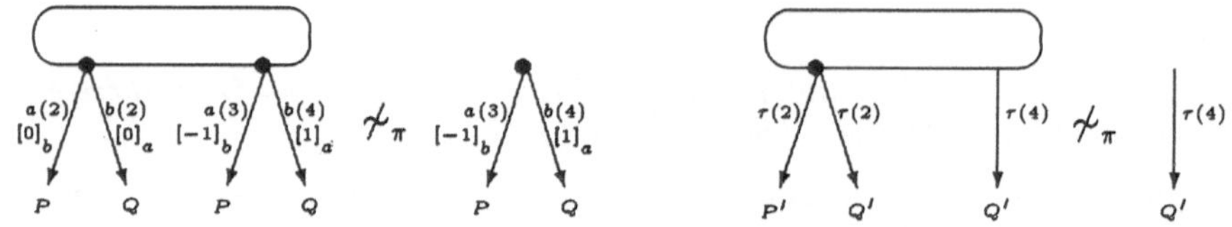

Figure 4: Inequivalent processes. The processes to the right are obtained by putting the processes to the left in the context $(\cdot)\backslash\{a, b\}$.

Proposition 13
$\sim_\pi$ is an equivalence relation.

Proposition 14
$\sim_\pi$ is a congruence with respect to prefixing, choice, and parallel composition, i.e., for any PC processes P, Q, R, and set of actions L: if $P \sim_\pi Q$ then $\alpha.P \sim_\pi \alpha.Q$, $P + R \sim_\pi Q + R$, and $P \,|_L\, R \sim_\pi Q \,|_L\, R$.

That is, $\sim_\pi$ is a congruence for all operators in PC except hiding. In Section 5 we give an example where $\sim_\pi$ is not preserved by the hiding operator. We also discuss possible remedies, aiming at defining an equivalence that coincides with the natural congruence. For $\sim_\pi$ we can only establish the following theorem, since $\sim_\pi$ is not a congruence with respect to hiding.

Theorem 15 (Semi-Characterization)
if $P \sim^c Q$ then $P \sim_\pi Q$.

4 Examples

In this section we give two examples to illustrate PC. But first we extend PC with *process constants* and the auxiliary operator $\oplus$.

Process constants are introduced in the syntax of PC by adding the clause $P ::= A$, where A is a process constant. For any constant A there is a *defining equation* written $A \stackrel{\text{def}}{=} P$, where $P \in \mathcal{P}roc$. In the semantics we add the rule **Con** given in Table 3. As usual (e.g. [Mil89]) process constants provide recursion since the defining equation for A may contain A itself.

The auxiliary operator $\oplus$ is denoting a choice between two transitions with incomparable priorities. We extend the syntax for PC with the clause $P ::= P \oplus P$. The semantics for $\oplus$ is given by the rule **Incomp$_+$** in Table 3.

Con:
$$\frac{Q \xrightarrow{\langle \mathcal{O}, \pi \rangle \alpha} Q'}{A \xrightarrow{\langle \mathcal{O}, \pi \rangle \alpha} Q'} \ (A \stackrel{\text{def}}{=} Q)$$

Incomp$_+$:
$$\frac{Q \xrightarrow{\langle \mathcal{O}, \pi \rangle \alpha} Q'}{P \oplus Q \xrightarrow{\langle \mathcal{O} \bullet r, \pi \rangle \alpha} Q' \ , \ Q \oplus P \xrightarrow{\langle \mathcal{O} \bullet l, \pi \rangle \alpha} Q'}$$

Table 3: Operational semantics for process constants and $\oplus$.

As a first example, consider a simple computer system consisting of a CPU and an external process (EP). The structure of the system is shown in Figure 5. When an interrupt is initiated by the external i-action, EP generates an

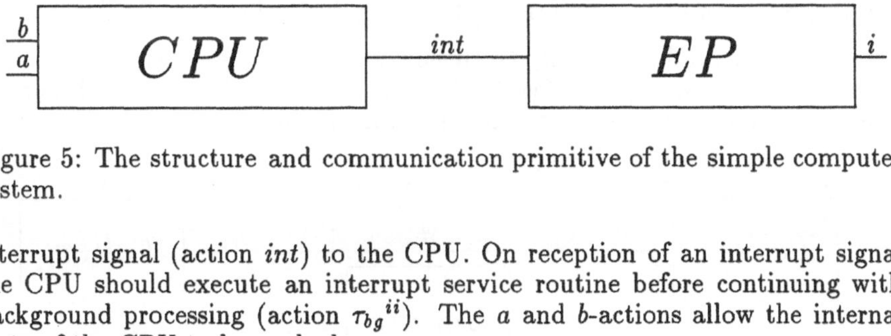

Figure 5: The structure and communication primitive of the simple computer system.

interrupt signal (action *int*) to the CPU. On reception of an interrupt signal the CPU should execute an interrupt service routine before continuing with background processing (action τ_{bg}[ii]). The a and b-actions allow the internal state of the CPU to be probed.

Specification of the components:

$$\begin{aligned} EP &\stackrel{\text{def}}{=} \tau_{idle}(0).EP + i(0).EP_1 \\ EP_1 &\stackrel{\text{def}}{=} int(1).EP \end{aligned}$$

$$\begin{aligned} CPU &\stackrel{\text{def}}{=} \tau_{bg}(1).CPU_1 + int(2).IS \\ CPU_1 &\stackrel{\text{def}}{=} a(0).CPU \\ IS &\stackrel{\text{def}}{=} b(0).CPU \end{aligned}$$

[ii] Here and in the following we will use annotated τ:s for clarification (e.g. τ_{bg} denotes that the τ is used to model background processing). Note that all τ's are equal, e.g. $\tau_a.0 \sim_\pi \tau_b.0$.

The complete system is then given by the following expression:

$$Sys \stackrel{\text{def}}{=} (EP \mid_{\{int\}} CPU) \backslash \{int\}$$

It is straightforward to establish that

$$\begin{array}{lcll}
Sys & \sim_\pi & \tau_{bg}(1).Sys'' \oplus \left(\tau_{idle}(0).Sys + i(0).Sys'\right) & \\
Sys' & \stackrel{\text{def}}{=} & \tau_{bg}(1).Sys^4 + \tau_{int}(3).Sys^3 & \sim_\pi \ \tau_{int}(3).Sys^3 \\
Sys'' & \stackrel{\text{def}}{=} & a(0).Sys \oplus \left(\tau_{idle}(0).Sys'' + i(1).Sys^4\right) & \\
Sys^3 & \stackrel{\text{def}}{=} & b(0).Sys \oplus \left(\tau_{idle}(0).Sys^3 + i(1).Sys^5\right) & \\
Sys^4 & \stackrel{\text{def}}{=} & a(0).Sys' & \\
Sys^5 & \stackrel{\text{def}}{=} & b(0).Sys' &
\end{array}$$

Note that in the term Sys', the subterm $\tau_{bg}(1).Sys^4$ is excluded by the subterm $\tau_{int}(3).Sys^3$. Also, the high priority subterms in Sys will not exclude the low priority one, since the priorities in the subterms are incomparable (a consequence of the $\oplus$-operator). Moreover, after an interrupt has occured (action i) the CPU cannot perform the τ_{bg}-action signifying that the CPU cannot perform any background processing while handling interrupts. This appears in Sys' as exclusion of the low priority term ($\tau_{bg}(1).Sys^4$).

As a second example we specify the behaviour of a simple system of automatic tellers. The system consists of two tellers and one server (depicted in Figure 6). The tellers normally work in a stand-alone mode, but when the local transac-

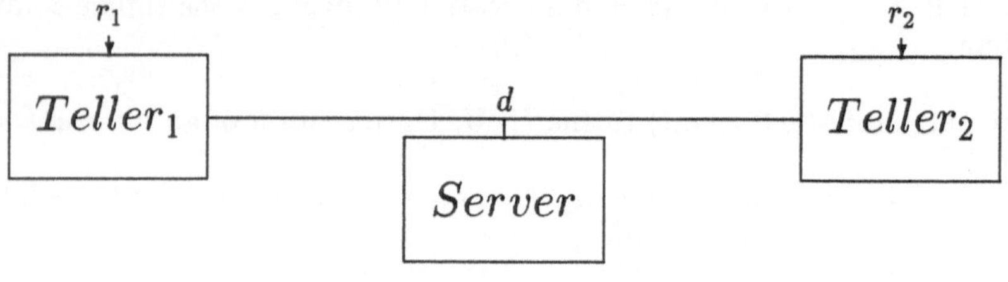

Figure 6: A system with two automatic tellers and one server.

tion buffers become full they must dump their buffers (action d) to the server before accepting further requests from customers (actions r_1 and r_2). To ensure consistency all tellers must participate in the dumping operation, i.e. the tellers cannot accept further request while dumping is in progress. The following is a PC specification of the components:

$$
\begin{aligned}
Teller_i &\stackrel{\text{def}}{=} r_i(4).T_i' + d(0).Teller_i \\
T_i' &\stackrel{\text{def}}{=} r_i(3).T_i'' + d(1).Teller_i \\
T_i'' &\stackrel{\text{def}}{=} d(2).Teller_i \\
\\
Server &\stackrel{\text{def}}{=} d(1).Server
\end{aligned}
$$

The complete teller-system is then given by the following expression:

$$TS^o \stackrel{\text{def}}{=} (Teller_1 \,|_{\{d\}}\, Server \,|_{\{d\}}\, Teller_2)\backslash\{d\}$$

It is possible to prove that TS^o is priority equivalent with the process S^o_{00}, given in Table 4.

If we place TS^o in an environment with two customers persistently attempting to request service from the tellers we get the following specification:

$$TS^c \stackrel{\text{def}}{=} \Big(U1 \,|_{\{r_1\}}\, TS^o \,|_{\{r_2\}}\, U2\Big)\backslash\{r_1, r_2\}$$

where

$$
\begin{aligned}
U1 &\stackrel{\text{def}}{=} r_1(0).U1 \\
U2 &\stackrel{\text{def}}{=} r_2(0).U2
\end{aligned}
$$

We can prove TS^c to be priority equivalent the process S^c_{00}, given in Table 4.

$S^o_{00} \stackrel{\text{def}}{=} \Big(r_1(4).S^o_{10} \oplus r_2(4).S^o_{01}\Big) + \tau_d(1).S^o_{00}$	$S^c_{00} \stackrel{\text{def}}{=} \tau_{r_1}(4).S^c_{10} \oplus \tau_{r_2}(4).S^c_{01}$
$S^o_{10} \stackrel{\text{def}}{=} \Big(r_1(3).S^o_{20} \oplus r_2(4).S^o_{11}\Big) + \tau_d(2).S^o_{00}$	$S^c_{10} \stackrel{\text{def}}{=} \tau_{r_1}(3).S^c_{20} \oplus \tau_{r_2}(4).S^c_{11}$
$S^o_{01} \stackrel{\text{def}}{=} \Big(r_1(4).S^o_{11} \oplus r_2(3).S^o_{02}\Big) + \tau_d(2).S^o_{00}$	$S^c_{01} \stackrel{\text{def}}{=} \tau_{r_1}(4).S^c_{11} \oplus \tau_{r_2}(3).S^c_{02}$
$S^o_{20} \stackrel{\text{def}}{=} r_2(4).S^o_{21} + \tau_d(3).S^o_{00}$	$S^c_{20} \stackrel{\text{def}}{=} \tau_{r_2}(4).D$
$S^o_{11} \stackrel{\text{def}}{=} \Big(r_1(3).S^o_{21} \oplus r_2(3).S^o_{12}\Big) + \tau_d(3).S^o_{00}$	$S^c_{11} \stackrel{\text{def}}{=} \Big(\tau_{r_1}(3).D \oplus \tau_{r_2}(3).D\Big) + \tau_d(3).S^c_{00}$
$S^o_{02} \stackrel{\text{def}}{=} r_1(4).S^o_{12} + \tau_d(3).S^o_{00}$	$S^c_{02} \stackrel{\text{def}}{=} \tau_{r_1}(4).D$
$S^o_{21} \stackrel{\text{def}}{=} r_2(3).S^o_{22} + \tau_d(4).S^o_{00}$	$D \stackrel{\text{def}}{=} \tau_d(4).S^c_{00}$
$S^o_{12} \stackrel{\text{def}}{=} r_1(3).S^o_{22} + \tau_d(4).S^o_{00}$	
$S^o_{22} \stackrel{\text{def}}{=} \tau_d(5).S^o_{00}$	

Table 4: Processes showing the behaviour of the open (left) and closed (right) specifications of the system of tellers.

In Table 4, S^o_{ij} denotes a process in the open system for which $Teller_1$'s transition buffer contains i transactions and $Teller_2$'s transaction buffer contains j

transactions, likewise in the closed system for S^c_{ij}. Note in Table 4, how comparable τ-transitions are excluded in the closed system, e.g. the τ_d-transition in S^o_{00} is not present in S^c_{00}. Note also that incomparable τ-transitions cannot be excluded even if they have different priorities (e.g. in Process S^c_{10}), and that S^o_{21} and S^o_{12} collapses to D in the closed system. Also, a state where both tellers have full transaction buffers (corresponding to Process S^o_{22}) is not reachable in the closed system.

5 Discussion

As mentioned in Section 3, $\sim_\pi$ is not a congruence with respect to the hiding operator. In this section we discuss alternative formulations of priority bisimulation equivalence aiming at finding an equivalence coinciding with the natural congruence. But first we present a counterexample showing that $\sim_\pi$ is not a congruence with respect to hiding. In the transition graphs, transitions originating from different ovals denote incomparable transitions. The counterexample in Figure 7 shows that if the two priority bisimilar processes S and T are put in the context $(\,\cdot\,)\backslash\{b,c\}$ the resulting processes will not be priority bisimilar. The $a(2)$-transition tr in $(S)\backslash\{b,c\}$ cannot be matched by any transition in $(T)\backslash\{b,c\}$, since $\mathcal{D}_{min}(tr,\tau) = \infty > 0 = \mathcal{D}_{min}(tr',\tau)$ for the only candidate a-transition tr' in $(T)\backslash\{b,c\}$.

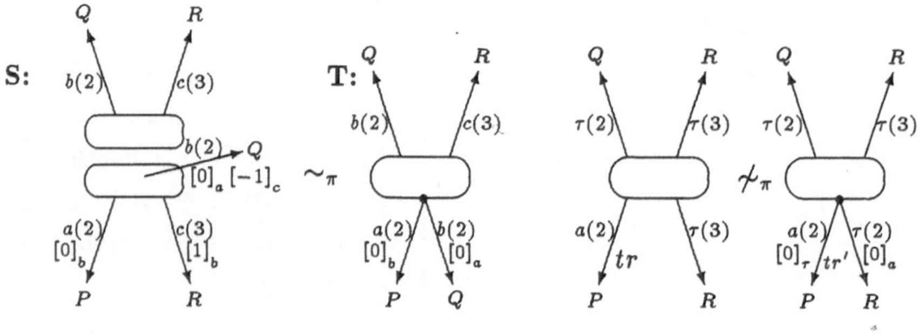

Figure 7: If the priority bisimilar processes to the left are put in the context $(\,\cdot\,)\backslash\{b,c\}$ the resulting processes (shown to the right) will not be priority bisimilar.

The following are some possible candidates for making $\sim_\pi$ a congruence for hiding as well:

1. Define $\mathcal{D}_{min}$ such that β-transitions with finite $\mathcal{D}_{min}$ in Definition 12 must be matched by finite $\mathcal{D}_{min}$, i.e., we redefine $\leq$ such that

   $$\forall \pi \in \mathcal{P}ri\,.\,\pi \nleq \infty \quad \text{and}\; \infty \leq \infty$$

   Let $\sim^1_\pi$ denote the induced priority bisimulation equivalence. Unfortunately, this modification is not sufficient to make priority bisimulation a congruence with respect to hiding. The counterexample in Figure 8 shows that if the two priority bisimilar processes S and T are put in the context $(\ \cdot\ )\backslash\{b,c\}$ the resulting processes will not be priority bisimilar. The $a(2)$-transition tr in $(S)\backslash\{b,c\}$ cannot be matched by any transition in $(T)\backslash\{b,c\}$, since $\mathcal{D}_{min}(tr,\tau) \not\leq \mathcal{D}_{min}(tr',\tau)$ for the only candidate a-transition (tr') in $(T)\backslash\{b,c\}$.

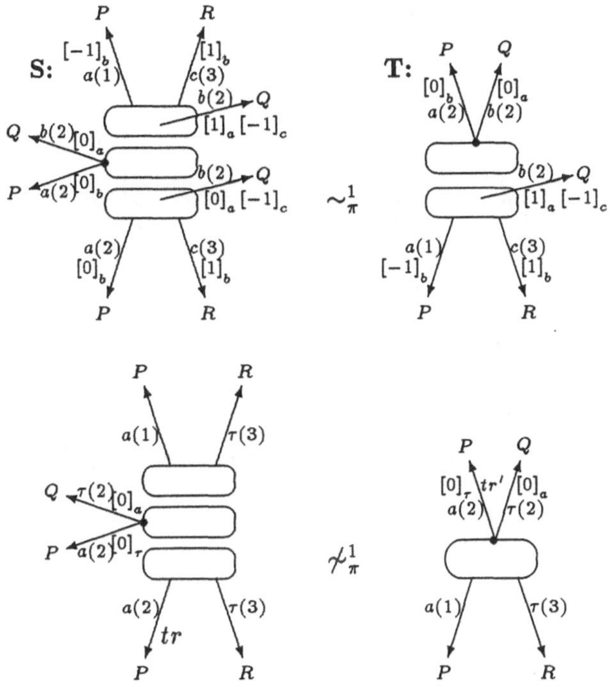

Figure 8: If the priority bisimilar processes S and T (top) are put in the context $(\ \cdot\ )\backslash\{b,c\}$ the resulting processes (bottom) will not be priority bisimilar.

2. In our second attempt we extend the $\mathcal{D}_{min}$ comparison scheme from 1 by recording information about indirect $\mathcal{D}_{min}$ dependencies of transitions. In the priority bisimulation equivalence $(\sim_\pi^2)$ we require the indirect dependencies to match. As an illustration, consider the processes S and T in Figure 9. When we investigate if the $c(3)$-transition in T is matched by one of the $c(4)$-transition in S we must check not only the $\mathcal{D}_{min}$ of comparable transitions, but also the $\mathcal{D}_{min}$ of transitions comparable to the comparable transitions. For the $c(3)$-transition in T we obtain the chain of $\mathcal{D}_{min}$: $[0]_b - [1]_a$, which is matched for instance by the chain $[1]_b - [2]_a$ associated to one of the $c(4)$-transitions in S.

   Unfortunately, the suggested modification is not sufficient to make priority bisimulation a congruence with respect to hiding. The counterexample in Figure 9 shows that if the two priority bisimilar processes S and T are put in the context $(\ \cdot\ )\backslash\{b,c\}$ the resulting processes will not be priority bisimilar, since there is no transition in $(T)\backslash\{b,c\}$ that can match the transition tr in $(S)\backslash\{b,c\}$.

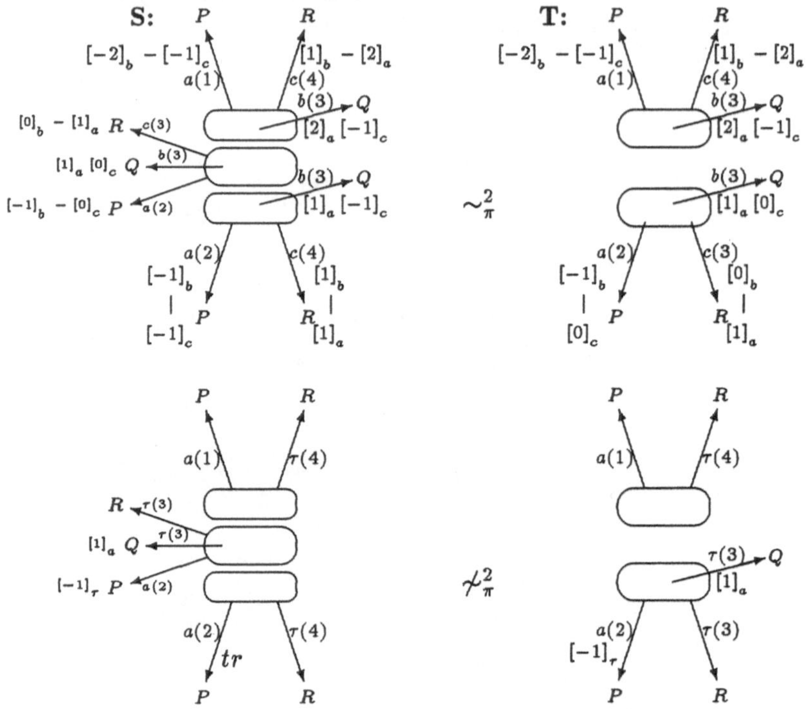

Figure 9: If the priority bisimilar processes S and T (top) are put in the context $(\cdot)\backslash\{b,c\}$ the resulting processes (bottom) will not be priority bisimilar.

3. In our next attempt we modify the semantics of the hiding operator by introducing new dependencies. Consider process S in Figure 7. The $a(2)$-transition has a comparable $b(2)$-transition. However, in a $(\cdot)\backslash\{b,c\}$-context the $\tau(2)$-transition emerging from the $b(2)$-transition is excluded by the $\tau(3)$-transition emerging from the $c(3)$-transition. The problem is that the $a(2)$-transition in S can be matched by an a-transition with a comparable b-transition that cannot be excluded by a c-transition in any context, e.g. consider the $a(2)$-transition in T. We remedy this by making the excluding transition (i.e. the $\tau(3)$-transition emerging from the $c(3)$-transition) comparable to the transitions which are comparable to the excluded transition (i.e. comparable to the $a(2)$-transition). Technically we make the following modification of the rule **hide**$_\tau$ in the operational semantics in Table 1:

$$\textbf{hide}_\tau : \quad \frac{P \xrightarrow{\langle \mathcal{O}, \pi \rangle \gamma} P'}{P \backslash L \xrightarrow{\langle \mathcal{O}', \pi \rangle \tau} P' \backslash L} \quad \begin{cases} \gamma \in L \cup \{\tau\} \\ \wedge \\ \pi = max \left\{ \pi' \mid \begin{array}{l} \pi' = max_\beta(P, \mathcal{O}) \wedge \\ \beta \in L \cup \{\tau\} \end{array} \right\} \\ \wedge \\ \mathcal{O}' = \mathcal{O} \bigcup_i \left\{ \mathcal{O}_i \mid \begin{array}{l} P \xrightarrow{\langle \mathcal{O}_i, \pi_i \rangle \beta} P_i \wedge \\ \beta \in L \cup \{\tau\} \wedge \\ \mathbf{Comp}(\mathcal{O}, \mathcal{O}_i) \wedge \\ \pi_i < \pi \end{array} \right\} \end{cases}$$

We use the definition of the bisimulation equivalence from Attempt 2. With this redefinition of semantics and priority bisimulation equivalence we claim the following:

Conjecture 16 (Charactarization) $\sim_\pi^{2'}$ *is a congruence for all PC operators, and*

$$P \sim^c Q \quad \textit{if and only if} \quad P \sim_\pi^{2'} Q.$$

4. In this attempt we modify the operational semantics by replacing **Comp** with the transitive closure of the **Comp**-relation. The priority bisimulation equivalence will be as defined in Section 3. Unfortunately, the processes in Figure 10 illustrate that $\sim_\pi$ is not a congruence for $|_L$. The $a(2)$ transition (tr) in U' cannot be matched by the $a(2)$ transition (tr') in V', since $\mathcal{D}_{min}(tr, b) > \mathcal{D}_{min}(tr', b)$.

5. Our final attempt will be a combination of attempts 1 and 4, i.e., we will use $\mathcal{D}_{min}$ as defined in Attempt 1 and a transitive **Comp**-relation as in Attempt 4. With this redefinition of semantics and priority bisimulation equivalence we claim the following:

Conjecture 17 (Charactarization) $\sim_\pi^{1'}$ *is a congruence with respect to all PC operators, and*

$$P \sim^c Q \quad \textit{if and only if} \quad P \sim_\pi^{1'Q} .$$

Note that the new equivalences (Case 3 and Case 5 above), which we claim both to characterize the natural congruence, are different in that they are defined for different operational semantics. In Case 5 a transitive **Comp** relation is used in the operational semantics, whereas the **Comp** relation in Case 3 is not transitive. Hovever, in Case 3 the transitivity (indirect dependencies) are captured by a transitive hiding operator and a transitive comparison of $\mathcal{D}_{min}$ values in the definition of the equivalence. Hence it seems, perhaps not surprisingly, that indirect dependencies must be taken into account (either in the semantics or in the equivalence) to obtain an equivalence coinciding with the natural congruence.

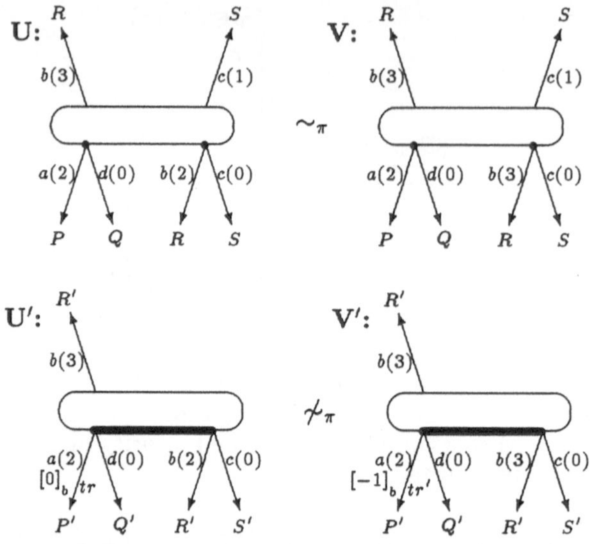

Figure 10: If the priority bisimilar processes U and V (top) are put in the context $(\cdot \mid_{\{c,d\}} c(0).\mathbf{0} + d(0).\mathbf{0})$ the resulting processes U' and V' (bottom) will not be priority bisimilar.

6 Conclusions

We have defined a process calculus, PC, where priorities are associated to transitions, such that the priorities of transitions originating from different processing elements are incomparable. We have defined a priority equivalence, which is a congruence with respect to all operators of PC except hiding. We have also showed that our equivalence is included in the natural strong congruence relation for PC. Further, we have suggested two possible candidates for a constructive definition of the natural congruence for PC.

The priority equivalence for PC makes a distinction between parallel and sequential behaviors. This relates our calculus to work on true concurrency, e.g. [Ace91, BCHK91, DDM88, BC88, FGM91].

The parallel composition operator ($\mid_L$) in PC expresses that its operands are executing on different processors. For modeling systems with several processes sharing one processor it would be useful to have another parallel composition operator expressing that its operands are executing on the same processor. Such an operator can easily be defined, since it only differs from the $\mid_L$ -operator in PC in that transitions originating from both its operands should be comparable. With such an operator it would be possible to express both sharing of (limited) resources and independent executions in the sense of [GL90].

Acknowledgments
We would like to thank Joachim Parrow for critical reading and clarifying discussions, Alan Jeffrey for drawing our attention to an error in an earlier

version of this paper, and Lars-åke Fredlund for critical reading.

This work was partially supported by the Swedish Board for Technical Development (ESPRIT/BRA projects 3096 SPEC and 3006 CONCUR). SICS is a non-profit research institute sponsored by the Swedish government and the following Swedish companies: Asea Brown Boveri, Nobeltech Systems AB, the Swedish Defense Material Administration, Ericsson, Swedish IBM, and the Swedish Telecommunication Administration.

References

[Ace91] L. Aceto. On relating concurrency and nondeterminism. In *Proc. Chalmers Workshop on Concurrency*, Båstad, Sweden, May 1991. Also available as Technical Report 6/89, University of Sussex, 1989.

[BBK86] J. Baeten, J. Bergstra, and J. Klop. Syntax and defining equations for an interrupt mechanism in process algebra. *Fundamenta Informaticae*, 9:127–168, 1986.

[BC88] G. Boudol and I. Castellani. Permutation of transitions: An event structure semantics for CCS and SCCS. In J.W. de Bakker, W.-P. de Roever, and G. Rozenberg, editors, *Linear Time, Branching Time and Partial Order in Logics and Models for Concurrency*, volume 354 of *Lecture Notes in Computer Science*, pages 411–427. Springer Verlag, 1988.

[BCHK91] G. Boudol, I. Castellani, M. Hennessy, and A. Kiehn. Observing localities. In *Proc. Chalmers Workshop on Concurrency*, Båstad, Sweden, May 1991. Also available as Technical Report 4/91, University of Sussex, 1991.

[Cam91] J. Camilleri. A conditional operator for CCS. In J. Baeten and J.-F. Groote, editors, *Proc. CONCUR'91 2nd International Conference on Concurrency Theory*, volume 527 of *Lecture Notes in Computer Science*, pages 142 - 156. Springer Verlag, August 1991.

[CH88] R. Cleaveland and M. Hennessy. Priorities in process algebra. In *Proc. 3rd IEEE Int. Symp. on Logic in Computer Science*, pages 193–202. IEEE Computer Society Press, 1988.

[CW91] J. Camilleri and G. Winskel. CCS with priority choice. In *Proc. 6th IEEE Int. Symp. on Logic in Computer Science*. IEEE Computer Society Press, July 1991.

[DDM88] P. Degano, R. De Nicola, and U. Montanari. Partial orderings descriptions and observations of nondeterministic concurrent processes. In J.W. de Bakker, W.-P. de Roever, and G. Rozenberg, editors, *Linear Time, Branching Time and Partial Order in Logics and Models for Concurrency*, volume 354 of *Lecture Notes in Computer Science*, pages 438–466. Springer Verlag, 1988.

[FGM91] G. Ferrari, R. Gorrieri, and U. Montanari. An extended expansion theorem. In S. Abramsky and T.S.E Mailbaum, editors, *TAPSOFT'91*, volume 494 of *Lecture Notes in Computer Science*, pages 29–48. Springer Verlag, 1991.

[GL90] R. Gerber and I. Lee. CCSR: a calculus for communicating shared resources. In J.C.M. Baeten and J.W. Klop, editors, *Proc. CONCUR'90 Theories of Concurrency: Unification and Extension*, volume 458 of *Lecture Notes in Computer Science*, pages 263 – 277. Springer Verlag, 1990.

[HO] H. Hansson and F. Orava. A calculus with incomparable priorities. Technical report, Swedish Institute of Computer Science. In preparation.

[Jef92] A. Jeffrey. Translating timed process algebra into prioritized process algebra. In J. Vytopil, editor, *Formal Techniques in Real-Time and Fault-Tolerant Systems*, volume 571 of *Lecture Notes in Computer Science*, pages 493–506. Springer Verlag, 1992.

[Mil80] R. Milner. *A Calculus of Communicating Systems*, volume 92 of *Lecture Notes in Computer Science*. Springer Verlag, 1980.

[Mil89] R. Milner. *Communication and Concurrency*. Prentice-Hall, 1989.

[Plo81] G. Plotkin. A structural approach to operational semantics. Technical report, DAIMI FN-19, Dept. of Computer Science, Aarhus University, Aarhus, Denmark, 1981.

[SS90] S. Smolka and B. Steffen. Priority as extremal probability. In J.C.M. Baeten and J.W. Klop, editors, *Proc. CONCUR'90 Theories of Concurrency: Unification and Extension*, volume 458 of *Lecture Notes in Computer Science*, pages 456–466. Springer Verlag, 1990.

[Tof90] C. Tofts. A synchronous calculus of relative frequency. In J.C.M. Baeten and J.W. Klop, editors, *Proc. CONCUR'90 Theories of Concurrency: Unification and Extension*, volume 458 of *Lecture Notes in Computer Science*, pages 467–480. Springer Verlag, 1990.

CCSR 92: Calculus for Communicating Shared Resources with Dynamic Priorities*

Patrice Brémond-Grégoire, Susan Davidson, Insup Lee
Department of Computer and Information Science,
University of Pennsylvania
Philadelphia, PA 19104-6389

Abstract

CCSR 92 is a process algebra which supports timed actions and instantaneous synchronization. In addition, it supports the notion of static and dynamic priorities, such as first-in-first-out and earliest deadline first. This paper describes the syntax and operational semantics of the language and a notion of compositionality of preemption. Compositionality allows us to eliminate possible behaviors without changing the potential effect of the process. We analyze several ways of combining the priorities of simultaneous actions that satisfy our compositionality requirement. Several examples of dynamic priority schemes are also shown to be compositional, while general dynamic priority schemes are not.

1 Introduction

Real-time distributed systems are of critical importance in our society. Errors in their design can often cost human lives, yet very few tools exist to model and prove properties of such systems. Our goal is to build an ensemble of language, methodology and tools to be used by system designers to formally specify and prove the correctness of distributed real-time systems. In this article, we concentrate on the language.

A characteristic of real-time systems is that their execution time is an important part of their correctness. A characteristic of distributed systems is that several things can happen at the same time. To be realistic in modeling real-time distributed systems, one must not only be able to model the notion of time, but also the fact that what can happen simultaneously is limited by the resources available to the system. This work addresses these two facets of formal specification of real-time systems in the context of a process algebra: CCSR 92 (Calculus for Communicating Shared Resources, vintage 1992).

We take the approach that systems evolve in discrete steps, and distinguish two kinds of steps: events and actions. Events are instantaneous and do not require resources. They are merely a named instant in time and are mainly used to achieve synchronization between processes. Actions take time and are

*This research was supported in part by ONR N00014-89-J-1131 and DARPA/NSF CCR90-14621.

bound to resources, thus modeling when work is being performed by the system. In order to arbitrate between several actions competing for the same resource, we also support the notion of priority, which yields a measure of the urgency of performing an action. Priority is also used to decide between several processes attempting to synchronize with the same event.

The CCSR 92 signature contains the usual prefix, choice, composition and recursion operators. It also includes operators to enforce synchronization and symbols such as deadlock and divergence. In the future we will incorporate constructs to express real-time constraints, delays and deadlines,

The operational semantics of CCSR 92 is defined in terms of two labeled transition systems, one for events and one for actions. Priority-less behavior is defined first; we then derive a notion of preemption using priorities, and enhance the transition systems accordingly. The notion of compositionality, by which process behaviors can be eliminated without changing the overall effect, is also introduced, and the preemption relation shown to be compositional.[1]

We then extend the notion of priority to dynamic priorities. Dynamic priorities are defined as a function of the history of the system, and can be used with events to model some common priority schemes such as first-in-first-out and least-slack-first. We also redefine the operational semantics of CCSR 92 to include the historical context. Since it can be shown that dynamic priorities are not in general compositional, we give a sufficient condition that ensures compositionality and show that our examples of priority schemes are compositional. We conclude by showing the difficulty of defining "local histories" to enforce compositionality.

2 A process algebra of actions

Our model assumes that there is a finite set of resources, which we call $\mathcal{R}$. A basic action is the usage of one of these resources, r, at a priority p, and for a finite amount of time, $t > 0$. We use the notation $\chi(t, p, r)$ for such an action. Each resource can execute one single action at any point in time, that is, resources can only be shared sequentially. Actions, however, can be composed with other actions to create more complex actions, representing the use of multiple resources. We will call $\mathcal{A}$ the set of all the actions (basic and composed); lowercase letters a, b, c range over actions.

We write $\rho(a)$ for the set of resources used by the action a. Similarly, we will write $\tau(a)$ for the time taken to execute action a.

Actions can be preordered by priorities. We say $a \succeq b$ if a has at least the same priority as b. For basic actions,

$$\chi(t, p, r) \succeq \chi(t', p', r') \iff r = r' \land p \geq p'.$$

Obviously, $\succeq$ restricted to basic actions is a preorder and the kernel of the preorder is the set of equivalence classes of actions having the same p and r but different t.

[1]For brevity, proofs have been omitted in this extended abstract, and can be found in the full paper (in preparation).

We also define a preemption relation. An action a is said to *preempt* an action b ($a \gg b$) when a choice between the two will always be made in favor of a. The preemption relation must be consistent with the priority relation, that is:

$$a \gg b \Longrightarrow a \succeq b.$$

For basic actions:

$$\chi(t,p,r) \gg \chi(t',p',r') \iff r = r' \wedge p > p'.$$

The extension of the preemption relation to general actions is deferred until later, when several possibilities are explored.

2.1 Basic signature

We start with a basic signature, Σ_b, for our algebra. The signature has been reduced to a minimum for clarity of presentation; we will add new operators as necessary to enhance the expressiveness of our language.

$$\begin{aligned}
\Sigma_b^0 &= \{\text{nil}\} \\
\Sigma_b^1 &= \{a : _ \mid a \in \mathcal{A}\} \\
\Sigma_b^2 &= \{_ + _, _ \| _\} \\
\Sigma_b^n &= \emptyset \text{ for } n \geq 3.
\end{aligned}$$

As usual, we write $T_{\Sigma_b}(X)$ to denote the set of all the syntactically finite terms that can be written using the operators of Σ_b and the countable set of variables X.

2.2 Operational semantics

The operational semantics of a term is defined as a labeled transition system:

$$\longrightarrow_{\mathcal{A}} \subseteq T_\Sigma(X) \times \mathcal{A} \times T_\Sigma(X).$$

Intuitively, an action transition corresponds to the execution of an action which takes some non-zero but finite amount of time and is externally observable. As usual, we write $p \xrightarrow{a}_{\mathcal{A}} q$ instead of $\langle p, a, q\rangle \in \longrightarrow_{\mathcal{A}}$. We will also use the notation $p \xrightarrow{a}_{\mathcal{A}}$ to mean: $\exists q$ such that $p \xrightarrow{a}_{\mathcal{A}} q$.

Nil. The process nil represents a deadlock, and has no behavior. More formally, there is no member of $\longrightarrow_{\mathcal{A}}$ with nil as its left component.

Prefix. The process $a : p$ has the following semantics:

$$a : p \xrightarrow{a}_{\mathcal{A}} p.$$

It corresponds to the immediate execution of the action a.

Choice. The operational semantics for choice is defined as follows:

$$p \xrightarrow{a}_{\mathcal{A}} p' \quad \Longrightarrow \quad p + q \xrightarrow{a}_{\mathcal{A}} p'$$
$$q \xrightarrow{a}_{\mathcal{A}} q' \quad \Longrightarrow \quad p + q \xrightarrow{a}_{\mathcal{A}} q'$$

In other words, the choice construct can execute as either of the component processes.

Parallel operator. The semantics of parallel is not as straightforward. It seems reasonable to think that if:

$$p \xrightarrow{a}_{\mathcal{A}} p' \wedge q \xrightarrow{b}_{\mathcal{A}} q'$$

then

$$p \| q \xrightarrow{f(a,b)}_{\mathcal{A}} g(a, b, p', q').$$

The question is what the functions "f" and "g" should be, and how $f(a, b)$ should compare, priority wise, with the other actions.

It seems natural to define the resources of $f(a, b)$ to be the union of the resources of a and b. Therefore:

$$\rho(f(a, b)) = \rho(a) \cup \rho(b).$$

However, the same resource should not be used twice at the same time because this would violate our very notion of resources. Therefore in order for $f(a, b)$ to exist, it must be the case that $\rho(a) \cap \rho(b) = \emptyset$.

We must now consider the timing of $f(a, b)$. If $\tau(a) = \tau(b)$, then the answer is straightforwardly $\tau(a)$. However, if the timings are different, say $\tau(a) < \tau(b)$, we must decompose the "longer" action b into two actions b' and b'' such that $\tau(b') = \tau(a)$, use b' in defining f and b'' in defining g.

For now, we assume that $\tau(a) = \tau(b)$, and define $f(a, b)$ as a composition operation between the two actions, noted $a * b$, with the following properties:

Property 2.1 (existence of composition) *For two actions a and b, $a * b$ exists if and only if:*

$$\tau(a) = \tau(b) \textit{ and } \rho(a) \cap \rho(b) = \emptyset$$

Property 2.2 (preservation of time) *The timing of $a * b$ is such that:*

$$\tau(a * b) = \tau(a) = \tau(b)$$

Property 2.3 (preservation of resources) *The set of resources of $a * b$ is such that:*

$$\rho(a * b) = \rho(a) \cup \rho(b)$$

Later we will give two possible interpretations of $a * b$ and show how to compare composed actions priority wise.

We can also write the first operational semantics rule for the parallel operator as:

$$\left.\begin{array}{l} p \xrightarrow{a}_{\mathcal{A}} p' \\ q \xrightarrow{b}_{\mathcal{A}} q' \\ a * b \ \textit{is defined} \end{array}\right\} \Longrightarrow p \| q \xrightarrow{a*b}_{\mathcal{A}} p' \| q'$$

To extend this operational semantics to the general case in which the timings are not equal, we must decompose the longer action. Thus if $\tau(a) < \tau(b)$, we decompose the action b into two actions b' and b'' such that $\tau(b') = \tau(a)$, $\tau(b') + \tau(b'') = \tau(b)$, and b' and b'' use the same resources as b, at the same priority. Here we write $b = \mathrm{seq}(b', b'')$ to indicate that b' and b'' form a sequential decomposition of b. Thus it seems natural to define that $b' : b'' : p$ is *operationally equivalent* to $b : p$. This justifies the following two transitions:

$$\left.\begin{array}{l} p \xrightarrow{a}_{\mathcal{A}} p' \\ q \xrightarrow{b}_{\mathcal{A}} q' \\ \rho(a) \cap \rho(b) = \emptyset \\ b = \mathrm{seq}(b', b'') \\ \tau(a) = \tau(b') \end{array}\right\} \Longrightarrow p \| q \xrightarrow{a*b'}_{\mathcal{A}} p' \| (b'' : q')$$

$$\left.\begin{array}{l} p \xrightarrow{a}_{\mathcal{A}} p' \\ q \xrightarrow{b}_{\mathcal{A}} q' \\ \rho(a) \cap \rho(b) = \emptyset \\ a = \mathrm{seq}(a', a'') \\ \tau(a') = \tau(b) \end{array}\right\} \Longrightarrow p \| q \xrightarrow{a'*b}_{\mathcal{A}} (a'' : p') \| q'$$

We are now left with two problems to resolve: what happens when $\rho(a) \cap \rho(b) \neq \emptyset$ and how does $a * b$ compare priority-wise with other actions? Let us delay the first problem until we talk about infinite processes and address the second one.

2.3 Prioritized transition system

To take into account the notion of priority, we define a subset of $\longrightarrow_{\mathcal{A}}$ which we call the prioritized transition system and denote as $\longrightarrow_{\pi\mathcal{A}}$. The prioritized transition system is defined from the un-prioritized transition system as follows:

$$p \xrightarrow{a}_{\pi\mathcal{A}} q \iff p \xrightarrow{a}_{\mathcal{A}} q \wedge \nexists a' . (p \xrightarrow{a'}_{\mathcal{A}} \wedge a' \succ\!\!\succ a)$$

It seems natural to require that preemption be compositional. That is if an action a preempts an action b in a process p, then in any context $C[\_]$, the behavior of $C[p]$ does not change if we eliminate b from the behavior of p.

To formally express this, we devise a notation to express the effect of a context on an action. For a given labeled transition system $\longrightarrow$ and a given context $C[\_]$, we define a function F as follows:

$$F(C[\_], a) = \{a' \mid \forall p . (p \xrightarrow{a} \Longrightarrow C[p] \xrightarrow{a'})\}$$

For example, if

$$C[\_] = (\_ \| c : Q + d : Q')$$
$$p \xrightarrow{a}, p \xrightarrow{b}$$

then $F(C[\_], a) = \{a * d, a * c\}$, $F(C[\_], b) = \{b * d, b * c\}$.

Definition 2.1 (Compositionality of preemption) *We say that a preemption relation is compositional with regard to an operational semantics $\rightarrow$ if for every unguarded context $C[\_]$, process p, and actions a and b such that:*

$$p \xrightarrow{a}_{\mathcal{A}} \wedge p \xrightarrow{b}_{\mathcal{A}} \wedge a \succ\!\!\succ b$$

it is true that:

$$\forall b' \in F(C[\_], b) \,.\, \exists a' \in F(C[\_], a) \,.\, a' \succ\!\!\succ b'$$

By unguarded context, we mean that the "hole" is not within the scope of a prefix operator.

Continuing with our example, $C[\_]$ is an unguarded context. If we assume $a \succ\!\!\succ b$, then the preemption relation is compositional if $a * c \succ\!\!\succ b * c$ and $a * d \succ\!\!\succ b * d$.

This notion of compositionality will guide us through the definition of the priority and preemption relations. There are two complementary ways to define the priority of composed actions: either a priority value is attached to each resource or a single priority value applies to all the resources. We are going to examine how each of these two ways can satisfy our notion of compositionality.

2.4 Distributed priority

The first definition of priority is based on the premise that basic actions using different resources have *totally unrelated* priorities. In this case, an action is made of a timing component and a set of pairs, each containing a resource and its priority, e.g., $a = \chi(t, \{\langle p_1, r_1\rangle, \langle p_2, r_2\rangle, \ldots\})$.

The composition operation can be defined by taking the union of the sets of prioritized resources:

$$\chi(t, R_a) * \chi(t, R_b) \quad = \quad \chi(t, R_a \cup R_b)$$

Note that this definition satisfies the properties of preservation of time and resources.

Let us define a family of functions, noted $\pi_r(a)$, returning the priority attached to resource r in the action a, that is:

$$\pi_r(\chi(t, R)) = p \text{ such that } \langle r, p\rangle \in R$$

We can define:

$$a \succeq b \quad \Longleftrightarrow \quad \forall r \in \rho(a), (r \in \rho(b) \wedge \pi_r(a) \geq \pi_r(b))$$

It is easy to see that this relation is reflexive and transitive and therefore $\succeq$ remains a pre-order. Note that the kernel of the pre-order is the equivalence classes of actions that have the same set of pairs $\langle p, r\rangle$ and different timings. This follows from the fact that actions can only be composed via the $*$ operation which does not produce any action that uses the same resource twice.

Lemma 2.1 $a \succeq b \Longrightarrow \rho(a) \subseteq \rho(b)$

We use the notation $a \cong_\pi b$ when $a \succeq b$ and $b \succeq a$.

The notion of preemption associated with this definition of priority is as follows:

$$a \succ\!\!\succ b \quad \Longleftrightarrow \quad a \succeq b \wedge (\exists r \in \rho(a) \,.\, \pi_r(a) > \pi_r(b))$$

We now need to prove that this definition satisfies our notion of compositionality. If we look at the operational semantics that we have defined so far, only the parallel operator can affect the actions. Since priorities are timing independent, we can avoid all the complex cases that require splitting actions, and concentrate on three actions $a' \cong_\pi a$, $b' \cong_\pi b$ and $c' \cong_\pi c$ and such that $\tau(a') = \tau(b') = \tau(c') = \min(\tau(a), \tau(b), \tau(c))$.

Compositionality follows directly from the following lemma:

Lemma 2.2

$$\left.\begin{array}{l} a \succ\!\!\succ b \\ \tau(a) = \tau(b) \\ b * c \;\; \textit{is defined} \end{array}\right\} \Longrightarrow a * c \textit{ is defined} \wedge a * c \succ\!\!\succ b * c$$

2.5 Combined priority

The other definition of priority is based on the premise that priority values are consistent between resources. In this case, we can define actions to be made of a timing value, a single priority and a set of resources, e.g., $\chi(t, p, R)$.

One can think of many different ways to combine actions that "make sense" in different circumstances. Two that come to mind immediately are adding the priorities or taking the average. However, not all simple combination functions like this are compositional. Take, for example, the max function; it fails to be compositional since although one action might have a higher priority than another one, if they are both combined with a third action with even higher priority, the resulting two actions have the same priority. Distributed priority is the compositional way to apply the max function.

It can be shown that a very simple condition is sufficient to ensure compositionality: the combination function, which we note "Comb(a, b)" must be symmetric and strictly monotonic. That is for all n, n', m:

$$\begin{array}{rcl} \mathrm{Comb}(m, n) & = & \mathrm{Comb}(n, m) \\ n > n' & \Longleftrightarrow & \mathrm{Comb}(n, m) > \mathrm{Comb}(n', m) \end{array}$$

Note that the max function fails this condition since Comb(n, m) may be equal to Comb(n', m).

The composition of two actions is therefore defined as:

$$\chi(t, p, R) * \chi(t, p', R') = \chi(t, \mathrm{Comb}(p, p'), R \cup R')$$

In other words, to compose two actions, provided they have the same timing component, we compose their priorities (with an adequate function) and take the union of their resources.

Let $\pi(a)$ be the priority component of the action a, that is:

$$\pi(\chi(t,p,R)) \quad = \quad p$$

The notion of priority associated with this is:

$$a \succeq b \quad \iff \quad \pi(a) \geq \pi(b) \wedge \rho(a) \subseteq \rho(b)$$

and the corresponding preemption is:

$$a \ggcurly b \quad \iff \quad a \succeq b \wedge \pi(a) > \pi(b)$$

Remark: requiring equality of resources in the above definition would lead to sparser but consistent priority and preemption relations.

This definition of preemption is compositional. As above we can limit ourselves to prove that composition preserves preemption.

Lemma 2.3

$$\left.\begin{array}{l} a \ggcurly b \\ \tau(a) = \tau(b) \\ b * c \ \ \textit{is defined} \end{array}\right\} \Longrightarrow a * c \ \textit{is defined} \wedge a * c \ggcurly b * c$$

3 Infinite processes

In this section, we expand the signature of the algebra to allow the definition of processes with infinite behavior, and define the operational semantics of recursive terms. We also introduce a new action, the "procrastination action", to formally cope with the case in which two parallel processes compete for the same resource.

3.1 Signature for recursive processes

We introduce a new family of unary operators, Rec , indexed by the set of variables X. Therefore the new signature, Σ_r, can be defined as follows:

$$\begin{array}{lcl} \Sigma_r^1 & = & \Sigma_b^1 \cup \{\text{Rec } x.\_ \mid x \in X\} \\ \Sigma_r^n & = & \Sigma_b^n \text{ for } n \neq 1 \end{array}$$

3.2 Operational semantics of recursive processes

The operational semantics of a recursion operation is defined by:

$$p \xrightarrow{a}_{\mathcal{A}} q \quad \Longrightarrow \quad \text{Rec } x.p \xrightarrow{a}_{\mathcal{A}} q[\text{Rec } x.p/x]$$

where $q[\text{Rec } x.p/x]$ is the standard notation for the substitution of Rec $x.p$ for each free occurrence of x in q.

From the above definition, it should be obvious that if

$$p \xrightarrow{a}_{\mathcal{A}} \text{ and } p \xrightarrow{b}_{\mathcal{A}} \text{ and } a \succ\!\!\succ b$$

then, regardless of the definition of "$\succ\!\!\succ$",

$$\text{Rec } x.p \xrightarrow{a}_{\mathcal{A}} \text{ and Rec } x.p \xrightarrow{b}_{\mathcal{A}} \text{ and } a \succ\!\!\succ b$$

therefore the two notions of preemption defined earlier remain compositional.

3.3 Interleaving and procrastination

Intuitively, if $p \xrightarrow{a}_{\mathcal{A}} p'$, $q \xrightarrow{b}_{\mathcal{A}} q'$ and $\rho(a) \cap \rho(b) \neq \emptyset$, we would like to specify that the behavior of $p||q$ interleaves the actions of both processes, and that the choice of which action is executed next is made according to their priorities. To achieve this, we introduce a new action δ, called the *procrastinating action*, with the following properties:

$$\begin{aligned} \delta * a = a * \delta &= a \text{ for all } a \in \mathcal{A} \\ \delta * \delta &= \delta \end{aligned}$$

If we write "$.p$" to express the fact that p can be delayed, the operational semantics of $.p$ should be:

$$\begin{aligned} p \xrightarrow{a}_{\mathcal{A}} p' \quad \Longrightarrow \quad & .p \xrightarrow{a}_{\mathcal{A}} p' \\ & .p \xrightarrow{\delta}_{\mathcal{A}} .p \end{aligned}$$

which is that of p augmented by a transition that lets p wait the time of another action before proceeding.

It should be obvious that this operational semantics is that of Rec $x.(\delta : x + p)$, and that "$.$" is therefore a macro to be expanded into the above expression — assuming, of course, that x is fresh and does not appear free in p. In the remainder of this paper we will use the "$.$" notation freely; furthermore, we will abbreviate "$a : .p$" into "$a.p$".

3.4 How to ensure progress

One of the possible traces of a process "$.p$" is an infinite sequence of δ. Obviously this is not a very desirable behavior in general, but it is required in some cases, for example, in the expression:

$$\text{Rec } x.(.a : x)||\text{Rec } y.(.b : y)$$

if $a \succ\!\!\succ b$ then the behavior of the right hand side process will be a continuous string of δ since it is preempted at every turn.

One might think of assigning δ a priority such that it would be preempted by any other action. Unfortunately this simple solution does not satisfy our requirement for compositionality. To see this, note that we would always have $a \succ\!\!\succ \delta$, but if $\rho(a) \cap \rho(b) \neq \emptyset$ then $b * a$ is not defined, while $b * \delta$ is always defined.

The root of the problem is that any process must be able to accommodate another process which is competing for the same resources and has a higher priority. To ensure that this situation cannot occur, we introduce an operator to reserve the rights to a set of resources for the exclusive use of one process. We call this the closure operator.

We augment, once again, the signature of our algebra with a family of operators, this time indexed by the powerset of $\mathcal{R}$ (the set of resources):

$$\begin{aligned} \Sigma_c^1 &= \Sigma_r^1 \cup \{[\_]_I \mid I \subseteq \mathcal{R}\} \\ \Sigma_c^n &= \Sigma_r^n \text{ for } n \neq 1 \end{aligned}$$

We define the operational semantics of closure as:

$$p \xrightarrow{a}_{\mathcal{A}} q \quad \Longrightarrow \quad [p]_I \xrightarrow{[a]_I}_{\mathcal{A}} [q]_I$$

All remains to be done is to define $[a]_I$ such that $\rho([a]_I) = \rho(a) \cup I$ to effectively close the resources.

Closure for combined priority.

$$\begin{aligned} [\chi(t, p, R)]_I &= \chi(t, p, R \cup I) \\ [\delta]_I &= \chi(*, 0, I) \end{aligned}$$

where "*" is a wild card that allows $\chi(*, 0, I)$ to assume the timing of any action it is composed with.

It is not hard to see that this definition is still compositional and that it satisfies our requirement of maximizing the use of resources.

Closure for distributed priority.

$$\begin{aligned} [\chi(t, R)]_I &= \chi(t, R \cup \{\langle 0, r\rangle \mid r \in I - \rho(\chi(t, R))\}) \\ [\delta]_I &= \chi(*, \{\langle 0, r\rangle \mid r \in I\}) \end{aligned}$$

where, again, "*" is a wild card that allow $\chi(*, 0, I)$ to assume the timing of any action it is composed with.

Unfortunately, this definition is not compositional and we have to re-visit our definition of distributed priority. To see this, observe that:

$$\chi(t, \{\langle 5, r_1\rangle\}) \succ\!\!\succ \chi(t, \{\langle 3, r_1\rangle, \langle 4, r_2\rangle\})$$

however, if we close them over r_2 we get:

$$\chi(t, \{\langle 5, r_1\rangle, \langle 0, r_2\rangle\}) \quad \not\succ\!\!\succ \quad \chi(t, \{\langle 3, r_1\rangle, \langle 4, r_2\rangle\})$$

To solve this problem, we have to redefine distributed priority as:

$$a \succeq b \quad \Longleftrightarrow \quad \rho(a) = \rho(b) \wedge \forall r \in \rho(a), \pi_r(a) \geq \pi_r(b)$$

The corresponding definition of preemption still stands:

$$a \succ\!\!\succ b \quad \Longleftrightarrow \quad a \succeq b \wedge (\exists r \in \rho(a) \,.\, \pi_r(a) > \pi_r(b))$$

4 Events and synchronization

In this section we introduce the notions of event and synchronization and define the corresponding operational semantics. For the sake of brevity, the details of the priority of events is omitted. Suffice it to say that the discussion would be very similar to the discussion of priority of actions. We also assume that events always have priority over actions since they do not consume any time.

We assume that there is a set of event names, which we call $\mathcal{L}$, We also use $\mathcal{E}$ to denote the set of all events and actions, and the variables e and f to range over $\mathcal{E}$. We call a basic event the occurrence of an event name at a given priority level.

Of course, multiple events can occur simultaneously — this is how synchronization is achieved. The composition of two events e and f, $e * f$, represents their simultaneous occurrence. While basic events have a single event name, composed events have several event names. We use the function $l : \mathcal{E} \rightarrow \mathbf{P}(\mathcal{L})$ to denote the set of event names of an event. As we did for actions, we will avoid giving a type to composed events and an interpretation to the composition function. For now we will rely on the following properties:

Property 4.1 (Composition of events) *Two events can be composed if and only if they have no event name in common:*

$$e * f \text{ is defined} \quad \Longleftrightarrow \quad l(e) \cap l(f) = \emptyset$$

Property 4.2 (Preservation of event names) *The set of event names are preserved by composition:*

$$l(e * f) \quad = \quad l(e) \cup l(f)$$

4.1 Signature of the algebra with events

We extend the signature Σ_c into a new signature Σ_e as follows:

$$\Sigma_e^1 \;=\; \Sigma_c^1 \cup \{e : \_ \mid e \in \mathcal{E}\} \cup \{\_\rfloor_E \mid E \subseteq \mathcal{L}\}$$

$$\Sigma_e^n \;=\; \Sigma_c^n \text{ for } n \neq 1$$

The family of prefix operators indexed over the set of events ($\{e : \_ \mid e \in \mathcal{E}\}$) is very similar to the family of prefix operators indexed over the set of actions; they both express sequencing. Note that we support the notion of "micro-sequencing" in which two events may occur successively but at the same *measurable* time. In this case we write $e_1 : e_2 : p$.

The family of operators $\{\_\rfloor_E \mid E \subseteq \mathcal{L}\}$ are called the *synchronization* operators. Each one defines a set of event names that must always occur simultaneously when they occur in the target process.

4.2 Operational semantics

In order to define the operational semantics of the expanded algebra, we make use of a second transition system "$\longrightarrow_{\mathcal{E}}$". The following set of rules specifies how to calculate "$\longrightarrow_{\mathcal{E}}$" and also how to calculate $\longrightarrow_{\mathcal{A}}$ for the new operators. These rules complement, rather than supersede, the rules that we have defined before.

Event prefixes. Event prefixes have an operational semantics very similar to that of action prefixes, the only difference being that they affect the $\longrightarrow_{\mathcal{E}}$ transitions.

$$e : p \xrightarrow{e}_{\mathcal{E}} p$$

This corresponds to the occurrence of the event e.

Synchronization. The semantics of the synchronization operator is defined as:

$$\left.\begin{array}{r} p \xrightarrow{e}_{\mathcal{E}} p' \\ E \subseteq l(e) \end{array}\right\} \implies p\lfloor_E \xrightarrow{e}_{\mathcal{E}} p'\lfloor_E$$

$$\left.\begin{array}{r} p \xrightarrow{e}_{\mathcal{E}} p' \\ E \cap l(e) = \emptyset \end{array}\right\} \implies p\lfloor_E \xrightarrow{e}_{\mathcal{E}} p'\lfloor_E$$

$$p \xrightarrow{a}_{\mathcal{A}} p' \implies p\lfloor_E \xrightarrow{a}_{\mathcal{A}} p'\lfloor_E$$

The first two rules formally specify, as we had hinted before, that the event names of E must occur together or not at all. For example, if $l(e) = \{e_{l_1}, e_{l_2}, e_{l_3}\}$ then:

$$
\begin{array}{lcll}
(e : p)\lfloor_{\{e_{l_1}, e_{l_2}\}} & \xrightarrow{e}_{\mathcal{E}} & p & \text{by the first rule} \\
(e : p)\lfloor_{\{e_{l_4}, e_{l_5}\}} & \xrightarrow{e}_{\mathcal{E}} & p & \text{by the second rule} \\
(e : p)\lfloor_{\{e_{l_3}, e_{l_6}\}} & \not\longrightarrow_{\mathcal{E}} & & \text{because no rule apply}
\end{array}
$$

The third rule specifies that the synchronization operators have no effect over actions.

There have been several ways of enforcing synchronization described in the literature. Our way is derived from Milner's [13, 11] hiding operator. In CCS, synchronized events are combined into generic τ events and synchronization of an event is enforced by pruning out transitions labeled with that event because they are un-synchronized. There are two problems with applying this technique to CCSR 92. The first one is that hiding priorities destroys compositionality. This can be solved in the context of combined priorities by assigning priorities to τ events, as done in [1] and [2], effectively creating a family of such events. However, this solution becomes very cumbersome when applied to distributed priorities. The second problem has to do with dynamic priorities; as we will see later, event names are heavily used in priority functions and should not be hidden.

Closure. The following rule formally states that none of the closure operators have any effect on the event transitions:

$$p \xrightarrow{e}_{\mathcal{E}} p' \implies [p]_I \xrightarrow{e}_{\mathcal{E}} [p']_I$$

Choice. As one might expect, the effect of the choice operator over the event transitions is similar to its effect over the action transitions:

$$\begin{array}{lcl} p \xrightarrow{e}_{\mathcal{E}} p' & \Longrightarrow & p+q \xrightarrow{e}_{\mathcal{E}} p' \\ q \xrightarrow{e}_{\mathcal{E}} q' & \Longrightarrow & p+q \xrightarrow{e}_{\mathcal{E}} q' \end{array}$$

Parallel. The operational semantics of the parallel operator over event transitions is somewhat different from what we have seen in the case of actions. This is because the element of time does not enter into the picture. We express this semantics by three rules:

$$\begin{array}{rcl} p \xrightarrow{e}_{\mathcal{E}} p' & \Longrightarrow & p\|q \xrightarrow{e}_{\mathcal{E}} p'\|q \\ q \xrightarrow{e}_{\mathcal{E}} q' & \Longrightarrow & p\|q \xrightarrow{e}_{\mathcal{E}} p\|q' \\ \left.\begin{array}{l} p \xrightarrow{e}_{\mathcal{E}} p' \\ q \xrightarrow{f}_{\mathcal{E}} q' \\ e * f \text{ is defined} \end{array}\right\} & \Longrightarrow & p\|q \xrightarrow{e*f}_{\mathcal{E}} p'\|q' \end{array}$$

Intuitively, if two processes in parallel are both ready for the occurrence of an event, those events can occur in any order. Furthermore, if the two events have disjoint sets of event names, they can also occur simultaneously. It is important to note that this operational semantics does not allow an event to occur simultaneously with an action.

4.3 Complete CCSR 92 signature

To summarize, the complete CCSR 92 signature is defined as follows:

$$\begin{array}{lcl} \Sigma^0 & = & \{\text{nil}\} \\ \Sigma^1 & = & \quad \{a : \_ \mid a \in \mathcal{A}\} \\ & & \cup \ \{e : \_ \mid e \in \mathcal{E}\} \\ & & \cup \ \{\text{Rec } x.\_ \mid x \in X\} \\ & & \cup \ \{[\_]_I \mid I \subseteq \mathcal{R}\} \\ & & \cup \ \{\_\!\downarrow_E \mid E \in \mathcal{Q}(\mathcal{L})\} \\ \Sigma^2 & = & \{\_ + \_, \_\|\_\} \\ \Sigma^n & = & \emptyset \text{ for } n \geq 3. \end{array}$$

5 Dynamic priorities

In the previous sections, we have assumed that priority was a constant. However, if we want to model priority schemes such as first-in-first-out, earliest deadline first or least slack, priorities must be able to change based on the system state. This implies that we have to keep a context from which priorities can be evaluated. We have decided to keep the history of the system as a context, i.e., the sequence of all the events and actions that have occurred. The reason for this choice is that it seems that all the information that one might want to use in a priority function can be derived from the past behavior of the system and the syntactic context which is static.

A history h is a string over $\mathcal{V} = \mathcal{E} \cup \mathcal{A}$, the combined vocabulary of actions and events. We sometimes also use the word trace instead of history. A dynamic priority is a function of h. Where we used to write $\pi(a), \pi_r(a), \pi(e)$ and $\pi_{e_l}(e)$, we will now write $\pi(a,h), \pi_r(a,h), \pi(e,h)$ and $\pi_{e_l}(e,h)$ respectively. The priority and preemption relations must also take history into account. We write $a \succeq_h b$ when a has a higher priority than b after a history h; this applies to distributed as well as composed priorities. Similarly we write $a \ggcurly_h b$ when a preempts b after a history h. We still use the unadorned symbols $\succeq$ and $\ggcurly$ when the relation is true for any history, which is consistent with our original definition.

So far, we have not specified the range of priorities and it did not matter as long as 0 was the lowest priority. In what follows we assume that the priority functions range over the positive real numbers augmented with a special constant $\perp$ which represents undefined priority. The undefined priority $\perp$ is not comparable with any priority; that is, for any constant priority p:

$$p \not\succeq \perp \wedge \perp \not\succeq p$$

5.1 Examples of dynamic priorities

We make use of a small set of functions to calculate a priority from a history. By definition, if any of the parameters of a function is undefined ($\perp$), then the result of the function is undefined as well.

- Last(e_l, h) gives the position of the last occurrence of an event with the label e_l in the history h. It is undefined if there is no such event.
- $h[i..j]$ is the substring of h starting at index i and ending at index j. If i is not specified, it is assumed to be the first element of h, if j is not specified, it is assumed to be the last.
- $\tau(h)$ The time taken for the execution of h, i.e. the sum of the timing of all actions of h. If any of the actions of h is δ, then the result is undefined.

All of our examples of dynamic priorities use the following process:

$$p \quad \stackrel{\Delta}{=} \quad e_{a_1} . a_1 : e_{a_2} . a_2 : e_{a_3} . a_3 \ldots$$

Note that the events e_{a_i} are not preceded by a colon and therefore no time can elapse between a_i and $e_{a_{i+1}}$. Intuitively, each e_{a_i} marks the earliest possible time of occurrence of the action a_i.

A first-in-first-out priority scheme is defined by the following priority function for each action a_i:

$$\pi(a_i, h) = \tau(h[\text{Last}(e_{a_i}, h)..])$$

The priority of a_i is the time elapsed since a_i was first enabled.

If p must complete its execution before a relative deadline d we can specify an earliest deadline first priority scheme with:

$$\pi(a_i, h) = \frac{1}{d - \tau(h[\text{Last}(e_{a_1}, h)..])}$$

We assume that there is a mechanism, not formalized here, to abort the process when the deadline expires and therefore to keep $\tau(h[e_{a_1}..]) < d$.

If d is the relative deadline and t_{a_i} is the execution time remaining after a_i (i.e. $sum_{j=i+1}^{n} t_j$, where $a_j = \chi(t_j, ...)$), in the absence of delay, the least slack first priority scheme can be specified by:

$$\pi(a_i, h) = \frac{1}{d - (\tau(h[\text{Last}(e_{a_1}, h)..]) + t_{a_i})}$$

5.2 Operational semantics with history

In order to formalize these notions, we now define an operational semantics that keeps track of the history of the system as a context. As before we define two labeled transition systems, replacing the terms by pairs: ⟨process, history⟩.

$$\begin{aligned} \longrightarrow_{\mathcal{A}} &\subseteq (T_\Sigma(X) \times \mathcal{V}^*) \times \mathcal{A} \times (T_\Sigma(X) \times \mathcal{V}^*) \\ \longrightarrow_{\mathcal{E}} &\subseteq (T_\Sigma(X) \times \mathcal{V}^*) \times \mathcal{E} \times (T_\Sigma(X) \times \mathcal{V}^*) \end{aligned}$$

To somewhat limit the introduction of new esoteric symbols, we overload the transition notation that we have used before. The generic variables α and β will range over $\mathcal{V}$; furthermore $\xrightarrow{\alpha}_{\mathcal{A}}$ or $\xrightarrow{\alpha}_{\pi\mathcal{A}}$ is meant when $\alpha \in \mathcal{A}$ and $\xrightarrow{\alpha}_{\mathcal{E}}$ or $\xrightarrow{\alpha}_{\pi\mathcal{E}}$ is meant when $\alpha \in \mathcal{E}$.

Prefix.

$$\langle \alpha : p, h\rangle \quad \xrightarrow{\alpha} \quad \langle p, h\alpha\rangle$$

Choice.

$$\begin{aligned} \langle p, h\rangle \xrightarrow{\alpha} \langle p', h\alpha\rangle \quad &\Longrightarrow \quad \langle p+q, h\rangle \xrightarrow{\alpha} \langle p', h\alpha\rangle \\ \langle q, h\rangle \xrightarrow{\alpha} \langle q', h\alpha\rangle \quad &\Longrightarrow \quad \langle p+q, h\rangle \xrightarrow{\alpha} \langle q', h\alpha\rangle \end{aligned}$$

Parallel.

$$\left.\begin{array}{l} \langle p, h\rangle \xrightarrow{a}_{\mathcal{A}} \langle p', ha\rangle \\ \langle q, h\rangle \xrightarrow{b}_{\mathcal{A}} \langle q', hb\rangle \\ a * b \textit{ is defined} \end{array}\right\} \quad \Longrightarrow \quad \langle p \| q, h\rangle \xrightarrow{a*b}_{\mathcal{A}} \langle p' \| q', h(a*b)\rangle$$

$$\left.\begin{array}{l} \langle p, h\rangle \xrightarrow{a}_{\mathcal{A}} \langle p', ha\rangle \\ \langle q, h\rangle \xrightarrow{b}_{\mathcal{A}} \langle q', hb\rangle \\ \rho(a) \cap \rho(b) = \emptyset \\ b = \text{seq}(b', b'') \\ \tau(a) = \tau(b') \end{array}\right\} \quad \Longrightarrow \quad \langle p \| q, h\rangle \xrightarrow{a*b'}_{\mathcal{A}} \langle p' \| (b'' : q'), h(a*b')\rangle$$

$$\left.\begin{array}{l}\langle p,h\rangle \xrightarrow{a}_{\mathcal{A}} \langle p',ha\rangle\\ \langle q,h\rangle \xrightarrow{b}_{\mathcal{A}} \langle q',hb\rangle\\ \rho(a)\cap\rho(b)=\emptyset\\ a=\mathrm{seq}(a',a'')\\ \tau(a')=\tau(b)\end{array}\right\} \implies \langle p\|q,h\rangle \xrightarrow{a'*b}_{\mathcal{A}} \langle (a'':p')\|q', h(a'*b)\rangle$$

$$\langle p,h\rangle \xrightarrow{e}_{\mathcal{E}} \langle p',he\rangle \implies \langle p\|q,h\rangle \xrightarrow{e}_{\mathcal{E}} \langle p'\|q,he\rangle$$

$$\langle q,h\rangle \xrightarrow{e}_{\mathcal{E}} \langle q',he\rangle \implies \langle p\|q,h\rangle \xrightarrow{e}_{\mathcal{E}} \langle p\|q',he\rangle$$

$$\left.\begin{array}{l}\langle p,h\rangle \xrightarrow{e}_{\mathcal{E}} \langle p',he\rangle\\ \langle q,h\rangle \xrightarrow{f}_{\mathcal{E}} \langle q',hf\rangle\\ e*f \ \textit{is defined}\end{array}\right\} \implies \langle p\|q,h\rangle \xrightarrow{e*f}_{\mathcal{E}} \langle p'\|q', h(e*f)\rangle$$

Recursion.

$$\langle p,h\rangle \xrightarrow{\alpha} \langle q,h\alpha\rangle \implies \langle \mathrm{Rec}\ x.p,h\rangle \xrightarrow{\alpha} \langle q[\mathrm{Rec}\ x.p/x],h\alpha\rangle$$

Closure.

$$\langle p,h\rangle \xrightarrow{a}_{\mathcal{A}} \langle q,ha\rangle \implies \langle [p]_I,h\rangle \xrightarrow{[a]_I}_{\mathcal{A}} \langle [q]_I, h[a]_I\rangle$$
$$\langle p,h\rangle \xrightarrow{e}_{\mathcal{E}} \langle q,he\rangle \implies \langle [p]_I,h\rangle \xrightarrow{e}_{\mathcal{E}} \langle [q]_I, he\rangle$$

Synchronization.

$$\left.\begin{array}{l}\langle p,h\rangle \xrightarrow{e}_{\mathcal{E}} \langle p',he\rangle\\ E\cap l(e)=\emptyset\end{array}\right\} \implies \langle p\lfloor_E,h\rangle \xrightarrow{e}_{\mathcal{E}} \langle p'\lfloor_E,he\rangle$$

$$\left.\begin{array}{l}\langle p,h\rangle \xrightarrow{e}_{\mathcal{E}} \langle p',he\rangle\\ E\subseteq l(e)\end{array}\right\} \implies \langle p\lfloor_E,h\rangle \xrightarrow{e}_{\mathcal{E}} \langle p'\lfloor_E,he\rangle$$

$$\langle p,h\rangle \xrightarrow{a}_{\mathcal{A}} \langle p',ha\rangle \implies \langle p\lfloor_E,h\rangle \xrightarrow{a}_{\mathcal{A}} \langle p'\lfloor_E,ha\rangle$$

5.3 Operational semantics extended to traces

The Greek letter ϵ is used to denote a trace of length 0 as well as an empty history.

From the prioritized action and event transition systems we derive a trace transition system as follows:

(1) $\quad \langle p,h\rangle \overset{\epsilon}{\Longrightarrow}_{\pi} \langle p,h\rangle$

(2) $\quad \left.\begin{array}{l}\langle p,h\rangle \xrightarrow{\alpha}_{\pi} \langle p',h\alpha\rangle\\ \langle p',h\alpha\rangle \overset{t}{\Longrightarrow}_{\pi} \langle p'',h\alpha t\rangle\end{array}\right\} \implies \langle p,h\rangle \overset{\alpha t}{\Longrightarrow}_{\pi} \langle p'',h\alpha t\rangle$

As before, we use the $\langle p, h\rangle \stackrel{t}{\Longrightarrow}_\pi$ to mean:

$\exists p'$ such that $\langle p, h\rangle \stackrel{t}{\Longrightarrow}_\pi \langle p', ht\rangle$

5.4 Compositionality and dynamic priorities

For constant priorities, our definition of compositionality was based on the effect of an *unguarded* context on the behavior of a process. In the case of dynamic priorities, we can no longer limit ourselves that way because prefix contexts is what builds history and history is now important. Therefore we need to re-define our notion of compositionality. For that, we first extend our F function, which denotes the possible effects of a context on a given step, to take histories into account:

$$F(C[\_], a, h) = \{a' \mid \forall p\,.\,\langle p, h\rangle \stackrel{a}{\longrightarrow} \Longrightarrow \langle C[p], h\rangle \stackrel{a'}{\longrightarrow}\}$$

Similarly, we need to denote the effect of a context on a history. For that, we define a function G which, given a context $C[\_]$, a history h and an unguarded context $C'[\_]$, returns the set of histories that can lead from C to C'.

$$\begin{aligned} G(C[\_], h, C'[\_]) \;&=\; \{h' \mid \forall p, p'\,. \\ &\quad (\langle \mathrm{p}, \epsilon\rangle \stackrel{\mathrm{h}}{\Longrightarrow}_\pi \langle \mathrm{p}, \mathrm{h}\rangle) \Longrightarrow (\langle \mathrm{C[p]}, \epsilon\rangle \stackrel{\mathrm{h'}}{\Longrightarrow}_\pi \langle \mathrm{C'[p']}, \mathrm{h'}\rangle)\} \end{aligned}$$

Definition 5.1 (compositionality with dynamic priorities) *We say that a preemption relation is compositional with respect to an operational semantics, $\rightarrow$, in presence of dynamic priorities if, for every steps α, β, every histories h, h', every context $C[\_]$ and every unguarded context $C'[\_]$ such that*

$$\alpha \succ\!\!\succ_h \beta,\ h' \in G(C[\_], h, C'[\_]),\ \textit{and}\ \beta' \in F(C'[\_], \beta, h'),$$

there is a step

$$\alpha' \in F(C'[\_], \alpha, h')\ \textit{such that}\ \alpha' \succ\!\!\succ_{h'} \beta'.$$

In other words, if an action preempts another one after every trace leading from one process to another process, this preemption relation remains regardless of the context; and in particular, regardless of what happened before or what happens in parallel.

It is interesting to note that a prerequisite for priorities to be both compositional and truly dynamic is that the range of the priority functions not be totally ordered. To see this, substitute ϵ for h in the definition above. It follows that for any pair of actions or events such that one preempt the other, the relation must remain true in any context; this, for all practical purposes, is static priority.

It is not hard to see that the compositionality condition is not satisfied in general. Take, for example, the following system:

$$\begin{aligned} p &= [(e\,.\,a\,.\,\mathrm{nil} \| f\,.\,b\,.\,\mathrm{nil})]_{\{r\}} \\ \rho(a) &= \rho(b) = \{r\} \\ \pi(a, h) &= \tau(h[\mathrm{Last}(e, h)..]) + 1 \text{ (modified first in first out)} \\ \pi(b, h) &= \tau(h[..\mathrm{Last}(e, h)]) \text{ (last in first out)} \end{aligned}$$

Let $p' = [(. a . \text{nil} \| . b . \text{nil})]_{\{\tau\}}$. The only possible traces from p to p' are: ef, fe, and $f * e$; that is the occurrence of both events in any order. For all three histories:

$$\pi(a, ef) = \pi(a, fe) = \pi(a, e * f) = 1$$
$$\pi(b, ef) = \pi(b, fe) = \pi(b, e * f) = 0$$

and therefore $a \not\succ_h b$.

However, if we use a context $C[\_] = c : \_$ with $\tau(c) = 2$; we obtain:

$$c : p \overset{ch}{\Longrightarrow}_\pi p'$$

where h is any of the three histories above. Now:

$$\pi(a, c\,ef) = \pi(a, cfe) = \pi(a, c\,(e * f)) = 1$$
$$\pi(b, c\,ef) = \pi(b, cfe) = \pi(b, c\,(e * f)) = 2$$

Therefore $a \not\succ_{ch} b$ and the scheme is not compositional.

5.5 A sufficient condition for compositionality

This condition requires that we extend the operators that we have defined over events and actions to traces. We do this in a straightforward way:

Closure. The closure of a trace is obtained by taking the closure of all the actions and leaving the events unchanged.

$$\begin{aligned} [\epsilon]_I &= \epsilon \\ [at]_I &= [a]_I[t]_I \\ [et]_I &= e[t]_I \end{aligned}$$

Note that the expression $[a]_I$ denotes the closure over actions that was defined earlier.

Synchronization. The synchronization of a trace is obtained by taking the synchronization of all the events and leaving all the actions unchanged. If any of the events cannot be synchronized, the trace cannot be synchronized and is considered undefined:

$$\epsilon\lfloor_E = \epsilon$$

$$(at)\lfloor_E = \begin{cases} a(t\lfloor_E) & \text{if } t\lfloor_E \text{ is defined} \\ \text{undefined} & \text{otherwise} \end{cases}$$

$$(et)\lfloor_E = \begin{cases} e(t\lfloor_E) & \text{if } t\lfloor_E \text{ is defined and } l(e) \cap E = \emptyset \\ e(t\lfloor_E) & \text{if } t\lfloor_E \text{ is defined and } l(e) \subseteq E \\ \text{undefined} & \text{otherwise} \end{cases}$$

Concatenation. This is the usual notion of string concatenation.

$$\begin{aligned} \epsilon t &= t \\ (at_1)t_2 &= a(t_1 t_2) \\ (et_1)t_2 &= e(t_1 t_2) \end{aligned}$$

Composition. This one is a little more tricky for several reasons. First we have to make sure action timings are consistent. When necessary we need to break an action into several consecutive actions to ensure we have compatible timings. Second we need to make sure that we account for all the possible ways of composing events; this results in a set of traces being generated. Therefore $t_1 * t_2$ is defined as a set of traces such that:

$$\begin{aligned} \epsilon * \epsilon &= \{\epsilon\} \\ et_1 * ft_2 &= \quad \{et \mid t \in t_1 * ft_2\} \\ &\cup \; \{ft \mid t \in et_1 * t_2\} \\ &\cup \; \{(e * f)t \mid (e * f) \textit{ is defined } \wedge t \in t_1 * t_2\} \\ et_1 * at_2 &= \{et \mid t \in t_1 * at_2\} \\ at_1 * et_2 &= \{et \mid t \in at_1 * t_2\} \\ at_1 * bt_2 &= \quad \{(a * b)t \mid \; (a * b) \textit{ is defined } \wedge t \in t_1 * t_2\} \\ &\cup \; \{(a' * b)t \mid \; (a' * b) \textit{ is defined } \wedge t \in a''t_1 * t_2 \\ &\qquad \wedge a = \mathrm{seq}(a', a'')\} \\ &\cup \; \{(a * b')t \mid \; (a * b') \textit{ is defined } \wedge t \in t_1 * b''t_2 \\ &\qquad \wedge b = \mathrm{seq}(b', b'')\} \end{aligned}$$

We can now define our necessary condition for compositionality:

Theorem 5.1 *If all the priority functions are such that whenever $\pi(\alpha, h)$ is defined, all the following conditions are true for all $h' \in \mathcal{V}^*, I \in \mathcal{R}$ and $E \subseteq \mathcal{Q}(\mathcal{L})$*

$$\begin{aligned} \pi(\alpha, [h]_I) &= \pi(\alpha, h) \\ h\downharpoonleft_E \textit{ is defined} &\Longrightarrow \pi(\alpha, h\downharpoonleft_E) = \pi(\alpha, h) \\ h'' \in h * h' &\Longrightarrow \pi(\alpha, h'') = \pi(\alpha, h) \\ \pi(\alpha, h'h) &= \pi(\alpha, h) \end{aligned}$$

then the associated preemption relation is compositional.

Our three examples satisfy the hypothesis of the above theorem and are therefore compositional *provided that event names are unique.* To see this, note that all three functions are entirely defined by the time elapsed since the *last* occurrence of a given event, and nothing that occurs in parallel or before that occurrence can change this priority once it is defined.

5.6 On local histories

The requirement that event names be unique limits our ability to compose arbitrary systems. One way to avoid this would be to define histories to be *local* to a "thread of execution". So, for example, the rule for parallel composition would be modeled on the following.

$$\left.\begin{array}{l}\langle p,h\rangle \xrightarrow{a}_{\mathcal{A}} \langle p',ha\rangle \\ \langle q,h'\rangle \xrightarrow{b}_{\mathcal{A}} \langle q',h'b\rangle \\ a*b \;\; \textit{is defined}\end{array}\right\} \Longrightarrow \langle\langle p,h\rangle \| \langle q,h'\rangle\rangle \xrightarrow{a*b}_{\mathcal{A}} \langle\langle p',ha\rangle \| \langle q',h'b\rangle\rangle$$

The problem with this approach is that the mathematical structure of this operational semantics is not well understood, in particular, this new type of labeled transition system is not a subset of:

$$(T_{\Sigma}(X) \times \mathcal{V}^*) \times \mathcal{A} \times (T_{\Sigma}(X) \times \mathcal{V}^*)$$

6 Conclusion and future area of research

We have presented a process algebra which combines the notions of real-time execution with continuous time, instantaneous synchronization, resource constraints and dynamic priorities. Our future efforts will concentrate on ways to define and prove equivalence between CCSR 92 processes.

Because of the continuous time assumption, it is not straightforward to apply bisimulation. We are investigating *continuous labeled transition systems* that can be derived from the transition systems defined in this article by simply adding an infinite set of transitions, corresponding to all possible ways of decomposing time intervals. Bisimulation can be applied to this kind of transition system and lead to a reasonable notion of equivalence.

We feel that the bisimulation equivalence is too fine because it differentiates between processes such as $abc + abd$ and $a(bc + bd)$ whose behavior is, for all practical purposes, indistinguishable. We are investigating ways to apply the notion of test equivalence defined by DeNicola and Hennessy [14]. The notion of priority affects thes testing relations in that, for example, the *may ordering* is no longer monotone, and not continuous.

This work is based on CCSR, developed by Gerber and Lee. We have expanded CCSR to include the notions of events, i.e., non resource-bound synchronization and dynamic priorities.

References

[1] R. Cleaveland and M. Hennessy. Priorities in Process Algebras. In *Proc. of IEEE Symposium on Logic in Computer Science*, 1988.

[2] R. Cleaveland and M. Hennessy. Priorities in Process Algebras. *Information and Computation*, 87:58–77, 1990.

[3] R. Gerber. *Communicating Shared Resrouces: A Model for Distributed Real-Time Systems.* PhD thesis, Department of Computer and Information Science, University of Pennsylvania, 1991.

[4] R. Gerber and I. Lee. Communicating Shared Resources: A Model for Distributed Real-Time Systems. In *Proc. 10th IEEE Real-Time Systems Symposium*, 1989.

[5] R. Gerber and I. Lee. The Formal Treatment of Priorities in Real-Time Computation. In *Proc. 6th IEEE Workshop on Real-Time Software and Operating Systems*, 1989.

[6] R. Gerber and I. Lee. A Proof System for Communicating Shared Resource. In *Proc. 11th IEEE Real-Time Systems Symposium*, 1990.

[7] R. Gerber and I. Lee. A Resource-Based Prioritized Bisimulation for Real-Time Systems. Technical Report MS-CIS-90-69, University of Pennsylvania, Department of Computer and Information Science, September 1990.

[8] R. Gerber and I. Lee. CCSR: A Calculus for Communicating Shared Resources. Technical Report MS-CIS-90-16, University of Pennsylvania, Department of Computer and Information Science, March 1990.

[9] R. Gerber and I. Lee. CCSR: A Calculus for Communicating Shared Resources. In *Proc. of CONCUR90, LNCS 458*, August 1990.

[10] R. Gerber and I. Lee. Specification and Analysis of Resource-Bound Real-Time Systems. In *Proceedings of REX Workshop on Real Time: Theory and Practice*, June 1991.

[11] M. Hennessy. *Algebraic Theory of Processes.* MIT Press Series in the Foundations of Computing. MIT Press, 1988.

[12] C.A.R. Hoare. Communicating sequential processes. *Communications of the ACM*, 21(8):666–676, August 1978.

[13] R. Milner. *Communication and Concurrency.* Prentice-Hall, 1989.

[14] R. De Nicola and M. Hennessy. Testing Equivalences for Processes. In *ICALP, LNCS 154*, pages 548–560, 1983.

Exception Handling in Process Algebra

F.S. de Boer * J. Coenen † R. Gerth‡

Eindhoven University of Technology

Department of Mathematics and Computing Science

P.O. Box 513, 5600 MB Eindhoven, The Netherlands

E-mail: {wsinfdb, wsinjosc, robg}@win.tue.nl

Abstract

We study exception handling as it occurs, e.g., in ADA, aiming at an algebraic characterization. We take Bergstra and Klop's Algebra of Communicating Processes (ACP) as our starting point and equationally define strong bisimulation for ACP extended with exception handling primitives. This theory is then applied to showing fault tolerance under an explicitly stated fault hypothesis of a system that is made more fault resilient by applying dynamic redundancy.

1 Introduction

Exception handling has received scant algebraic treatment. In fact, [HH87] and [Dix83] are the only papers that we are aware of that touch on this topic. In [HH87], the *interrupt* construct of [Hoa85], $P\hat{}Q$, is utilized to express recovery from errors or exceptions. In [Dix83] the term exception is used for Hoare's interrupt construct. The construct satisfies the following SOS rules: $P\hat{}Q \xrightarrow{a} P'\hat{}Q$ provided $P \xrightarrow{a} P'$ and $P\hat{}Q \xrightarrow{a} Q'$ just in case $Q \xrightarrow{a} Q'$. I.e., execution of P can always be interrupted by the first action of Q. If † is a symbol standing for an error, then associating an *error handler* Q for this error to process P is done by $P\hat{}(\dagger \to Q)$: if one assumes that P does not generate †, then in a process $\dagger \,||\, (P\hat{}(\dagger \to Q))$, in which the leftmost process specifies the *error hypothesis* that at most one error may occur while P or Q is executing, the handler Q can interrupt P only if the error actually occurs (Hoare's parallel operator imposes synchronization on common actions.)

Admirably though this approach fits their purpose, we feel there is room for improvement. Let us write $P \hookrightarrow Q$ for associating an exception handler Q to a process P. Now, what should this construct satisfy? Authoritative answers to this question can be found in [Ada83, Cri85, LS90]:

1. In a process $(P \hookrightarrow Q) \hookrightarrow Q$ only the inner handler Q should be activated by an exception during execution of P so that a second exception occurring while Q is active can still be caught by the outer handler.

---

*NWO/SION project "Research and Education in Computer Science (REX)."

†NWO/SION project "Fault Tolerance: Paradigms, Models, Logics, Construction."

‡ESPRIT project: "Building Correct Reactive Systems (REACT)."

2. If the process P in $P \hookrightarrow Q$ raises an exception, then Q ought to handle this exception if it can.

Now, $P\hat{}(\dagger \to Q)$ as an implementation of $P \hookrightarrow Q$ fails on both counts: exception handlers need not be invoked innermost first; if P can raises the error $\dagger$ in $\dagger \,||\, (P\hat{}(\dagger \to Q))$ then it can not be handled by Q, and therefore it ought to be handled by a parallel process.

Another issue that we address is the algebraic treatment of *failure* due to unrecoverable errors. Such failures we want to be visible.

In this paper we propose to extend aprACP [BK84][1] to aprACP$_E$ with an exception handling construct that does satisfy the three earlier conditions. Furthermore, a recovery operator is introduced, which interprets the occurrence of certain actions as errors. Recovery from errors is described in terms of the synchronization merge of aprACP, i.e., the occurrence of an error requires synchronization with a corresponding action, a so-called handler, otherwise a failure occurs which gives rise to uncontrolled behaviour of the system. One of the main difficulties of a proper algebraic treatment of failure is the asymmetry between errors and their corresponding handlers: an error is an autonomous action whereas handlers are only activated when an error occurs. This asymmetry is analogous to the one between asynchronous send and receive actions, and between synchronous put and get actions [BW90]. However there is an important difference between the occurrence of an error, on the one hand, and synchronous put and asynchronous send actions, on the other hand: an error *has* to synchronize with a corresponding handler, otherwise a failure occurs, whereas a synchronous put or asynchronous send action is completely autonomous.

The language and an axiomatization of strong bisimulation is presented in Section 2. In Section 3 we turn to an example due to Peleska [Pel91]. A simple transformational process P is made more fault resilient by putting two copies of P in an arbitration protocol. The fault resilient version equals P under the fault hypothesis that the time interval between errors is large enough. Some conclusions are presented in Section 4.

2 aprACP$_E$ and its Axiomatization

2.1 Language and SOS

Let A be an alphabet of actions. We have $a, b, \ldots \in A$. We assume two more actions disjoint from A: δ and $\bot$; the former denoting inaction and the latter indicating the occurrence of an unrecoverable exception. Elements of $A \cup \{\delta, \bot\}$ are denoted by $\alpha, \beta, \ldots$.

The grammar in Table 1 specifies the syntax of process terms of aprACP$_E$ (we treat the left-merge $\cdot \,⌊\!⌊\, \cdot$ and the communication merge $\cdot|\cdot$ of aprACP as auxiliary operators.) By convention, prefixing $(\alpha\cdot)$ binds strongest, then comes $\hookrightarrow$, then $+$ and finally $||$. The behaviour of aprACP$_E$-terms is described in Table 2. Some of the SOS rules have negative premises, so there is a question of well-definedness. However, the rules are all in GSOS format, hence stratifiable, so that they define a proper transition relation on the process terms [Gro90].

[1] I.e., ACP with action prefixing instead of sequential composition.

x	$:=$	$\delta \mid \perp \mid \alpha{\cdot}x$	$(\alpha \in A \cup \{\delta, \perp\})$
	$\mid$	$x + y \mid x \,\|\, y$	
	$\mid$	$x \hookrightarrow y$	
	$\mid$	$\partial_H(x)$	$(H \subseteq A)$
	$\mid$	$\mathcal{R}_H(x)$	$(H \subseteq A)$

Table 1: Process terms

We have the following definition of bisimulation:

2–1 DEFINITION (Bisimulation). *Two processes x and y are bisimilar, notation: $x \simeq y$, if and only if there exists a relation R on processes such that $x \; R \; y$ and whenever $x' \; R \; y'$ then for every $a \in A$ if $x' \xrightarrow{a} x''$ then $y' \xrightarrow{a} y''$ for some y'' such that $x'' \; R \; y''$ and vice versa; and also if $x' \xrightarrow{\perp} x''$ then $y' \xrightarrow{\perp} y''$ for some y'' and vice versa.*

It should be noted that in the above definition in case of the occurrence of a failure the resulting processes are *not* required to be bisimilar. As a consequence the behaviour of a process becomes uncontrollable after the occurrence of a failure.

Since the SOS rules are obviously well-founded, we obtain that bisimulation is a congruence for the operators in Table 1 (Theorem 4.4 in [Gro91].)

The rules for $\perp$, $\alpha{\cdot}x$ and $+$ are as should be expected. In aprACP, the parallel operator is modelled as interleaving plus synchronization, where synchronization is described in terms of a communication function $\cdot|\cdot \in Act \cup \{\delta, \perp\} \times Act \cup \{\delta, \perp\} \to Act \cup \{\delta, \perp\}$. The encapsulation $\partial_H(\cdot)$ prohibits any action in H to occur and, hence, is similar to the CCS restriction [Mil89]. The process $x \hookrightarrow y$ resembles Hoare's $x \widehat{\ } y$ in that y may interrupt x anytime, but is dissimilar w.r.t. one essential point: control may only transfer to y through executing an initial action of y that x *cannot* perform. E.g., the process $a{\cdot}\delta \hookrightarrow a{\cdot}b{\cdot}\delta$ admits only one sequence of transitions:

$$a{\cdot}\delta \hookrightarrow a{\cdot}b{\cdot}\delta \xrightarrow{a} \delta \hookrightarrow a{\cdot}b{\cdot}\delta \xrightarrow{a} b{\cdot}\delta \xrightarrow{b} \delta \; .$$

We stress that in $x \hookrightarrow y$, activation of y is not subject to any other constraints. At the end of this subsection we shall see how to enforce that exception handlers can be activated by the occurrence of an error only. Given an action a the set of handlers of a, i.e., those actions b such that $a|b \in A$, is denoted by $a^{\downarrow}$. We assume that $a^{\downarrow} \cap b^{\downarrow} = \emptyset$ if $a \neq b$, i.e. a recovery action b can recover a particular type of exception actions only (viz. the unique action a such that $b \in a^{\downarrow}$.) The recovery operator $\mathcal{R}_H(\cdot)$ interprets the execution of an action $a \in H$ as the occurrence of an error which raises an exception handled by an

		$+$ and $\parallel$ are commutative
Fail	$\bot \xrightarrow{\bot} \delta$	$\bot\cdot x \xrightarrow{\bot} \delta$
Prefixing	$a\cdot x \xrightarrow{a} x$	
Choice	$\dfrac{x \xrightarrow{\alpha} z}{x+y \xrightarrow{\alpha} z}$	
Merge	$\dfrac{x \xrightarrow{\alpha} z}{x \parallel y \xrightarrow{\alpha} z \parallel y}$	$\dfrac{x \xrightarrow{\alpha} u \quad y \xrightarrow{\beta} v}{x \parallel y \xrightarrow{\alpha\|\beta} u \parallel v}$
Exception Handler	$\dfrac{x \xrightarrow{\alpha} z}{x \hookrightarrow y \xrightarrow{\alpha} z \hookrightarrow y}$	$\dfrac{x \stackrel{\alpha}{\not\longrightarrow} \quad y \xrightarrow{\alpha} z}{x \hookrightarrow y \xrightarrow{\alpha} z}$
Encapsulation	$\dfrac{x \xrightarrow{\alpha} y \ (\alpha \notin H)}{\partial_H(x) \xrightarrow{\alpha} \partial_H(y)}$	
Recovery	$\dfrac{x \xrightarrow{a} y \quad y \xrightarrow{b} z \quad (a \in H,\ b \in a^{\downarrow})}{\mathcal{R}_H(x) \xrightarrow{a\|b} \mathcal{R}_H(z)}$	
	$\dfrac{x \xrightarrow{a} y \quad y \stackrel{b}{\not\longrightarrow} \quad (a \in H, \forall_b(b \in a^{\downarrow}))}{\mathcal{R}_H(x) \xrightarrow{\bot} \delta}$	$\dfrac{x \xrightarrow{\alpha} y \quad (\alpha \notin H)}{\mathcal{R}_H(x) \xrightarrow{\alpha} \mathcal{R}_H(y)}$

Table 2: SOS rules for $\mathsf{aprACP_E}$

action in $a^{\downarrow}$. The result of the recovery of an error generated by an action a by a handler b for a is indicated by $a|b$. Unrecovered actions are indicated by $\bot$. As an example consider the process $\mathcal{R}_{\{a\}}(x)$, where $x = (a\cdot\delta \hookrightarrow b\cdot y) \parallel b\cdot z$, with b a handler of a. The occurrence of a is interpreted as an error, which can be handled by either $b\cdot y$ or $b\cdot z$: we have both

$$\mathcal{R}_{\{a\}}(x) \xrightarrow{a|b} y \parallel b\cdot z$$

and

$$\mathcal{R}_{\{a\}}(x) \xrightarrow{a|b} (\delta \hookrightarrow b\cdot y) \parallel z\ .$$

Thus we see that an error generated by a process is broadcasted so that it may raise an exception in any (but only one) process of the system. In this way, fault hypotheses, which are used to specify relative to what fault scenarios the systems is fault resilient, can be described as a parallel process. (see Section 3 for an example.) The scope within which an error or exception must be caught is determined by the recovery operator. E.g., in the process $\mathcal{R}_{\{a\}}(a{\cdot}\delta \hookrightarrow b{\cdot}y) \parallel b{\cdot}z$, the error a can only be caught by the handler $b{\cdot}y$. The autonomous character of an error can be best illustrated by the following example: consider the process $x = a{\cdot}\delta + a{\cdot}\delta \hookrightarrow b{\cdot}y$. Then

$$\mathcal{R}_{\{a\}}(x) \xrightarrow{\perp} \delta$$

because the error generated by the left summand cannot be recovered. So, once an error occurs other alternatives are disregarded.

An exception handler that is activated only if an error occurs can now be modelled as

$$\partial_{\{b\}} \circ \mathcal{R}_{\{a\}}(p \hookrightarrow b{\cdot}q)\,,$$

with $b \in a^{\perp}$ $(a|b \neq b)$. Thus we model exception handling analogously to the ACP treatment of concurrency: first, freely generate all potentially possible executions and then restrict this set to the actual ones.

2.2 The Axiom System

The axiomatization is an extension of the usual axiomatization of aprACP which consists of all the axioms of Table 4 and Table 3 but for the axioms concerning $\perp$: $\perp{\cdot}x = \perp$ and $\alpha|\perp = \perp$.

$x + y$	$=$	$y + x$
$(x + y) + z$	$=$	$x + (y + z)$
$x + x$	$=$	x
$x + \delta$	$=$	x
$\delta{\cdot}x$	$=$	δ
$\perp{\cdot}x$	$=$	$\perp$

Table 3: $\mathsf{aprBPA}_{\delta,\perp}$ axiomatization

Tables 5 and 6 extend these axioms. The combined set of equations is also denoted as $\mathsf{aprACP_E}$. The way $\cdot \hookrightarrow \cdot$ is axiomatized is analogous to that of the merge. We introduce auxiliary operators that force the left (right) process to move first, thus allowing choices to be resolved. Two auxiliary operators are needed here because $\cdot \hookrightarrow \cdot$ is not commutative. As an example, consider the

$x \parallel y$	$=$	$x \mathbin{\lfloor\!\lfloor} y + y \mathbin{\lfloor\!\lfloor} x + x \mid y$	
$(\alpha{\cdot}x) \mathbin{\lfloor\!\lfloor} y$	$=$	$\alpha{\cdot}(x \parallel y)$	
$(x+y) \mathbin{\lfloor\!\lfloor} z$	$=$	$x \mathbin{\lfloor\!\lfloor} z + y \mathbin{\lfloor\!\lfloor} z$	
$\alpha{\cdot}x \mid \beta{\cdot}y$	$=$	$(\alpha \mid \beta){\cdot}(x \parallel y)$	
$(x+y) \mid z$	$=$	$x \mid z + y \mid z$	
$x \mid (y+z)$	$=$	$x \mid y + x \mid z$	
$\alpha \mid \beta$	$=$	$\beta \mid \alpha$	
$\alpha \mid (\beta \mid \gamma)$	$=$	$(\alpha \mid \beta) \mid \gamma$	
$\alpha \mid \delta$	$=$	δ	
$\alpha \mid \bot$	$=$	$\bot$	$\alpha \neq \delta$
$\partial_H(a{\cdot}x)$	$=$	δ	$a \in H$
$\partial_H(\alpha{\cdot}x)$	$=$	$\alpha{\cdot}\partial_H(x)$	$\alpha \notin H$
$\partial_H(x+y)$	$=$	$\partial_H(x) + \partial_H(y)$	

Table 4: Merge and encapsulation

following derivation:

$$
\begin{aligned}
& (a{\cdot}\delta \hookrightarrow b{\cdot}y) \hookrightarrow b{\cdot}z \\
= \; & (a{\cdot}\delta \mathrel{\twoheadleftarrow\!\!\rhook} b{\cdot}y + a{\cdot}\delta \mathrel{\lhook\!\!\twoheadrightarrow} b{\cdot}y) \hookrightarrow b{\cdot}z \\
= \; & (a(\delta \hookrightarrow b{\cdot}y) + (a{\cdot}\delta + \delta) \mathrel{\lhook\!\!\twoheadrightarrow} b{\cdot}y) \hookrightarrow b{\cdot}z \\
= \; & (a(\delta \mathrel{\twoheadleftarrow\!\!\rhook} b{\cdot}y + \delta \mathrel{\lhook\!\!\twoheadrightarrow} b{\cdot}y) + \delta \mathrel{\lhook\!\!\twoheadrightarrow} b{\cdot}y) \hookrightarrow b{\cdot}z \\
= \; & (a(\delta + b{\cdot}y) + b{\cdot}y) \hookrightarrow b{\cdot}z \\
= \; & (a{\cdot}b{\cdot}y + b{\cdot}y) \hookrightarrow b{\cdot}z \\
= \; & (a{\cdot}b{\cdot}y + b{\cdot}y) \mathrel{\twoheadleftarrow\!\!\rhook} b{\cdot}z + (a{\cdot}b{\cdot}y + b{\cdot}y) \mathrel{\lhook\!\!\twoheadrightarrow} b{\cdot}z \\
= \; & (a{\cdot}b{\cdot}y \mathrel{\twoheadleftarrow\!\!\rhook} b{\cdot}z + b{\cdot}y \mathrel{\twoheadleftarrow\!\!\rhook} b{\cdot}z) + (a{\cdot}b{\cdot}y + b{\cdot}y) \mathrel{\lhook\!\!\twoheadrightarrow} b{\cdot}z \\
= \; & (a(b{\cdot}y \hookrightarrow b{\cdot}z) + b{\cdot}(y \hookrightarrow b{\cdot}z)) + \delta \\
= \; & a(b{\cdot}y \mathrel{\twoheadleftarrow\!\!\rhook} b{\cdot}z + b{\cdot}y \mathrel{\lhook\!\!\twoheadrightarrow} b{\cdot}z) + b{\cdot}(y \hookrightarrow b{\cdot}z) \\
= \; & a{\cdot}b(y \hookrightarrow b{\cdot}z) + \delta + b(y \hookrightarrow b{\cdot}z) \\
= \; & a{\cdot}b(y \hookrightarrow b{\cdot}z) + b(y \hookrightarrow b{\cdot}z)
\end{aligned}
$$

The left summand of the conclusion of this derivation describes the situation that after a the handler $b{\cdot}y$ is activated, whereas the right summand describes the immediate activation of the handler $b{\cdot}y$. Note that the handler $b{\cdot}z$ can only be activated after $b{\cdot}y$.

The recovery operator resembles the state operator [BW90]. After having seen an $a \in H$ action, the operator changes its behavior: the operator $\mathcal{R}^a_H$ searches for a handler for a, which then is transformed into $a|b$, in case such a handler cannot be found $\bot$ is delivered.

$x \hookrightarrow y$	$=$	$x \twoheadleftarrow\joinrel\rhook\, y + x \lhook\joinrel\twoheadrightarrow y$	
$(x+y) \twoheadleftarrow\joinrel\rhook\, z$	$=$	$x \twoheadleftarrow\joinrel\rhook\, z + y \twoheadleftarrow\joinrel\rhook\, z$	
$\alpha{\cdot}x \twoheadleftarrow\joinrel\rhook\, z$	$=$	$\alpha{\cdot}(x \hookrightarrow z)$	
$\delta \twoheadleftarrow\joinrel\rhook\, x$	$=$	δ	
$\perp \twoheadleftarrow\joinrel\rhook\, x$	$=$	$\perp$	
$x \lhook\joinrel\twoheadrightarrow (y+z)$	$=$	$(x \lhook\joinrel\twoheadrightarrow y) + (x \lhook\joinrel\twoheadrightarrow z)$	
$(\alpha{\cdot}x + y) \lhook\joinrel\twoheadrightarrow \beta{\cdot}z$	$=$	δ	$\alpha = \beta$
$(\alpha{\cdot}x + y) \lhook\joinrel\twoheadrightarrow \beta{\cdot}z$	$=$	$y \lhook\joinrel\twoheadrightarrow \beta{\cdot}z$	$\alpha \neq \beta$
$\delta \lhook\joinrel\twoheadrightarrow x$	$=$	x	

Table 5: Exception handler

2.2.1 Soundness

To prove soundness of the axiom system we define a model for the language which associates with each process a labelled transition system. A labelled transition system is a triple $(S, A, \longrightarrow)$ consisting of a set of states S, a set of labels A, and a transition relation $\longrightarrow\ \subseteq S \times A \times S$. We will use a representation of transition systems in non-well-founded set theory ([Acz88].) The techniques underlying the model construction are taken from [Rut92]. One of the advantages of this new approach is that the model is defined directly in terms of the SOS. Given that the SOS is well-founded and in GSOS format so that bisimulation is a congruence, we then can prove soundness of the axioms *without* having to define explicitly the semantic operators corresponding to the operators of the language.

First observe that we can associate SOS rules to the auxiliary operators that were used; see Table 7. These rules, too, are in GSOS format and are well-founded. In other words, bisimulation is a congruence for all operators, auxiliary or not.

2–2 DEFINITION. *Let P be the largest class satisfying*
$P = \mathcal{P}(A_\perp \times P)$.
(Here $A_\perp = A \cup \{\perp\}$.)

Formally, P is obtained as the largest fixed-point of the class operator Φ that assigns to every class X the class $\mathcal{P}(A_\perp \times X)$, i.e., the class of all *subsets* of $A_\perp \times X$ (see [Rut92].)

2–3 DEFINITION. *Let $M \in \mathsf{aprACP_E} \rightarrow P$ be defined as follows:*

$$M(x) = \{\langle \alpha, M(y)\rangle \mid x \xrightarrow{\alpha} y\}$$

This recursive definition can be justified by an application of the *solution lemma* according to which systems of equations of a certain class have a unique solution in non-well-founded set theory [Acz88]. We have the following theorem:

$\mathcal{R}_H(\alpha{\cdot}x)$	$=$	$\alpha{\cdot}\mathcal{R}_H(x)$	$\alpha \notin H$
$\mathcal{R}_H(a{\cdot}x)$	$=$	$\mathcal{R}_H^a(x)$	$a \in H$
$\mathcal{R}_H(x+y)$	$=$	$\mathcal{R}_H(x)+\mathcal{R}_H(y)$	
$\mathcal{R}_H^a(\delta)$	$=$	$\perp$	
$\mathcal{R}_H^a(b{\cdot}x)$	$=$	$(a\|b){\cdot}\mathcal{R}_H(x)$	$b \in a^{\downarrow}$
$\mathcal{R}_H^a(\alpha{\cdot}x)$	$=$	$\perp$	$\alpha \notin a^{\downarrow}$
$\mathcal{R}_H^a(\perp+x)$	$=$	$\perp+\mathcal{R}_H^a(x)$	
$\mathcal{R}_H^a(b{\cdot}x+y)$	$=$	$\mathcal{R}_H^a(y)$	$b \notin a^{\downarrow}$
$\mathcal{R}_H^a(b{\cdot}x+c{\cdot}y+z)$	$=$	$(a\|b){\cdot}\mathcal{R}_H^a(x)+\mathcal{R}_H^a(c{\cdot}y+z)$	$b,c \in a^{\downarrow}$

Table 6: Recovery

2–4 THEOREM. *For any processes x and y we have*
$x \simeq y \Leftrightarrow M(x) = M(y)$.

For a proof of this theorem we refer to [Rut92]. Since we know that $\simeq$ is a congruence, it suffices for proving soundness to show that for any axiom $x = y$ we have $M(x) = M(y)$.

2–5 THEOREM (Soundness). *For any two processes x and y*
$\mathsf{aprACP_E} \vdash x = y$ implies $x \simeq y$

As explained above we need now only to inspect the individual equations. We treat the following case: $x \hookrightarrow y = x \twoheadleftarrow\!\!\!\hookleftarrow y + x \hookrightarrow\!\!\!\!\twoheadrightarrow y$.

$$\begin{aligned}
& M(x \hookrightarrow y) \\
= \;& \{\langle\alpha, M(z)\rangle \mid x \hookrightarrow y \xrightarrow{\alpha} z\} \\
= \;& \{\langle\alpha, M(x' \hookrightarrow y)\rangle \mid x \xrightarrow{\alpha} x'\} \cup \{\langle\alpha, M(y')\rangle \mid x \stackrel{\alpha}{\not\longrightarrow},\ y \xrightarrow{\alpha} y'\} \\
= \;& M(x \twoheadleftarrow\!\!\!\hookleftarrow y) \cup M(x \hookrightarrow\!\!\!\!\twoheadrightarrow y) \\
= \;& M(x \twoheadleftarrow\!\!\!\hookleftarrow y + x \hookrightarrow\!\!\!\!\twoheadrightarrow y)
\end{aligned}$$

Finally we note that every guarded recursive (process) equation has a unique solution in P (see [Rut92].) For example, the equation $x = a{\cdot}x$ is interpreted in non-well-founded set theory as $x = \{\langle a, x\rangle\}$. Let $\pi(x)$ be the unique solution of x. Since P is the largest class satisfying the equation used for its definition we have that $\pi(x) \in P$.

2.2.2 Completeness

We first prove an elimination lemma. Let, $\mathsf{aprBPA}_{\delta,\perp}$ be the axiom system of Table 3 for basic processes, i.e., processes formulated in the signature $\{\delta, a{\cdot}, \perp{\cdot}, \cdot + \cdot \mid a \in A\}$.

Left Handler	$\dfrac{x \xrightarrow{\alpha} y}{x \twoheadleftarrow\!\!\rhook z \xrightarrow{\alpha} y \hookrightarrow z}$	
Right Handler	$\dfrac{y \xrightarrow{\alpha} z \quad x \overset{\alpha}{\not\longrightarrow}}{x \lhook\!\!\twoheadrightarrow y \xrightarrow{\alpha} z}$	
Handler-Search	$\dfrac{x \xrightarrow{b} y \quad (b \in a^{\downarrow})}{\mathcal{R}^{a}_{H}(x) \xrightarrow{a\mid b} \mathcal{R}_{H}(y)}$	$\dfrac{x \overset{b}{\not\longrightarrow} z \quad (b \in a^{\downarrow})}{\mathcal{R}^{a}_{H}(x) \xrightarrow{\perp} \delta}$

Table 7: SOS rules for the auxiliary operators

2–6 Lemma (Elimination). *For any* $x \in \mathsf{aprACP_E}$ *there is a basic process* y *such that* $\mathsf{aprACP_E} \vdash x = y$.

Using the axioms we can eliminate all the operators but prefixing and choice starting from the innermost one and "working our way up".

Completeness then follows from the completeness of $\mathsf{aprBPA}_{\delta,\perp}$ (which is a trivial extension of aprBPA_{δ}) and the above soundness theorem: let $x \simeq y$, according to the above lemma there exists basic processes x' and y' such that $\mathsf{aprACP_E} \vdash x = x'$ and $\mathsf{aprACP_E} \vdash y = y'$. By the soundness theorem we have that $x \simeq x'$ and $y \simeq y'$, so $x' \simeq y'$. From the completeness of $\mathsf{aprBPA}_{\delta,\perp}$ it then follows that $\mathsf{aprBPA}_{\delta,\perp} \vdash x' = y'$, and thus $\mathsf{aprACP_E} \vdash x = y$.

3 A distributed fault-tolerant system

In order to achieve a higher degree of reliability, a fault-tolerant system must exploit some form of redundancy. In the example below, which is very much inspired by Peleska's fault-tolerant system [Pel91], a system P is duplicated and embedded in a protocol that ensures correct behaviour despite the presence of faults. We have modified Peleska's original system in order to preserve correctness under a larger class of fault scenario's. Nevertheless we will refer to this system as Peleska's system.

In this section each exception can be recovered by one action only, i.e. $|a^{\downarrow}| = 1$ for all $a \in A$. Therefore, in this section we simply write $a^{\downarrow}$ instead of the unique action b that recovers $a^{\uparrow}$, for all $a^{\uparrow} \in a$.

Peleska's system is build around two duplicates P_1 and P_2 of the basic system P in table 8. The basic system inputs a value x on channel a and then computes the value $\varphi(x)$, which is outputted on channel b. It is assumed that all values are within a *finite* data domain D, and that $\varphi : D \mapsto D$ is a function on D. Any system that satisfies the equation of the basic system, can be systematically transformed into a more resilient system T that is *weakly* bisimilar with P; i.e., a system that is bisimilar if we abstract from internal

$P = \sum_{x \in D} a(x) \cdot b(\varphi(x)) \cdot P$

Table 8: Basic system

actions. This implies that for many applications one may simply replace P by system T.

The transformed system T consists of six components RP_1, RP_2, Q_1, Q_2, RR_1, and RR_2 (see Figure 1.) Components RP_1 and RP_2 are *restartable* (see e.g. [Pra84, Pra87, HH87].) A restartable system can be defined with the exception handler of aprACP$_E$. For example, in Table 9 a restartable version RP of the basic system P is defined. The restartable system RP behaves like the

$RP = P \hookrightarrow p^{\downarrow} \cdot RP$

Table 9: A restartable system

basic system P, until the exception $p^{\uparrow}$ is raised after which it is restarted. Components RP_1 and RP_2 are defined slightly different, because they operate in a master-slave configuration. Initially RP_1 is the active master until a fault is detected in P_1 (signalled by exception $z_1{}^{\uparrow}$.) Upon detection of the fault RP_1 is de-activated and RP_2 takes over, thereby switching the rôle of master and slave. If a fault is detected in P_2 (signalled by exception $z_2{}^{\uparrow}$) process RP_1 takes over again. As a matter of fact this is an example of a *dynamic redundant system* in which RP_1 and RP_2 alternately function as *hot-standby components*. The philosophy of such a system is that if faults don't occur too frequently — i.e. the *Mean Time Between Failures* (MTBF) is sufficiently larger than the *Mean Time To Repair* (MTTR) — the de-activated faulty component can be replaced while the other duplicate is operational.

To ensure that no data will be lost if the currently active component P_i ($i \in \{1, 2\}$) crashes, the stand-by component should receive a copy of the input data whenever the active component receives an input. Because the stand-by component is not active this might result in a deadlock. For this reason each RP_i is connected to a component RR_i. RR_i is a restartable component with a core process R_i. Processes R_i simultaneously accept the input data and then offer it to their corresponding component RP_i. Because RP_i may not be active, process RR_i might deadlock. For this reason the other process R_j — which is guaranteed to succeed because only one of the components RP_i can be de-activated at the time — sends a reset signal (f or g) after it has forwarded the result obtained from RP_j. Component RR_i restarts R_i when the exception handler $r_i{}^{\downarrow}$ is activated. The exception handler $r_i{}^{\downarrow}$ is synchronized with the exception handler $p_i{}^{\downarrow}$, and therefore triggered by exception $z_i{}^{\uparrow}$. If P_i crashes after accepting an input of RR_i, but before resetting the other component RR_j, it should be willing to accept a reset signal before restarting in order to avoid deadlock.

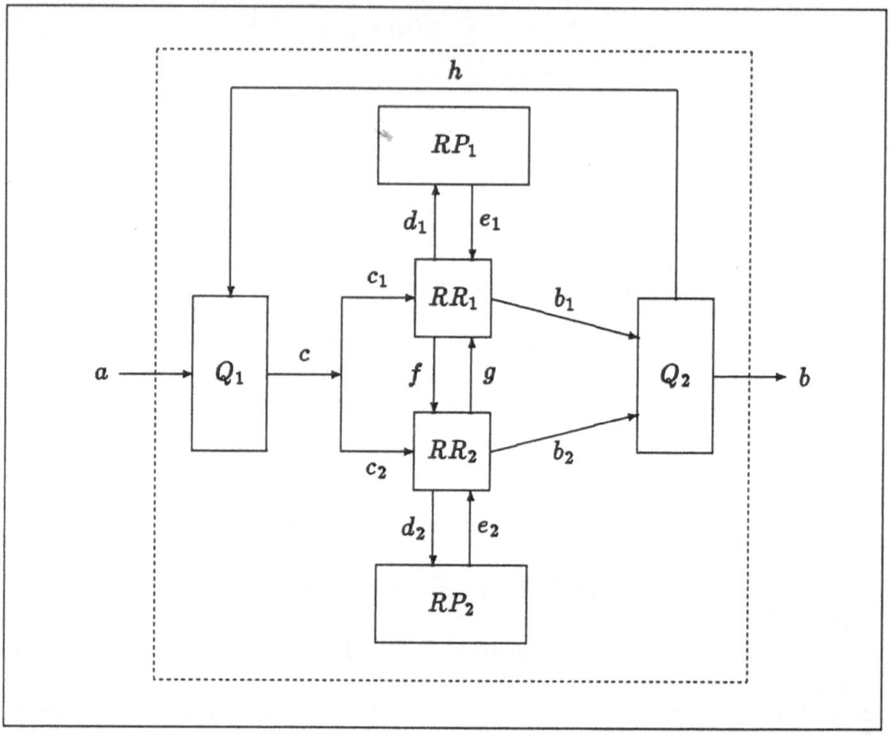

Figure 1: Fault-tolerant system T

There is still one problem to be resolved. In case a component P_i crashes just after RR_i has forwarded the result, but before RR_i has send the reset signal to the other component RR_j, RP_j becomes active and RR_j will forward its output also. To avoid such duplicate outputs an additional layer consisting of components Q_1 and Q_2 is included. Components Q_1 and Q_2 execute an alternating-bit protocol. Process Q_1 receives its inputs from the input channel a of the system T. Upon receipt of a message, it adds an extra bit to the message and forwards it to both components RR_1 and RR_2. A component RR_i removes the additional bit before passing the message to RP_i, but re-appends it again before forwarding the output messages from RP_i. Component Q_2 waits for a message of one of the components R_i. Upon receipt of a message the extra bit is inspected and removed. If the extra bit has the expected value then Q_2 outputs the message, to channel b of T and sends a signal h to Q_1. If the extra bit does not has the expected value then the received message is simply discarded. The signal h, which is not present in Peleska's original example, informs component Q_1 that it may accept a new input. Peleska's original system has a buffer capacity due to internal communications. This results in a communication latency which allows the transformed system T to input more than one message before giving an output message. For this reason Peleska's original transformed system is not weakly bisimilar with the basic system P.

The specification of the transformed system T and its components is listed in Table 10. The synchronization function is defined in Table 11, and the encapsulation set H is defined in Table 13.

T	$=$	$\partial_H((P_1 \hookrightarrow p_1{}^{\downarrow}\cdot RP_1) \parallel RP_2 \parallel Q_1(0) \parallel Q_2(0) \parallel RR_1 \parallel RR_2)$
P_1	$=$	$\sum_{x\in D} d_1(x)\cdot e_1(\varphi(x))\cdot P_1$
P_2	$=$	$\sum_{x\in D} d_2(x)\cdot e_2(\varphi(x))\cdot P_2$
RP_1	$=$	$(\delta \hookrightarrow p_2{}^{\downarrow}\cdot P_1) \hookrightarrow p_1{}^{\downarrow}\cdot RP_1$
RP_2	$=$	$(\delta \hookrightarrow p_1{}^{\downarrow}\cdot P_2) \hookrightarrow p_2{}^{\downarrow}\cdot RP_2$
$Q_1(n)$	$=$	$\sum_{x\in D} a(x)\cdot c(x,n)\cdot h\cdot Q_1(1-n)$
$Q_2(n)$	$=$	$\sum_{\substack{x\in D\\ i\in\{1,2\}}} (b_i(x,n)\cdot b(x)\cdot h\cdot Q_2(1-n) + b_i(x,1-n)\cdot Q_2(n))$
R_1	$=$	$\sum_{\substack{x\in D\\ n\in\{0,1\}}} c_1(x,n)\cdot(d_1(x)\cdot(\sum_{y\in D} e_1(y)\cdot b_1(y,n)\cdot f\cdot R_1) + g\cdot R_1)$
R_2	$=$	$\sum_{\substack{x\in D\\ n\in\{0,1\}}} c_2(x,n)\cdot(d_2(x)\cdot(\sum_{y\in D} e_2(y)\cdot b_2(y,n)\cdot g\cdot R_2) + f\cdot R_2)$
RR_1	$=$	$R_1 \hookrightarrow r_1{}^{\downarrow}\cdot(g\cdot RR_1 + RR_1)$
RR_2	$=$	$R_2 \hookrightarrow r_2{}^{\downarrow}\cdot(f\cdot RR_2 + RR_2)$

Table 10: Specification of T and it's components

$b_i(x,n)\vert b_i(x,n)$	$=$	$cb_i(x,n)$	$f\vert f$	$=$	cf
$c(x,n)\vert c(x,n)$	$=$	$cc(x,n)$	$g\vert g$	$=$	cg
$c_1(x,n)\vert c_2(x,n)$	$=$	$c(x,n)$	$h\vert h$	$=$	ch
$d_i(x)\vert d_i(x)$	$=$	$cd_i(x)$	$p_i{}^{\downarrow}\vert p_i{}^{\downarrow}$	$=$	$r_i{}^{\downarrow}$
$e_i(x)\vert e_i(x)$	$=$	$ce_i(x)$	$r_i{}^{\downarrow}\vert r_i{}^{\downarrow}$	$=$	$z_i{}^{\downarrow}$
			$z_i{}^{\uparrow}\vert z_i{}^{\downarrow}$	$=$	$z_i{}^{\Downarrow}$
$i \in \{1,2\}, n \in \{0,1\}, x \in D$					

Table 11: Synchronization function

Of course, no system can be guaranteed to function correctly in arbitrary conditions. Therefore we have to make some assumptions about occurrences of faults in a *fault hypothesis*. It then suffices to prove correctness of a system with respect to the fault hypothesis. In aprACP$_E$ a fault hypothesis can be modelled as a process. As such one may think of the fault hypothesis as Cristian's *adverse environment* [Cri85]. To prove correctness of the system T with respect to a fault hypothesis modelled by process FH, we have to verify the property in Table 12 ($\simeq_\tau$ denotes weak bisimulation and τ_I renames the action in I as τ.) The set I of internal moves and the set J of actions that must be recovered

$$P \quad \simeq_\tau \quad \tau_I \circ \mathcal{R}_J(T \,\|\, FH)$$

Table 12: Proof obligation

are defined in Table 13. In order to prove the property in Table 12 we need additional axioms for weak bisimulation and hiding. These axioms are included in Table 14 and their justification can be found in e.g. [BW90]. Note that aprACP$_E$ allows P and $\mathcal{R}_J(T \,\|\, FH)$ to be reduced to aprBPA$_{\delta,\perp}$-terms and the axioms in Table 14 suffice for weak bisimulation on aprBPA$_{\delta,\perp}$.

H	$=$	$\{b_i(x,n), c(x,n), c_i(x,n), d_i(x), e_i(x), f, g, h, p_i^{\downarrow}, r_i^{\downarrow}\}$
I	$=$	$\{cb_i(x,n), cc(x,n), cd_i(x), ce_i(x), cf, cg, ch, z_i^{\Downarrow}\}$
J	$=$	$\{z_i^{\uparrow}\}$
$i \in \{1,2\}, n \in \{0,1\}, x \in D$		

Table 13: Encapsulation, hide, and recovery set

$a{\cdot}\tau{\cdot}x$	$=$	$a{\cdot}x$	
$\tau{\cdot}x + x$	$=$	$\tau{\cdot}x$	
$a{\cdot}(\tau{\cdot}x + y)$	$=$	$a{\cdot}(\tau{\cdot}x + y) + a{\cdot}x$	
$\tau_I(\delta)$	$=$	δ	
$\tau_I(\perp)$	$=$	$\perp$	
$\tau_I(a{\cdot}x)$	$=$	$\tau{\cdot}\tau_I(x)$	$a \in I$
$\tau_I(a{\cdot}x)$	$=$	$a{\cdot}\tau_I(x)$	$a \notin I$
$\tau_I(x + y)$	$=$	$\tau_I(x) + \tau_I(y)$	

Table 14: Axioms for weak bisimulation and hiding

Peleska's system is weakly bisimilar with the basic system P for the trivial fault hypothesis δ, which means that the *normal behaviour* of T satisfies the property in Table 12. Peleska's original system ([Pel91]) can tolerate a single failure of one of its basic components P_1 or P_2, which is expressed by the fault hypothesis $FH = z_1{}^{\uparrow}\cdot\delta + z_2{}^{\uparrow}\cdot\delta$. The system we present can tolerate any number of faults of P_1 and P_2 provided the interval between consecutive faults is large enough. This is modelled by synchronizing the fault hypothesis with the feedback signal h (see Table 15.) The corresponding proof obligation is also given in that Table (It is assumed that $ch|ch = cch$.)

FH	$=$	$z_1{}^{\uparrow}\cdot ch\cdot FH + z_2{}^{\uparrow}\cdot ch\cdot FH + ch\cdot FH$	Fault hypothesis
P	$\simeq_\tau$	$\tau_{I\cup\{cch\}} \circ \mathcal{R}_J \circ \partial_{\{ch\}}(T \parallel FH)$	Proof obligation

Table 15: Proof obligation for extended fault hypothesis

4 Conclusions

We have defined an exception handling primitive and recovery operator that have properties that are more in line with what is found in the fault tolerance literature [Cri85]; specifically, handlers are invoked innermost out and handlers can only become active through the occurrence of an error. We have developed an algebraic theory for these operators based on ACP [BK84]. We choose ACP because it is well developed, uniform theory. However, nothing stands in the way of developing a similar theory based on CCS [Mil89] or TCSP [Hoa85]. We have used this theory to specify a generalization of a fault resilient system of Peleska's [Pel91]. Finally, we want to note our use of non-well founded sets [Acz88] to construct models for our axiomatization. The standard method in ACP is to use the *process graph model*. In this model, elements are bisimulation equivalence classes of graphs and this fact makes the process graph model more difficult to use than the concrete model we introduce in this paper in which bisimilar process terms map onto the same element in the model, which element is straightforwardly determined by the SOS.

Future work includes further working out the example towards a formal proof of weak bisimilarity and extending the theory. We need to investigate more closely the connection of our theory with others such as the one in [Pra87]. Another question is whether a process algebra with prioritized actions [BW90] can be used to model exception handling. We already have some preliminary results. Finally, we want to extend our axiomatization to congruences coarser than bisimulation; specifically to maximal trace congruence.

Acknowledgement

We would like to thank Jos Baeten for his helpful comments.

References

[Acz88] P. Aczel. *Non-well-founded sets.* Number 14 in CSLI Lecture Notes. 1988.

[Ada83] American National Standards Institute, Inc. *The Programming Language Ada Reference Manual.* LNCS 155, Springer-Verlag 1983.

[BK84] J.A. Bergstra & J.W. Klop. *Process Algebra for Synchronous Communication.* Information and Control 60:109–137, 1984.

[BW90] J.C.M. Baeten & W.P. Weijland. *Process Algebra.* Cambridge Tracts in Theoretical Computer Science, Vol **18**, 1990.

[Cri85] F. Cristian. *A Rigorous Approach to Fault-Tolerant Programming.* IEEE Transactions on Software Engineering 11:23–31, 1985.

[Dix83] T.I. Dix. *Exceptions and Interrupts in CSP.* Science of Computer Programming 3:189–204, 1983.

[Gro90] J.F. Groote. *Transition System Specifications with Negative Premises.* Proc. CONCUR '90, LNCS 443 pp. 332-341, 1990.

[Gro91] J.F. Groote *Process Algebra and Structured Operational Semantics.* PhD Thesis Centre for Mathematics and Computer Science, University of Amsterdam, 1991.

[Hoa85] C.A.R. Hoare. *Communicating Sequential Processes.* Prentice-Hall 1985.

[HH87] He Jifeng & C.A.R. Hoare. *Algebraic Specification and Proof of a Distributed Recovery Algorithm.* Distributed Computing 2:1–12, 1987.

[LS90] K. Lodaya & R.K. Shyamasundar. *Proof Theory for Exception Handling in a Tasking Environment.* Acta Informatica 28:7–42, 1990.

[Mil89] R. Milner. *Communication and Concurrency.* Prentice-Hall 1989.

[Pel91] J. Peleska. *Design and Verification of Fault Tolerant Systems with CSP.* Distributed Computing 5:95–106, 1991.

[Pra84] K.V.S. Prasad. *Specification and Proof of a Simple Fault Tolerant System in CCS.* Internal Report CSR-1178-84, Department of Computer Science, University of Edinburgh, 1984.

[Pra87] K.V.S. Prasad. *Combinators and Bisimulation Proofs for Restartable Systems.* PhD Thesis University of Edinburgh, 1987.

[Rut92] J.J.M.M. Rutten. *Processes as terms: non-well-founded models for bisimulation.* Technical report CWI CS-R9211, also to appear in Mathematical Structures in Computer Science.

Session 3

Observation Trees
(Extended Abstract)

Pierpaolo Degano
Dipartimento di Matematica, Università di Parma
&
Dipartimento di Informatica, Università di Pisa

Rocco De Nicola
Dipartimento di Scienze dell'Informazione, Università di Roma "La Sapienza"

Ugo Montanari
Dipartimento di Informatica, Università di Pisa

Synopsis. *Four ways of structuring experiments over transition systems as observation trees are introduced. The nodes of the trees are labelled by the observations of the corresponding computations. The trees are compared via bisimulations and it is shown that various observational equivalences proposed in the literature can be recast in a general experimental setting. This permits assessing the different equivalences and putting forward criteria for choosing among them. It also throws light on the different assumptions made on observations by the different equivalences and permits a deeper understanding of the differences between, e.g., pomset and history preserving bisimulation, two well studied equivalences for true concurrency.*

1. Introduction

A successful approach to describing distributed concurrent systems consists of defining their operational semantics through transition systems and introducing equivalence relations to identify those systems which exhibit the same behaviour up to details that are considered inessential. Many of these equivalences rely on Park's notion of bisimulation [Par81].

Checking whether two systems are equivalent can be restated as the following game, involving also an observer: either of the two systems performs a sequence of transitions; the observer extracts from this sequence an *observation* that records only those aspects that are relevant to the actual comparison; then the observer waits for the other system to perform (possibly empty) another sequence of transitions with the same observation. If so, the first step of the game is successful, and the equiva-

Work partially supported by ESPRIT Basic Research Action n° 3011 CEDISYS and by Progetto Finalizzato Informatica e Calcolo Parallelo. The second author has also been supported by Istituto di Elaborazione dell'Informazione of CNR at Pisa.

lence check goes on from the states reached by the two systems after the transitions. If it is not possible to set apart systems by means of these experiments then they are indistinguishable.

The literature reports many equivalences based on the above paradigm. They mainly differ on the way nondeterminism is taken into account and on what is observed out of experiments. Common observations are sequences of labels of transitions, possibly enriched with additional structure, e.g., representing the causal relations among the performed activities, or providing information about the sites where these occurred. In a companion paper [DDM92], we propose a unifying framework for comparing most of the bisimulation-based equivalences that have been introduced both in the interleaving and the non interleaving approaches to concurrency. Our proposal is based on the following four step procedure, each step of which can be investigated in isolation, regardless of choices made in the others.

1) Define, e. g. in a syntax driven way, *elementary transitions* which describe the immediate evolutions of the system from every state.
2) Obtain *experiments* of systems (from a given initial state) and give them a *tree structure*, by structuring system computations, i.e., paths in the transition system.
3) Introduce *observations* over experiments, to abstract from unwanted details, and decorate the tree above with observations, so obtaining what we call an *observation tree*.
4) Compare labelled trees (e.g., via bisimulations) to determine which terms have an *equivalent behaviour* according to the introduced observations.

The focus of [DDM92] is on systems equivalences, i.e., on item (4) above. Equivalences, which are parametric with respect to the way trees are labelled, are defined on observation trees, both operationally and axiomatically. Observation trees are labelled on their *nodes*, that *are* the experiments made on systems. In general, the label of a node is *not an action*, or whatever is obtained by observing a *single* transition, rather it is the observation of the whole experiment represented by the node. Observations may be of any kind, e.g., sequences of actions, partial orderings, localities, and the like. To prove whether any two finite systems, described through an operational semantics, are equivalent, one has only to build the suitable observation trees and then use the axiomatization of the preferred congruence, specialized on the actual observation.

The way in which experiments are organized as trees, i.e., item (2) above, has been neglected in [DDM92] and will be our main concern here. Four different ways of structuring computations as trees will be introduced. This enables us to recast within our framework many of the bisimulation based semantics discussed in the literature (among which [Mil80 and 83, AB84, vGV87, DDM87, BC88, vGG89, DD89, DDM90]), and to combine them in new, more powerful ways.

We will still assume elementary transitions (item (1) above) as given, possibly defined in the SOS style (like in [Mil89] for interleaving-based descriptions and in [DDM85], [DDM88], [DDM90] for more detailed ones). Similarly we will not specify how observations are actually extracted from computations to label trees (item (3) above). For this, we refer the reader to [DDM90] where instances of observations of CCS computations are defined.

Four Ways of Structuring Experiments

Assume a transition system as given, with elementary transitions denoted by ϑ. The simplest experiment which can be made on a system consists of firing an elementary transition at a time. If all the experiments are of this kind, the resulting tree has as nodes *sequences* of experiments of the form $\vartheta_1 ; \vartheta_2 ;...; \vartheta_n$, where ";" denotes the *sequential composition* of experiments (as usual, we have that $\vartheta_i ; \vartheta_{i+1}$ is defined only when the target of ϑ_i and the source of ϑ_{i+1} coincide). Obviously, when these experiments are structured as trees, the node $\vartheta_1 ; \vartheta_2 ; ...; \vartheta_n$ precedes every node $\vartheta_1 ; \vartheta_2 ;...; \vartheta_n ; \vartheta$ where ϑ is any elementary transition. The observation trees obtained in this way will be a *tree of kind 0* where $\vartheta_1 ; \vartheta_2 ; ...; \vartheta_n \leq_0 \vartheta_1 ; \vartheta_2 ; ...; \vartheta_n ; \vartheta$.

The experiments outlined above are somewhat too basic; often, in checking systems equivalence one would like to let them run for a while and not stop them after any elementary step. Consider as an example those weak equivalences that ignore the occurrence of some "internal" activities. In order to deal with these more general situations, sequences of transitions (i.e., *computations* or paths in the transition system) can be considered as experiments. As any other experiment, they occur as a whole and cannot be interrupted, namely they are *atomic*. As usual, there will be the empty computation, "single-step" computations (each made of a single transition), and longer computations of the form $\vartheta_1 \bullet \vartheta_2 \bullet ... \bullet \vartheta_n$, where "$\bullet$" denotes the *concatenation* c • c' of computations (again, we have that this is defined only if the target of c coincides with the source of c'). The new trees have now as nodes *sequences* of computations of the form $\xi = \vartheta_{11} \bullet ... \bullet \vartheta_{1m_1} ; \vartheta_{21} \bullet ... \bullet \vartheta_{2m_2} ; ... ; \vartheta_{n1} \bullet ... \bullet \vartheta_{nm_n}$, where ϑ_{ij} are elementary transitions here considered as simple computations consisting of a single transition.

Within this new experimental setting, a node ξ precedes any other node of the form $\xi ; \xi'$, where ξ' is a sequence of experiments; the observation trees obtained in this way will be *tree of kind* 2 and we will have $\xi \leq_2 \xi ; \xi'$.

By considering a degenerate case of the above, where only atomic observations are allowed, or, alternatively, sequential composition of experiments and concatenation of computations are collapsed, it is possible to get trees of *kind 1*. The nodes of the trees of kind 1 are just long computations and not sequences of experiments. In this case we have that a node c precedes the nodes c' • c" (notationally, $c \leq_1 c' \bullet c''$).

We would like to remark that although sequential composition of experiments and concatenation of computations are distinct in general, in the literature their difference is not fully appreciated and they are identified almost always. This is possibly due to the fact that in the classical interleaving approach to concurrency they lead to the same distinctions. It is obvious that, when the two notions of sequential composition of experiments and concatenation of computations do coincide, we obtain trees of kind 1.

Our list of observation trees ends with those obtained by unioning the orderings of type 1 with those of type 2. The new trees have sequences of computations as nodes and are partially ordered by the union of $\leq_2$ and (the obvious extension to sequences of computations of) $\leq_1$. The observation trees obtained in this way will be *tree of kind 3* and we will have $\xi \leq_3 \xi'$ iff $\xi \leq_1 \xi'$ or $\xi \leq_2 \xi'$. This type of trees has never been used in the literature, although, as we will see later, they provide a sharper tool for analyzing concurrent systems.

Two sequential compositions of experiments

The introduction of two distinct operators, the sequential composition of experiments ";" and the concatenation of computations "•", may seem unnecessarily complex and obscure. We will try to convince the reader, by means of an example, that such a distinction is indeed necessary, and that it has been already used implicitly in the literature.

Consider the following system (described by a CCS term)

$$\alpha \cdot \gamma \cdot nil \mid \beta \cdot nil$$

consisting of two concurrent sub-systems, one performing action α followed by γ, the other action β. This system may perform, among others, the following two-step computation

$$c_1 = \alpha \cdot \gamma \cdot nil \mid \beta \cdot nil \rightarrow \gamma \cdot nil \mid \beta \cdot nil \bullet \gamma \cdot nil \mid \beta \cdot nil \rightarrow \gamma \cdot nil \mid nil$$

and, after it, the single-step computation

$$c_2 = \gamma \cdot nil \mid nil \rightarrow nil \mid nil.$$

If one is interested in the causal relation between the actions of these computations, and partially ordered multisets (*pomsets*) observations have been chosen, we have the association reported below:

$$o(c_1) = \alpha \text{ PAR } \beta \quad \text{and} \quad o(c_2) = \gamma,$$

where α PAR β is meant to be the pomset with two concurrent events, one labelled by α and the other by β, and γ the pomset with one event, labelled by γ.

Let us now consider the *sequence* of computations

$$\xi = c_1;c_2 = (\alpha \cdot \gamma \cdot nil \mid \beta \cdot nil \rightarrow \gamma \cdot nil \mid \beta \cdot nil \bullet \gamma \cdot nil \mid \beta \cdot nil \rightarrow \gamma \cdot nil \mid nil) \; ; \; (\gamma \cdot nil \mid nil \rightarrow nil \mid nil).$$

Following [BC88], [DDM90] or [Bro86] the observation of ξ amounts to some sequentialization "⨾" of the observations of c_1 and c_2, i.e.,

$$o(\xi) = (\alpha \text{ PAR } \beta) \fatsemi \gamma$$

Had we identified ";" and "•", we would get a computation that constitutes the *single* experiment

$$c = c_1 \; ; \; c_2 = \alpha \cdot \gamma \cdot nil \mid \beta \cdot nil \rightarrow \gamma \cdot nil \mid \beta \cdot nil \bullet \gamma \cdot nil \mid \beta \cdot nil \rightarrow \gamma \cdot nil \mid nil \bullet \gamma \cdot nil \mid nil \rightarrow nil \mid nil.$$

Its observation (according to [BC88, DDM90]) is

$$o(c) = (\alpha\gamma) \text{ PAR } \beta$$

where $(\alpha\gamma)$ PAR β is meant to be the pomset with three events, α causing β, and γ concurrent with both.

Clearly, the two observations $o(\xi)$ and $o(c)$ cannot be the same, for reasons of symmetry, no matter how the sequentialization operation "⨾" between pomsets is defined. And indeed the system has been experimented in two different ways: in the first one on its ability to perform, possibly concurrently, two events and then another one; in the second way on the possibility it has of executing a sequence of two actions concurrently with a third one. This remark will support our choice of considering "⨾" as a free operation of sequentialization. Therefore, as shown in the partial ordering approach, it makes sense to consider the experiments ξ and c as different and to let them give rise to two different nodes in a tree.

Observing a simple labelled event structure

Four different kinds of trees of experiments can thus be built for a given transition system. In order to check whether two such trees are equivalent, each node will be labelled by a specific observation, e.g., a sequence of actions, a pomset or the like. As anticipated in the example above, we will define the observation of a sequence of computations as the free sequentialization of their observations, i.e., $o(\xi\ ;\ \xi') = o(\xi)\ ;\ o(\xi')$.

Below, we give an example within the partial ordering approach to concurrency. Consider the labelled event structure (les) [Win87] of Fig. 1, where symbol # and the downwards arrow denote conflict and causal dependency, respectively. The computations of the les are the left closed and conflict free subsets of its events. As observations of (sequences of) experiments we take (sequences of) weak pomsets, i.e., pomsets where events labelled by the invisible action τ have been eliminated (graphically, labelled dots • denote event occurrences).

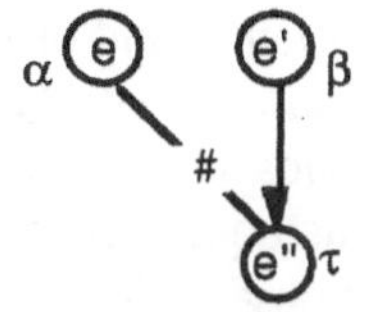

Figure 1: A Labelled Event Structure

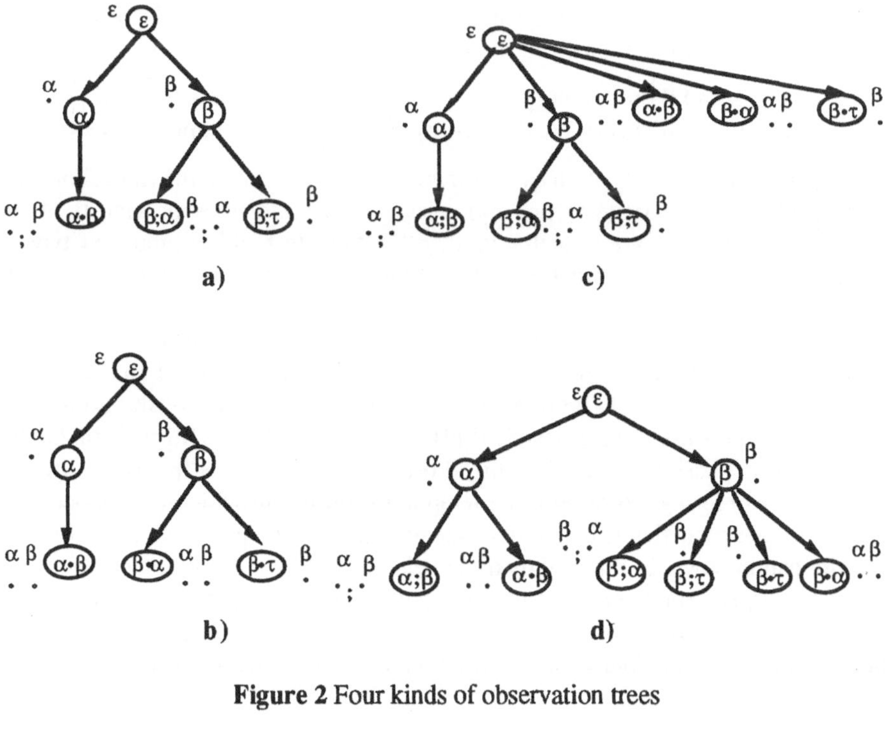

Figure 2 Four kinds of observation trees

The tree of kind 0 is in Fig. 2.a; that of kind 1 is in Fig. 2.b. Fig. 2.c shows the tree of kind 2, in which the experiment $\alpha;\beta$, consisting of the atomic transition α followed by the atomic transition β, is different from $\alpha \bullet \beta$, consisting of a single atomic step, and has a different observation. The tree of kind 3 (Fig. 2.d), is a combination of those of type 1 and 2. Its nodes, i.e., the experiments, are the same as those of the tree of type 2, while the ordering between the nodes is obtained as the union of the two orderings.

Observations drive the choice of the tree

Two groups of observations will be considered in the paper, regardless of their expressive power. The relationships between them and the different kinds of trees will then be studied, thus providing a rationale for choosing one kind of tree or another. Indeed, the discriminative power of an equivalence does not depend only on the way it deals with nondeterminism and on the expressive power of the chosen observations, but also on the chosen strategy to build the observation tree. At one extreme of the spectrum "observations / kind of trees" there are the classical interleaving observations, namely sequences of (labels of) actions. It is immediate verifying that any of the four kinds is as good as the others in making use of bisimulation to determine systems equivalence— string juxtaposition *is* the free operation of sequentialization ; of observations. The example observation of $\alpha{\cdot}\gamma{\cdot}$nil | $\beta{\cdot}$nil above shows that this is not the case when pomsets are chosen as observations and thus, in this case different kinds of trees lead to different equivalences.

Observations can be characterized in terms of two intuitive properties which are based on how they can be sequentialized. Let c_1, c_2, c_1' and c_2' be sequences of transitions, i.e., computations. We say that an observation function o is:

- *incremental* if $o(c_1) = o(c_1')$ and $o(c_2) = o(c_2')$ imply $o(c_1 \bullet c_2) = o(c_1' \bullet c_2')$.
- *cancellative* if $o(c_1) = o(c_1')$ and $o(c_1 \bullet c_2') = o(c_1' \bullet c_2')$ imply $o(c_2) = o(c_2')$.

In the paper we will see that, in case of incremental observation, the right choice for testing whether two systems are equivalent, is to build trees of kind 0; if the observation functions are cancellative then the trees to build are those of type 1; finally if the observation functions enjoy both properties all four kind of trees do collapse.

In our opinion, the fact that sequences of actions are both incremental and cancellative, has hindered the difference between the four kind of trees we have presented. Examples of incremental but not cancellative observations are concatenable concurrent histories [FMM91a] and causal streams [FMM91b]. On the contrary, it is possible to see that mixed ordering observations (pomsets enriched with a total ordering expressing the temporal ordering in which events occurred [DDM90]) are cancellative but not incremental. Thus in all these cases no increase in discrimination power is obtained by introducing tree of kind 3. Interestingly enough, pomsets, the structure more frequently used for describing causality preserving semantics of concurrent systems, are neither incremental nor cancellative; thus, with these observations, the four kinds of trees lead to four different semantics.

Embedding other semantics

By a suitable combination of kinds of trees and type of observation, many equivalences studied in the literature are captured.

It is not difficult to see that interleaving observations permit recovering the observational equivalence of [HM85, Mil80], whichever kind of observation trees is used. The difference between the kinds of trees is ineffective also with multiset observations, i.e., sequences of multisets of actions; these observations together with strong and weak bisimulation induce the same identifications as the semantics for SCCS and Meije of [Mil83] and [AB84], respectively.

When pomset observations are used, the differences between the kinds of trees come into play. The trees of kind 1 are the same trees introduced in [DDM87] and called NMS. When coupled with strong bisimulation pomset observations and trees of kind 1 give rise to the po-bisimulation of [DDM87], also studied under the name of weak history preserving bisimulation in [vGG89]. The trees of kind 2 permit recovering pomset bisimulation as described, e.g., in [BC88, vGV87, DDM90]. The trees of kind 3, a combination of those of type 1 and 2, give rise to another semantics which is certainly naturally characterized and has not yet been investigated in the literature. Some examples will be discussed in the paper.

The history preserving bisimulation of [RT88], the mixed ordering one of [DDM89] and the causal tree one of [DD89] are obtained when mixed ordering observations are used. Roughly, mixed orderings are pomsets, with an additional total ordering that keeps track of the temporal ordering in which actions took place.

Also equivalences that take into account the spatial distribution of processes are captured by an adequate combination of proper kind of trees and of observations. Indeed, [MY92] proves that the location equivalence of [BCHK92] is recovered by using trees of kind 1 and suitable location observations, that essentially associate to each event also a label describing the place where it occurred.
The two tables of Figures 3 and 4 show, by referring to the relevant papers, which ones of the strong observational equivalences (i.e. equivalence which do not disregard τ's) proposed in the literature fall within our general framework.

We would like to remark that we concentrate here only on strong observations and thus on the induced strong bisimulations. For details about weak and branching bisimulation and about a new relation called jumping bisimulation, we refer the reader to [DDM92]. Indeed, the issue of taking into account or ignoring invisible moves is completely orthogonal to the discussion about the role of the four kinds of trees with respect to cancellative or incremental observations.

We end this section by briefly mentioning proved trees [DP92], a model tightly related to observation trees, as shown also by Figure 4. A proved tree is built from a transition system defined in SOS style, and each of its arcs, representing the occurrence of a transition, is labelled by the proof of the transition itself. As done here, proved trees are then observed yielding different models.

observations → / orderings on computations ↓	interleaving	multiset
$\leq_1$ and $\leq_2$ (and $\leq_3$)	Strong obs. equiv [Mil80]	Strong bisimul. [Mil83, AB84]

Figure 3

observations → / orderings on computations ↓	partial ordering	mixed ordering	locality observations
$\leq_1$	NMS - po equiv. [DDM87] Weak history preserving bisim. [vGG89]	Mo bisim. [DDM89] Causal bisim. [DD89], [DP92] History preserving bisim. [RT88]	Location equiv. [BCHK92], [DP92], [MY92]
$\leq_2$	Pomset bisim. [BC88], [vGV87] Po observ. equiv. [DDM90]	Mo observ. equiv. [DDM90]	

Figure 4

2. Background

Here, we recall from [DDM92] the basic definitions and results on Observation Trees. As discussed in the Introduction, the nodes of an observation tree represent (sequences of) experiments made on systems and the label of a node gives a description of the experiment, i.e., a particular *observation* of it, taken from a given domain.

2.1 Definition (*Observation Structures*)
Given a set D of *observations*, an *observation structure* is a triple $\langle S, \rightarrow, o \rangle$, where

- S is a set of *nodes*, ranged over by r, s, possibly indexed;
- $\rightarrow \subseteq S \times S$ is a relation on nodes called *transition relation*; and
- o: $S \rightarrow D$ is an *observation function* mapping nodes into observations.

2.2 Definition (*Observation Tees*)
Given a partial ordering $(D, \preccurlyeq)$ of observations, an *observation tree* is an observation structure $\langle S, \rightarrow, o \rangle$, where:

(i) relation $\rightarrow$ is acyclic;
(ii) $r \rightarrow s$ and $r' \rightarrow s$ implies $r = r'$; and
(iii) $r \rightarrow s$ implies $o(r) \preccurlyeq o(s)$. ♦

The next two definitions, introduce the notion of strong bisimulation over observation trees that will be used to obtain strong observational congruence and a syntax for describing finite observation structures.

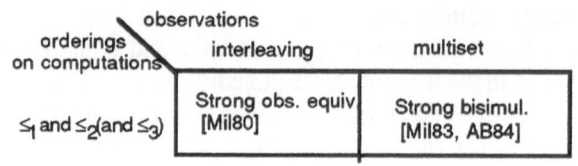

2.3 Definition (*strong bisimulation and equivalence*)
Given an observation structure ‹S, →, o›, a symmetric binary relation R on S such that r R s implies o(r) = o(s) is a *strong bisimulation*

if r R s and r → r' imply that there exists s' with s → s' and r' R s'.

The maximal *strong equivalence* will be denoted by ≈. ♦

2.4 Definition (*language for Observation Structures*)
Let D be a set of observations ranged over by $\mathfrak{a}, \mathfrak{b}$, ...; and let $* \notin D$ be a distinguished element. An *observation term* over D is E : x, where

• E is a term generated by the following grammar $E ::= NIL \mid \mathfrak{a}(E) \mid E \oplus E$

and

• $x \in D \cup \{*\}$ is the type of E given by the following equational type formulae [MSS90]

$NIL : * \quad E : x \Rightarrow \mathfrak{a}(E) : \mathfrak{a} \qquad E_1 : \mathfrak{a} \text{ and } E_2 : \mathfrak{a} \Rightarrow E_1 \oplus E_2 : \mathfrak{a}.$

We will use P, Q, R,... to denote observation terms, and resort to $P^{\mathfrak{a}}, Q^{\mathfrak{a}}, R^{\mathfrak{a}}$, ... when they are typed by $\mathfrak{a} \in D$, i.e., they are different from NIL. ♦

2.5 Definition (*from observation terms to the observation structure TOS*)
The observation structure TOS = ‹S, →, o› generated by observation terms is obtained by letting

• the *set of states* S be the set of observation terms $P^{\mathfrak{a}}$ typed on D;
• the *transition relation* $P^{\mathfrak{a}} \to Q^{\mathfrak{a}}$ be the smallest relation such that:

(i) $\mathfrak{b}(P^{\mathfrak{a}}) \to P^{\mathfrak{a}}$
(ii) $P^{\mathfrak{a}} \to Q^{\mathfrak{b}} \Rightarrow P^{\mathfrak{a}} \oplus R^{\mathfrak{a}} \to Q^{\mathfrak{b}}$
(iii) $R^{\mathfrak{a}} \to Q^{\mathfrak{b}} \Rightarrow P^{\mathfrak{a}} \oplus R^{\mathfrak{a}} \to Q^{\mathfrak{b}}$;

• the *observation function* o defined as $o(P^{\mathfrak{a}}) = \mathfrak{a}$. ♦

As it might be expected, it is possible to establish a correspondence between observation terms and finite observation trees. In order to associate a tree with a term, consider the corresponding state of TOS, and unfold TOS itself from it. If there is no corresponding state, i.e., if the term is NIL, the corresponding tree is empty. Alternatively, the operations of observation terms can be carried over observation trees. Clearly, NIL is the tree with no nodes; $\mathfrak{a}(P)$ adds to the tree of P a new root labelled by $\mathfrak{a}$ (provided that $\mathfrak{a}$ is smaller than, or equal to the label of the root of P); and $P^{\mathfrak{a}} \oplus Q^{\mathfrak{a}}$ coalesces the roots of the trees of $P^{\mathfrak{a}}$ and $Q^{\mathfrak{a}}$ (provided that the roots have the same label). Of course, different terms can represent the same tree. Indeed, finite observation trees are isomorphic to observation terms, up to commutativity, associativity and absorption of identity of ⊕ (see axiom A3 in Table 1).

The above definition states what is the observation structure to be associated with observation terms. We just give the flavour of the *viceversa* by the following term of type ε (observations are enclosed in square brackets, and typing has been omitted, for the sake of readability)

$$[\varepsilon]([\bullet^{\alpha}]([\bullet^{\alpha}\bullet^{\beta}](NIL))) \oplus [\varepsilon]([\bullet^{\beta}]([\bullet^{\alpha}\bullet^{\beta}](NIL)) \oplus [\bullet^{\beta}]([\bullet^{\beta}](NIL]))).$$

that represents the observation tree of Figure 2.b).

The following theorem states that strong equivalence is actually a congruence, i.e., it is preserved by the operations defined on the nodes of TOS. Next, we report a set of laws that are sound and complete for strong congruence.

2.6 Theorem (*$\approx$ is a congruence over the algebra of observation terms*)
Let P_1 and P_2 be observation terms, then

$P_1 \approx P_2$ implies $\mathfrak{a}(P_1) \approx \mathfrak{a}(P_2)$ and $P \oplus P_1 \approx P \oplus P_2$ and $P_1 \oplus P \approx P_2 \oplus P$. ♦

2.7 Theorem (*soundness and completeness*)
Let $\mathcal{A} = \{A1, A2, A3, A4\}$ be the set of axioms of Table 1; and let $P^{\mathfrak{a}}$ and $Q^{\mathfrak{a}}$ be observation terms. Then, $P^{\mathfrak{a}} \approx Q^{\mathfrak{a}}$ if and only if $\mathcal{A} \vdash P^{\mathfrak{a}} = Q^{\mathfrak{a}}$. ♦

(A1)	$P^{\mathfrak{a}} \oplus Q^{\mathfrak{a}} = Q^{\mathfrak{a}} \oplus P^{\mathfrak{a}}$	(A2)	$(P^{\mathfrak{a}} \oplus Q^{\mathfrak{a}}) \oplus R^{\mathfrak{a}} = P^{\mathfrak{a}} \oplus (Q^{\mathfrak{a}} \oplus R^{\mathfrak{a}})$
(A3)	$P^{\mathfrak{a}} \oplus \mathfrak{a}(NIL) = P^{\mathfrak{a}}$	(A4)	$P^{\mathfrak{a}} \oplus P^{\mathfrak{a}} = P^{\mathfrak{a}}$

Table 1

3. From Transition Systems to Observation Trees

Here we will show how observation trees can be associated to transition systems.

3.1. Definition (*transition systems*)
A *transition system* is a four-tuple $\langle Q,T,\partial_0,\partial_1\rangle$ where Q is a set of *states*, T is a set of *transitions* and ∂_0,∂_1: T $\rightarrow$ Q are two functions yielding *source* and *target* of any transitions ϑ. As usual, (finite) *computations* are defined as sequences of transitions $c = \vartheta_1\vartheta_2\ldots\vartheta_n$, $n \geq 1$, such that $\forall i$, $\partial_0(\vartheta_{i+1}) = \partial_1(\vartheta_i)$, plus an empty computation εq for every state q. Functions ∂_0 and ∂_1 are obviously extended to computations, and an associative operation of *concatenation* $c_1 \bullet c_2$ is defined whenever $\partial_1(c_1) = \partial_0(c_2)$, with empty computations being the unities.

3.2. Definition (*observable transition systems*)
An *observable transition system* TS = $\langle Q,T,\partial_0,\partial_1,o\rangle$ is a transition system $\langle Q,T,\partial_0,\partial_1\rangle$, plus a monotone *observation function* o from its computations to a partial ordering D of observations. We require that o identifies all unities, i.e. $o(\varepsilon q) = \varepsilon$; moreover we define the observation function o to be:

- *incremental* if $o(c_1) = o(c_1') \wedge o(c_2) = o(c_2') \Rightarrow o(c_1 \bullet c_2) = o(c_1' \bullet c_2')$, when defined.
- *cancellative* if $o(c_1) = o(c_1') \wedge o(c_1 \bullet c_2) = o(c_1' \bullet c_2') \Rightarrow o(c_2) = o(c_2')$. ♦

3.3. Definition (*experiments and their observations*)
Let TS = ‹Q,T,∂_0,∂_1,o› be an observable transition system.

(i) An *experiment* ξ of TS is either a computation or a sequence of experiments $\xi = \xi_1 ; \xi_2 ; ... ; \xi_n$, where ";" is an associative operation of *sequentialization* $\xi_1 ; \xi_2$, defined whenever $\partial_1(\xi_1) = \partial_0(\xi_2)$, with the empty computations as unities (∂_0 and ∂_1 being the obvious generalizations to experiments).
(ii) Similarly, the observation of an experiment, or the *outcome*, is a sequence of observations $\Omega = \omega_1 ; \omega_2 ; ... ; \omega_n$. The observation function o is homomorphically extended by letting $o(\xi_1 ; \xi_2) = o(\xi_1) ; o(\xi_2)$ and by identifying ε with the empty sequence of observations.
(iii) Concatenation • is generalized by associativity to the case of nonempty experiments: $(\xi_1 ; \xi_2) \bullet \xi_3 = \xi_1 ; \xi_2 \bullet \xi_3$ and $\xi_1 \bullet (\xi_2 ; \xi_3) = \xi_1 \bullet \xi_2 ; \xi_3$, where ξ_1, ξ_2 and ξ_3 are nonempty.

In the sequel, we will assume that concatenation binds tighter than sequentialization and will thus write $c_1 \bullet c_2 ; \xi$ instead of $(c_1 \bullet c_2); \xi$. ♦

3.4. Property (*standard forms of experiments*)
An experiment ξ has the following standard form:
$\xi = \vartheta_{11}\bullet...\bullet \vartheta_{1m_1} ; \vartheta_{21}\bullet...\bullet\vartheta_{2m_2} ;... ; \vartheta_{n1}\bullet...\bullet\vartheta_{nm_n}$, $\vartheta_{ij}\in T$, $0\leq i\leq n$, $1\leq j\leq m_i$. ♦

3.5. Property (*; is more discriminating than • for incremental observations, less for cancellative*)

(i) Given a transition system with an incremental observation function o, we have: $o(\xi_1) = o(\xi_1')$ and $o(\xi_1;\xi_2) = o(\xi_1';\xi_2')$ imply $o(\xi_1\bullet\xi_2) = o(\xi_1'\bullet\xi_2')$.
(ii) Given a transition system with a cancellative observation function o, we have: $o(\xi_1) = o(\xi_1')$ and $o(\xi_1\bullet\xi_2) = o(\xi_1'\bullet\xi_2')$ imply $o(\xi_1;\xi_2) = o(\xi_1';\xi_2')$. ♦

The following definition introduces four partial orderings on experiments that will be used to define the four different kinds of observation trees discussed above. They are closely related to the discussion of the Introduction. Below, rather than working with different sequences of transitions we will define orderings only on *experiments*, i.e. on sequences of computations. Thus, we will obtain the orderings on computations and on sequences of elementary transitions as special cases.

3.5. Definition (*orderings on experiments*)
Let TS = ‹Q,T,∂_0,∂_1,o› be an observable transition system. We define the following four partial orderings on its experiments:

0) $\xi_1 \leq_0 \xi_2$ if there exists a transition $\vartheta \in T$ such that $\xi_1; \vartheta = \xi_2$;
1) $\xi_1 \leq_1 \xi_2$ if there exists an experiment ξ such that $\xi_1 \bullet \xi = \xi_2$;
2) $\xi_1 \leq_2 \xi_2$ if there exists an experiment ξ such that $\xi_1 ; \xi = \xi_2$;
3) $\xi_1 \leq_3 \xi_2$ if $\xi_1 \leq_1 \xi_2$ or $\xi_1 \leq_2 \xi_2$. ♦

3.6. Definition (*four kinds of observation trees*)
Let TS = ‹Q,T,∂_0,∂_1,o› be an observable transition system, and let $q\in Q$. We define the following four kinds of *observation trees associated to* q, written $t_i(q)$.

0) $t_0(q)$ is an observation tree of *kind 0*, and has as nodes the experiments ϑ_1 ; ϑ_2 ; ...; ϑ_n from q made of sequences of transitions, ordered by $\leq_0$;
1) $t_1(q)$ is an observation tree of *kind 1* the nodes of which are experiments consisting of a *single computation* c from q (i.e., with $q = \partial_0(c)$), ordered by $\leq_1$;
2) $t_2(q)$ is an observation tree of *kind 2* the nodes of which are the experiments ξ from q, ordered by $\leq_2$;
3) $t_3(q)$ is an observation tree of *kind 3* the nodes of which are the experiments ξ from q, ordered by $\leq_3$. ♦

The theorem below describes some general results about the relationships between the four different kinds of observation trees under different assumptions on the observation function. Obviously, in all cases the trees of kind 3 are more (or equally) discriminating than the other two.

3.7. Theorem (*relationships between the three orderings on observation trees*)
Let TS = $\langle Q,T,\partial_0,\partial_1,o\rangle$, q', q"$\in$ T, and ξ_1, ξ'_1, ξ_2 and ξ'_2 be experiments of TS. Then

(i) $t_3(q') \approx t_3(q") \Rightarrow t_i(q') \approx t_i(q")$ with $i \in \{0,1,2\}$;
(ii) $t_2(q') \approx t_2(q") \Rightarrow t_0(q') \approx t_0(q")$
(iii) if o is *incremental* then
$t_0(q') \approx t_0(q") \Leftrightarrow t_2(q') \approx t_2(q") \Leftrightarrow t_3(q') \approx t_3(q") \Rightarrow t_1(q') \approx t_1(q")$
(iv) if o is *cancellative* then
$t_1(q') \approx t_1(q") \Rightarrow t_i(q') \approx t_i(q")$ with $i \in \{0,2,3\}$.
(v) if o is both *incremental* and *cancellative* then
$t_0(q') \approx t_0(q") \Leftrightarrow t_1(q') \approx t_1(q") \Leftrightarrow t_2(q') \approx t_2(q") \Leftrightarrow t_3(q') \approx t_3(q")$. ♦

The above theorem implies that whenever the observation function is incremental, the tree of kind 0 is the right choice. When it is cancellative, trees of kind 1 are at least as discriminating as the others. It is immediate to see that the interleaving observation function (yielding strings of actions, with or without τ's) is both incremental and cancellative. An interesting case is when the observation function does not satisfy the conditions of either ii) or iii): we will see below that this happens for pomset observations. In such a case a weak observational equivalence based on the fourth kind of trees is actually more discriminating than one based on either the first or the second.

We briefly mention how known semantics are captured by our framework. The proofs of theorems 3.8 and 3.10 can be found in [Con92]. There the claimed results are also generalized to the observational equivalences of [DDM87, 89, 90] and [DD90] that ignore internal moves.

3.8. Theorem (*truly concurrent equivalences for CCS*)
Let TS = $\langle Q,T,\partial_0,\partial_1,o\rangle$ be the truly concurrent (intentional) semantics of CCS of [DDM90], with q', q" $\in$ Q, and let o be the observation function that yields pomsets. Let $\approx_{NMS\text{-}po}$ be the strong NMS-equivalence of [DDM87] and let $\approx_{po}$ be the strong observational equivalence of [DDM90] or the pomset observational equivalence of [BC88]. Then,

- $t_1(q') \approx t_1(q") \Leftrightarrow q' \approx_{NMS\text{-}po} q"$;
- $t_2(q') \approx t_2(q") \Leftrightarrow q' \approx_{po} q"$. ♦

The next theorem, due to [MY92], shows that the trees of kind 1 with locality observations capture the location equivalence of [BCHK92]

3.9. Theorem (*location equivalences for CCS*)
Let TS = $\langle Q,T,\partial_0,\partial_1,o\rangle$ be the spatial transition system of CCS of [MY92], with q', q" $\in$ Q, and let o be the observation function that yields locality observations. Let $\approx_{loc}$ be the location equivalence of [BCHK92]. Then,

- $t_1(q') \approx t_1(q'') \Leftrightarrow q' \approx_{loc} q''$. ♦

Since location equivalence, $\approx_{loc}$, coincides with the locational equivalence defined on proved trees in [DP92], a trivial consequence of the above theorem is that the latter is recovered by trees of kind 1 with locality observations. Also the equivalence proposed by [Kie91] that combines the causal and location approaches can be recast in our parametric framework, by joining the relevant observations [MY92]. Two more equivalences are shown to be expressible *via* observation trees by the following theorem.

3.10. Theorem (*equivalences for Event Structures*)
Let TS = $\langle Q,T,\partial_0,\partial_1,o\rangle$ be the truly concurrent (intentional) semantics of CCS of [DDM90], with q', q" $\in$ Q, and let o be the observation function that yields pomsets. Let $\approx_{NMS\text{-}po}$ be the strong NMS-equivalence of [DDM87] and let $\approx_{po}$ be the strong observational equivalence of [DDM90] or the pomset observational equivalence of [BC88]. Then,

- $t_1(q') \approx t_1(q'') \Leftrightarrow q' \approx_{NMS\text{-}po} q''$;
- $t_2(q') \approx t_2(q'') \Leftrightarrow q' \approx_{po} q''$. ♦

We have also that in virtue of [Ace92] the result of the latter theorem can be extended to the Behavioural Structures equivalence of [RT88], called history preserving equivalence in [vGG89]. We conclude this section by exhibiting two pairs of les's which show that the partial ordering equivalences induced by orderings $\leq_1$ and $\leq_2$ are incomparable. This calls for further investigation on the instantations to specific observations of the equivalence induced by tree of kind 3. This equivalence has never been encountered in the literature, but arises naturally and meet the rationale which is behind the notion of observational equivalence.

The two les's in Fig. 5 are not strong equivalent according to $\leq_2$ because the choice of the leftmost α event in the first les prevents to perform an α causing a β afterwards, while in the second les this is always possible, no matter which α is chosen.

Similarly, the two les's in Fig. 6 are not strong equivalent according to $\leq_1$ because if in the second les the rightmost α is chosen, the only successive observation will be α causing β, while in the first les there is still a choice between α concurrent with β, and α causing either β or γ.

According to Theorem 3.7 i), the pairs of les's in Fig. 5 and Fig. 6 will be both discriminated by the partial ordering equivalence induced by $\leq_3$.

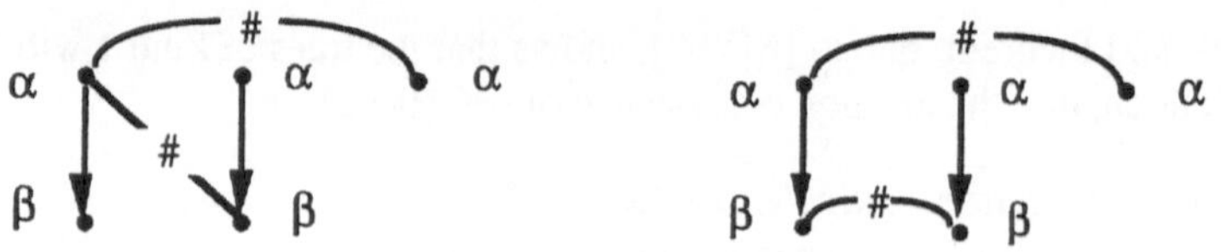

Figure 5

Two les's which are strong observational equivalent with po observations and ordering $\leq_1$ ($\approx_{NMS\text{-}po}$) and are **not** strong observational equivalent with po observations and ordering $\leq_2$ ($\approx_{po}$)

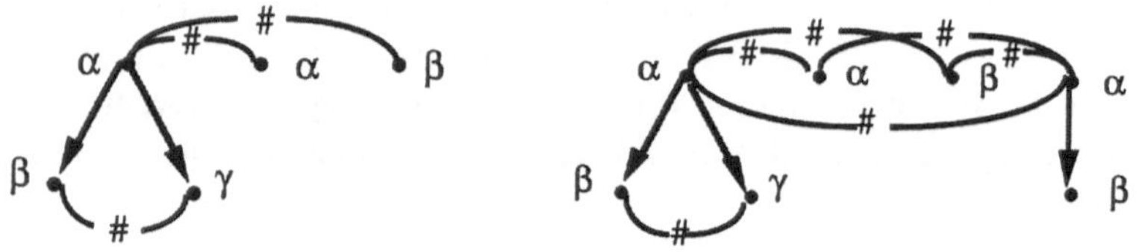

Figure 6

Two les's which are strong observational equivalent with po observations and ordering $\leq_2$ ($\approx_{po}$) but are **not** strong observational equivalent with po observations and ordering $\leq_1$ ($\approx_{NMS\text{-}po}$)

References

[AB84] Austry,D. and Boudol,G. Algèbre de Processus et Synchronization, *Theoret. Comput. Sci.*, **30**, 1, (1984), 91-131.

[Ace92] Aceto,L. History Preserving, Causal and Mixed-Ordering Equivalence over Stable Event Structures, To appear in *Fundamenta Informaticae*.

[BC88] Boudol,G. and Castellani,I. On the Semantics of Concurrency: Partial Orders and Transition Systems, *Theoret. Comput. Sci.*, **59**, 1,2, (North Holland, 1988), 25-84.

[BCHK92] Boudol, G., Castellani, I., Hennessy, M., Kiehn, A., A Theory of Processes with Localities, CONCUR'92 (W. R. Cleaveland, ed.), LNCS **630**, Springer-Verlag, 1992, pp. 108-122.

[Bro86] Broy,M. Process Semantics of Communicating Concurrent Programs. Internal Report Universität Passau, MIP - 8602, 1986.

[Con92] Conte,P.G. Confronti tra semantiche ad ordinamenti parziali per sistemi concorrenti, Tesi di Laurea in Scienze dell'Informazione, Università di Pisa, 1992.

[DD89] Darondeau,Ph. and Degano,P. Causal Trees, ICALP '89 (Ausiello,G. et al., eds) LNCS **372**, Springer-Verlag, 1989, pp. 234-248.

[DDM85] Degano,P., De Nicola,R. and Montanari,U. Partial ordering derivations for CCS, Proc. 5th Int. Conf. Fundamentals of Computation Theory, Cottbus (DDR), 1985. Lecture Notes in Computer Science, Vol. 199, Springer, Berlin, pp. 520-533.

[DDM87] Degano,P., De Nicola,R. and Montanari,U. Observational Equivalences for Concurrency Models, in *Formal Description of Programming Concepts* III (M. Wirsing, ed.), North Holland (1987), pp. 105-132.

[DDM88] Degano, P., De Nicola, R. e Montanari, U. A distributed operational semantics for CCS based on C/E systems, *Acta Informatica*, 26 (1988), 59-91.

[DDM89] Degano,P., De Nicola,R. and Montanari,U. Partial Ordering Description of Nondeterministic Concurrent Systems, in *Linear Time, Branching Time and Partial Orderd in Logic and Models for Concurrency*, LNCS **354**, Springer-Verlag, (J. De Bakker et al, eds.) 1989, 438-466.

[DDM90] Degano,P., De Nicola,R. and Montanari,U. A Partial Ordering Semantics for CCS, *Theoret. Comput. Sci.*, **75**, (1990), 223-262.

[DDM92] Degano,P., De Nicola,R. and Montanari,U. Universal Axioms for Bisimulations, Technical Report TR-9/92, Dipartimento di Informatica, Università di Pisa. *Theoret. Comput. Sci.*, to appear.

[DP92] Degano,P. and Priami,C. Proved Trees, ICALP '92 (W. Kuich eds) LNCS **623**, Springer-Verlag, 1992, pp. 629-640.

[FMM91a] Ferrari,G.L., Montanari,U. and Mowbray,M. On Causality Observed Incrementally, Finally, *Proc. CAAP'91*, LNCS **430**, Springer-Verlag, 1991.

[FMM91b] Ferrari G. L., Montanari, U. and Mowbray, M. Causal Streams: Tracing Causality in Distributed Systems (Extended Abstract), January 1991, submitted for publication.

[vGG89] van Glabbeek,R. Goltz,U. Equivalence Notions for Concurrent Systems and Refinement of Actions. *Proc. MFCS '89*, LNCS **379**, Springer-Verlag, 1989.

[vGV87] van Glabbeek,R. and Vaandrager,F.W. Petri Net Models for Algebraic Theories of Concurrency. *Proc. PARLE* (J.W. de Bakker, et al., eds.) LNCS **259**, Springer-Verlag, 1987, pp. 224-242.

[HM85] Hennessy,M. and Milner,R. Algebraic Laws for Nondeterminism and Concurrency, *Journal of ACM*, **32** (1985), pp. 137-161.

[Kie91] Kiehn,A. Local and Global Causes, Rep. Tech. Univ. Munchen 342/23/91 A, 1991.

[Mil80] Milner,R. *A Calculus of Communicating Systems*. LNCS **92**, Springer-Verlag, 1980.

[Mil83] Milner,R. Calculi for Synchrony and Asynchrony, *Theoret. Comput. Sci.*, **25**, (1983), 267-310.

[Mil89] Milner,R. *Communication and Concurrency*. Prentice Hall International, London 1989.

[MSS90] Manca,V., Salibra,A. and Scollo,G. Equational Type Logic, *Theoret. Comput. Sci.*, **77**, (1990), 131-159.

[MY92] Montanari,U. and Yankelevich,D. A Parametric Approach to Localities, ICALP '92 (W. Kuich, ed.) LNCS **623**, (Springer-Verlag, 1992), pp. 617-628.

[Par81] Park,D. Concurrency and Automata on Infinite Sequences, in Proc. GI, LNCS **104**, Springer-Verlag, 1981, pp. 167-183.

[RT88] Rabinovich,A. and Trakhtenbrot,B. Nets of Processes, *Fundamenta Informaticae* **11 (4)**, 1988, North Holland, 357-404.

[Win87] Winskel,G. Event Structures, in *Advances in Petri Nets 1987*, (G. Rozenberg, ed.) LNCS **266**, Springer-Verlag, 1987, pp. 196-223.

Computing Ready Simulations Efficiently

Bard Bloom*
Department of Computer Science, Cornell University
Ithaca, New York, USA
bard@cs.cornell.edu

Robert Paige†
Department of Computer Science, New York University
New York, New York, USA
paige@cs.nyu.edu

December 1, 1992

Abstract

Ready simulation is the finest fully abstract notion of process equivalence in the CCS setting. We give an $O(mn + n^2)$-time algorithm for deciding when n-state, m-transition processes are ready similar; a substantial improvement on the $\Theta(mn^6)$ algorithm presented in [4].

1 Introduction

1.1 Motivation from Process Algebra

In process algebras such as Milner's CCS [19], a process P is an entity capable of repeatedly participating in uninterpreted atomic actions a. In practice, events such as "the y key on the keyboard is pressed," "the processor sends a message on channel 83," and "circuit component 412 experiences a signal on its input wire" are modeled as atomic actions. The basic operational notion in such languages is $P \xrightarrow{a} P'$, indicating that P is capable of performing action a and thereafter behaving like P'. In general, processes are nondeterministic; P may have several alternative possible behaviors after performing a: that is, $P \xrightarrow{a} P'$ for several P''s.

One of the basic problems in this area is to find good notions of *process equivalence*. For example, a process that simply does an a and then stops ought to be the same as one that has several different ways of doing a's and then stopping, since one a or stopped process is the same as another. A wide variety of notions of equivalence have been proposed, *e.g.*[15, 17, 12, 3, 13, 14, 24], appropriate for different kinds of process algebras and conceptual settings.

---

*Supported by NSF grant (CCR-9003441)
†Supported by ONR grant (N00014-90-J-1890) and AFOSR grant (AFOSR-91-0308)

Process equivalences can be partially ordered by *fineness*: finer notions make more distinctions between processes; coarser ones consider more processes identical. Both fine and coarse notions are useful in theory and practice. For theory, fine notions give more detailed insight into the precise structure of concurrency; coarse notions can give insight into the nature of a particular programming language. In practice, fine notions allow one to prove extremely strong theorems about processes: *e.g.*, prove that a process meets its specification extremely precisely and will behave properly in all reasonable (and, generally, most unreasonable) environments. Coarse notions can only prove weaker theorems — *e.g.*, that the program is correct in composition with other programs in the same language — but such theorems are usually good enough, and (as they are weaker) are more likely to be true. A notion of equivalence is best when it is neither too fine nor too coarse; that is, when it is *fully abstract* for the language in question: when two processes are semantically equivalent if and only if they are indistinguishable in the language: *e.g.*, iff they produce the same sequences of actions in all contexts in the language (*i.e.*, they produce visibly the same results). In general, coarse notions are introduced because they are fully abstract for some language of interest.

In this study, we consider two of the finest (most discriminating) notions: ready simulation and (strong) bisimulation. Bisimulation [18, 19, 2, 3] is generally regarded as the finest usable notion of process equivalence in this setting. If processes P and Q are bisimilar, written $P \leftrightarrow Q$, then at all future times, they have exactly the same set of nondeterministic choices available.

Bisimulation is a meaningful notion in many settings beyond process algebra (*e.g.*, [20, 16]). It admits several powerful proof methods, and protocol-verification environments based on bisimulation have been built [10]. Furthermore, bisimulation of n-state, m-transition processes can be computed in $O(m \lg n)$ time [23], making verification of even relatively large protocols a viable possibility.[1]

However, bisimulation is too fine in general. Bisimulation pays great attention to exactly *when* decisions were made. For example, consider a *lossy queue*: a communication channel that accepts messages on one end, and nondeterministically either delivers them at the other end or loses them. (While this is not a data structure many programs are likely to build intentionally, it is a respectable approximation of many physical communication mechanisms, and thus is a reasonable thing to appear in protocol analyses.) There are several versions of the lossy queue. If queue Q_1 can only lose messages when they are enqueued, and queue Q_2 can lose messages on enqueueing or dequeueing, then Q_1 and Q_2 are not bisimilar. However, they behave identically in all CCS programs. Thus it seems reasonable to wish to consider them the same. Ready simulation is a slight coarsening of bisimulation that does identify them, and, more generally, it ignores irrelevant differences in decision times.

In [4, 5, 6] we formalize the concept of a "CCS-like language" as a GSOS language; that is, one whose rules are defined in a style that generalizes CCS's style. Using this definition, we show that bisimulation is not fully abstract for any CCS-like language; indeed, in the lossy queue example Q_1 and Q_2 are indistinguishable in all CCS-like languages.

[1]Note that n is the number of states in the process, rather than the size of the process. Even for restricted calculi, state spaces may be exponentially larger than code. This is a fundamental weakness of all state-space exploration methods.

Based on that analysis, we proposed the notion of *ready simulation*, which is the finest notion of process equivalence that is fully abstract for *some* CCS-like language. That is,

1. if two processes are ready similar, then one may be replaced by the other freely in any program written in a CCS-like language; and

2. if two processes are not ready similar, then there is reason to consider them different; namely a CCS-like language and a program in it in which the two will produce visibly different results.

See [5, 14, 16]. for more details. Ready simulation is thus more appropriate for CCS-like languages for foundational reasons. The theory and practice of bisimulation are well-developed.

The mathematics of ready simulation is straightforward. Let A be the (finite) alphabet of action symbols. Processes are labelled transition systems $P = \left\langle \Sigma_P, \dot{\rightarrow}_P, i_P \right\rangle$, where Σ_P is a finite set of states, $i_P \in \Sigma_P$ is the *initial state* and $\dot{\rightarrow}_P$ is a *transition relation*, a subset of $\Sigma_P \times A \times \Sigma_P$. Lower-case letters range over states. Transitions are written in infix: $p \xrightarrow{a} p'$. We omit the subscript on the transition relation $\xrightarrow{a}$ for both processes' transition relation, and write $p \xrightarrow{a}$ if $\exists p'.p \xrightarrow{a} p'$. We write $p \xrightarrow{a} p' \in P$ iff $p \xrightarrow{a} p'$ is a labelled transition of process P.

Let P and Q be finite-state processes. A relation $\leq$ is a *ready simulation relation* if, whenever $p \in \Sigma_P$, $q \in \Sigma_Q$, and $p \leq q$, then:

1. For every p' such that $p \xrightarrow{a} p'$, there is some q' such that

$$q \xrightarrow{a} q' \text{ and } p' \leq q' \tag{1}$$

2. p and q have precisely the same set of initial actions: for all a,

$$p \xrightarrow{a} \text{ iff } q \xrightarrow{a}. \tag{2}$$

State p is said to be ready similar to state q if $p \leq q$. P is a ready simulation approximation of Q, $P \sqsubseteq Q$, if such a relation $\leq$ exists where $i_P \leq i_Q$; P and Q are ready similar, $P \rightleftharpoons Q$, if $P \sqsubseteq Q \sqsubseteq P$.

This study expands the practical side of the theory of ready simulation by giving an algorithm that computes ready simulations in time $O(n^2 + nm)$, where $n = \max(|\Sigma_P|, |\Sigma_Q|)$, and m is the total number of transitions in P and Q. The previous best time was $\Theta(mn^6)$ presented in [4]. An $O(mn^2)$ algorithm for a related problem is presented in [11].

2 Algorithm Derivation

2.1 Methodology

We will present the new algorithm as the end-product of a semi-formal derivation. The methodology underlying this derivation integrates formal specification, algorithm design, proof, and analysis within a unified framework. We begin by presenting a formal specification of the function to be computed, and map this function by successive meaning preserving program transformations into an efficient implemention. Each transformation being selected is either an obvious simplification or is guided by complexity considerations. Surprisingly, the step-by-step description below corresponds closely to the systematic way in which the new algorithm was discovered. In using this syntax-directed approach no inspired decisions (called 'eureka' steps by Burstall and Darlington[7]) were needed.

2.2 Mathematical Preliminaries

Before going through the derivation, we first need to present a few definitions, formal tools and convenient notations. A partially ordered set X has the *descending chain condition* if every descending chain is finite. Given a function $f : X \to X$, we define $f^0(x) = x$ and $f^{n+1}(x) = f(f^n(x))$. If X has a partial order, then $\text{gfp}(x = f(x))$ denotes the greatest solution to the equation $x = f(x)$, *viz.* the greatest fixed point of f.

We will use the fixed point theorem and dominated convergence argument just below as tools for proving correctness of a formal specification and taking the initial step in the algorithm derivation:

Theorem 2.1 (Tarski): *Given a semilattice $(L, \sqcap, 1)$, where $\sqcap$ is meet, 1 is the greatest element, and L has a descending chain condition; and given a monotone function $f : L \to L$, then*

$$(\max Y | Y \leq f(Y)) = \text{gfp}(Y = f(Y)) = f^i(1) \tag{3}$$

for some finite i.

Theorem 2.2 (Dominated Convergence [8]): *Given a semilattice $(L, \sqcap, 1)$ as in Theorem 2.1, and any sequence s_i such that*

1. $s_0 = 1$
2. $s_{i+1} = s_i$ *if* $s_i = f(s_i)$
3. $s_i > s_{i+1} \geq f(s_i)$ *otherwise.*

Then s_i converges to $s_k = \text{gfp}(Y = f(Y))$ for some finite k.

Corollary 2.3 *If S is a finite set, $L = \wp(S)$ is the powerset of S, $\sqcap$ is intersection, then $\text{gfp}(Y = f(Y))$ can be computed by the following code:*

```
Y := S;
while exists x ∈ (Y − f(Y)) loop
    Y := Y - {x}
end loop
```

Proof: The code assigns the values s_i to Y in succession, where $s_0 = S$, $s_{i+1} = s_i$ if $s_i = f(s_i)$, and $s_{i+1} \in \{s_i - \{x\} : x \in s_i - f(s_i)\}$ otherwise. It is clear by induction on i that $f^i(S) \subseteq f(s_i) \subseteq s_{i+1} \subseteq s_i \subseteq S$ for all i. The corollary is thus a simple consequence of the dominated convergence theorem. $\triangle\!\!\!\!+$

We use the following notation. Let R and R' be binary relations and S be a set. Then:

$$\begin{aligned}
\mathsf{domain}(R) &= \{x|\exists y.\langle x,y\rangle \in R\} \\
\mathsf{range}(R) &= \{y|\exists x.\langle x,y\rangle \in R\} \\
R \circ R' &= \{\langle x,z\rangle|\exists y.\langle x,y\rangle \in R \wedge \langle y,z\rangle \in R'\} \\
R^{-1} &= \{\langle y,x\rangle : \langle x,y\rangle \in R\} \\
R\{x\} &= \{y : \langle x,y\rangle \in R\} \\
R[S] &= \{y : \exists x \in S.\langle x,y\rangle \in R\} = \bigcup_{x\in S} R\{x\}
\end{aligned}$$

We use these most often for the transition relations $R_a = \left\{\langle p,p'\rangle : p \xrightarrow{a} p'\right\}$. Specifically, note that:

$$p' \in R_a\{p\} \quad \text{iff} \quad p \xrightarrow{a} p' \quad \text{iff} \quad p \in R_a^{-1}\{p'\} \tag{4}$$

3 Formal Specification

In order to decide whether one process is a ready simulation approximation of another, it is useful to observe that whenever there exists a ready simulation relation $\leq$, there also exists a unique maximum one. Consequently, to decide whether two process are ready similar, we will compute the largest ready simulation relations in both directions, and then test whether the initial states of these processes are ready similar in both directions.

Let $P = \left\langle \Sigma_P, \dot{\rightarrow}_P, i_P \right\rangle$ and $Q = \left\langle \Sigma_P, \dot{\rightarrow}_Q, i_Q \right\rangle$ be two processes. The largest relation satisfying property (2) in our definition of ready simulation can be specified by,

$$\begin{aligned}
&E_1 = \max E \subseteq \Sigma_P \times \Sigma_Q \text{ such that} \\
&\left(\forall \langle p,q\rangle \in E|\left\{a : q \xrightarrow{a} q' \in Q\right\} = \left\{a : p \xrightarrow{a} p' \in P\right\}\right)
\end{aligned} \tag{5}$$

which is a convenient initial approximation that overestimates the largest ready simulation relation.

We calculate E_1 by the the following method. Scan the edges in Q and P in linear time to obtain the sets $\left\{a : q \xrightarrow{a} q' \in Q\right\}$ for each state q in Q, and $\left\{a : p \xrightarrow{a} p' \in P\right\}$ for each state p in P. Next, partition the states of $\Sigma_P \cup \Sigma_Q$ into blocks of states each having the same set of labels (computed in the previous pass). The time to form this partition takes time linear in the sum of the labels in all the label sets by using multiset discrimination [9].

This partition ER_1 is a compact representation of E_1; *i.e.*,

$$E_1 = \bigcup\{(B \cap \Sigma_P) \times (B \cap \Sigma_Q) : B \in ER_1\} \tag{6}$$

Recall that R_a is the transition map for symbol a: $R_a = \left\{\langle r,r'\rangle | r \xrightarrow{a} r'\right\}$. Then the following is a specification for the largest ready simulation relation when one exists and the empty set otherwise:

$$\max E \subseteq E_1 \text{ such that } (\forall \langle p,q\rangle \in E, \forall a \in A, \forall \langle p,p'\rangle \in R_a, \exists \langle q,q'\rangle \in R_a | \langle p',q'\rangle \in E) \tag{7}$$

Theorem 3.1 *Specification (7) is well-defined;* i.e., *there is always a unique maximum relation E satisfying the constraints.*

Proof: Use the transformation $(\forall x \in S | \Phi(x)) \longrightarrow (S = \{x \in S | \Phi(x)\})$ to obtain the following specification equivalent to (7):

$$\max E \subseteq E_1 \text{ such that } E = \{\langle p,q\rangle \in E | (\forall a \in A, \forall p' \in R_a\{p\}, \exists q' \in R_a\{q\} | q' \in E\{p'\})\} \tag{8}$$

The predicate $\exists q' \in R_a\{q\} | q' \in E\{p'\}$ can be simplified to $R_a\{q\} \cap E\{p'\} \neq \emptyset$, which is further reduced to $p' \in E^{-1}[R_a\{q\}]$. Consequently, we have the predicate

$$\forall p' \in R_a\{p\} | p' \in E^{-1}[R_a\{q\}] \tag{9}$$

which can be turned into the simpler but equivalent form

$$R_a\{p\} \subseteq E^{-1}[R_a\{q\}] \tag{10}$$

The ready simulation specification that results from these simplifications is,

$$\max E \subseteq E_1 | (E = \{\langle p,q\rangle \in E | (\forall a \in A | R_a\{p\} \subseteq E^{-1}[R_a\{q\}])\}) \tag{11}$$

Since the function

$$F(E) = \{\langle p,q\rangle \in E | (\forall a \in A | R_a\{p\} \subseteq E^{-1}[R_a\{q\}])\} \tag{12}$$

is monotone in E, and since $(\wp(E_1), \cap, E_1)$ is a semilattice, we see that specification (7) is well-defined and equivalent to gfp$(E = F(E))$ by Theorem 2.1.
$\triangle$
$+$

4 Derivation of a Prototype Algorithm

Our derivation will proceed in two stages. First we derive a prototype algorithm that could be readily implemented in a high level programming language such as SETL[26, 25]. Next, we use this prototype as the starting point for a derivation of a much more complex and efficient algorithm that could be implemented with somewhat greater effort in a conventional lower level language.

4.1 Dominated Convergence

According to Theorem 2.1 we can compute the largest ready simulation relation by $F^i(E_1)$ for some integer i, but the high degree of redundancy makes this approach too inefficient. Fortunately, we can use Corollary 2.3 to compute gfp$(E = F(E))$ more efficiently by performing the code shown in Figure 1. In the following two sections we will derive an efficient prototype algorithm from this initial program.

```
E := E_1
while exists ⟨p,q⟩ ∈ (E − F(E)) loop
    E := E - {⟨p,q⟩}
end while
```

Figure 1: First approximation to ready simulation algorithm

4.2 Invariant Maintenance

In this section we will recognize computational bottlenecks appearing in Figure 1, and remove them by maintaining program invariants. The method has been called finite differencing by Paige [21] and 'strengthening invariants' by Dijkstra and Gries.

Our first goal is to maintain the invariant $W = E - F(E)$ appearing within the predicate of the while-loop. However, it is useful to first simplify expression $E - F(E)$ into the equivalent expression

$$\{\langle p,q\rangle \in E | (\exists a \in A | R_a\{p\} - E^{-1}[R_a\{q\}] \neq \varnothing)\} \tag{13}$$

which can be rewritten more conveniently,

$$\{\langle p,q\rangle \in E | (\exists a \in A | R_a\{p\} - (E^{-1} \circ R_a)\{q\} \neq \varnothing)\} \tag{14}$$

We will keep track of the subset W of E containing those pairs of states that cannot be ready similar, because they violate condition (1). In order to maintain the invariant

$$W = \{\langle p,q\rangle \in E | (\exists a \in A | R_a\{p\} - (E^{-1} \circ R_a)\{q\} \neq \varnothing)\} \tag{15}$$

we will maintain the value of its subexpressions (from innermost to outermost) in program variables as invariants also. The first such invariant is the relational composition

$$T1(a) = E^{-1} \circ R_a = \{\langle y,x\rangle : \exists w.\langle y,w\rangle \in R_a, x \in E^{-1}\{w\}\} \tag{16}$$

Assume that invariant (16) for *T1(a)* holds on entry to the while-loop of Figure 1. This invariant is re-established when it is spoiled by the update

$$E := E - \{\langle p,q\rangle\} \tag{17}$$

inside the loop by updating *T1(a)* by the code in Figure 2 just before the change to E.

The code in Figure 2 can be performed more efficiently by also maintaining the invariants,

$$T2(a) = \{\langle\langle y,x\rangle,w\rangle : \langle x,w\rangle \in R_a, y \in E^{-1}\{w\}\} \tag{18}$$

and

$$T3 = \{\langle q,a\rangle : a \in A, q \in \mathsf{range}(R_a)\} \tag{19}$$

```
for a in A such that R_a^{-1}{q} ≠ ∅ loop
    for qq in R_a^{-1}{q} loop
        if |{w ∈ R_a{qq}|p ∈ E^{-1}{w}}| = 1 then
            T1(a){qq} := T1(a){qq} - {p}
        end if
    end loop
end loop
```

Figure 2: Updating *T1(a)*

```
for a in T3{q} loop
    for qq in R_a^{-1}{q} loop
        if |T2(a){⟨p, qq⟩}| = 1 then
            T1(a){qq} := T1(a){qq} - {p}
        end if
        T2(a){⟨p, qq⟩} := T2(a){⟨p, qq⟩} - {q}
    end loop
end loop
```

Figure 3: Maintaining invariants for *T1(a)* and *T2(a)*

That is, just prior to the change in E, we can maintain both invariants *T1(a)* and *T2(a)* together by executing the code in Figure 3

Now we can consider how to maintain the top-level invariant,

$$W = \{\langle p, q\rangle \in E | (\exists a \in A | R_a\{p\} - T1(a)\{q\} \neq \varnothing)\} \tag{20}$$

relative to changes in both E and *T1(a)*. Just before pair $\langle p, q\rangle$ (which was originally selected arbitrarily from W) is removed from E, we can update W by executing

$$W := W - \{\langle p, q\rangle\}$$

Just prior to the change[2]

$T1(a)\{qq\} := T1(a)\{qq\} - \{p\}$

we can update W by executing,

[2] This assignment is to be interpreted as updating the representation of a set of pairs.

```
for pp in R_a^{-1}{p} loop
    if ⟨pp, qq⟩ ∈ (E − W) then
        W := W ∪ {⟨pp, qq⟩}
    end if
end loop
```

which can be made more efficient by maintaining the additional invariant,

$$WC = E - W \tag{21}$$

Observe that invariant *WC* and *W* can be maintained collectively by executing the following code just before the modification to *T1(a)*,

```
for pp in R_a^{-1}{p} loop
    if ⟨pp, qq⟩ in WC then
        WC := WC - {⟨pp, qq⟩}
        W := W ∪ {⟨pp, qq⟩}
    end if
end loop
```

WC does not have to be updated just prior to the element deletion from E, since the deleted element belonged to *W* and not *WC*.

Putting the preceding code fragments together, and noting that *T1(a)* is never used, that a set of reference counts

$$T2count(a)(p, qq) = |T2(a)\{\langle p, qq\rangle\}|$$

can replace the stored sets *T2(a)*$\{\langle p, qq\rangle\}$, and that $WC = E$ on exit from the while-loop, we obtain the prototype code in Figure 4.

To make an initial performance analysis of the code in Figure 4, we can assume that all element additions, element deletions, membership tests, and arbitrary selection operations (*e.g.*, the exists predicate) take unit time, and searching through a set X takes time linear in the size of X. Certainly, universal hashing [6] will achieve these times in a probabilistic sense.

Under these assumptions, it is apparent that the time bounds depend entirely on the total number of for-loop iterations through $R_a^{-1}\{q\}$ and through $R_a^{-1}\{p\}$. Since each pair $\langle p, q\rangle$ can be selected only once from W, the number of times that $R_a^{-1}\{q\}$ can be searched for fixed q is at most n. Hence, the total iteration count for this loop is nm. Since the predicate $T2count(a)(p, qq) = 1$ will be true at most once for a pair $\langle p, qq\rangle$, we know that the number of times that $R_a^{-1}\{p\}$ can be searched for fixed p is most n. Thus, the total iteration count for this loop is also nm.

4.3 Establishing Invariants

To complete the derivation of a prototype algorithm, we need to discuss how to establish the invariants for *T1(a)*, *T2count*, *T3*, *W*, and *WC* on entry to the while-loop, where E_1 is assigned to E. The method to establish these invariants involves loop fusion similar to the approach found in [22].

```
while exists ⟨p,q⟩ in W loop
    for a in T3{q} loop
        for qq in R_a^{-1}{q} loop
            if T2count(a)(p, qq) = 1 then
                for pp ∈ R_a^{-1}{p} loop
                    if ⟨pp, qq⟩ in WC then
                        WC := WC - {⟨pp, qq⟩}
                        W := W ∪ {⟨pp, qq⟩}
                    end if
                end loop
            end if
            T2count(a)(p, qq) := T2count(a)(p, qq) − 1
        end loop
    end loop
    W := W - {⟨p,q⟩}
end while
(* WC is the desired ready simuation relation.  *)
```

Figure 4: Prototype code

We can establish *T3* by executing

```
T3 := ∅
for a ∈ A loop
    for q ∈ domain(R_a^{-1}) loop
        T3{q} := T3{q} ∪ {a}
    end loop
end loop
```

T3 can then be used to establish *T1(a)* and *T2count* by the code in Figure 5.
In order to establish

$$W = \{\langle x,y\rangle \in E | (\exists a \in A | R_a\{x\} - T1(a)\{y\} \neq \varnothing)\}, \tag{22}$$

we will make use of invariants,

$$wcount2(a)(x,y) = |R_a\{x\} - T1(a)\{y\}|, \tag{23}$$

for all $\langle x,y\rangle \in E$ such that $R_a\{x\} \neq \varnothing$,

$$T4 = \{\langle x,a\rangle : a \in A, x \in \mathrm{domain}(R_a)\}, \tag{24}$$

and

$$wcount(x,y) = \sum_{a \in T4\{x\}} wcount2(a)(x,y). \tag{25}$$

```
T2count := ∅
T1 := ∅
for ⟨p,q⟩ ∈ E₁ loop
    for a ∈ T3{q} loop
        for qq ∈ R_a^{-1}{q} loop
            if T2count(a)(p, qq) = 0 then
                T1(a){qq} := T1(a){qq} ∪ {p}
            end if
            T2count(a)(p, qq) := T2count(a)(p, qq) + 1
        end loop
    end loop
end loop
```

Figure 5: Code to establish *T1(a)* and *T2count*

```
T3 := ∅
T4 := ∅
for a ∈ A loop
    for q ∈ range(R_a) loop
        T3{q} := T3{q} ∪ {a}
    end loop
    for q ∈ domain(R_a) loop
        T4{q} := T4{q} ∪ {a}
    end loop
end loop
```

Figure 6: Code to establish *T3* and *T4*

Thus, $W = \{\langle x,y\rangle \in E | wcount(x,y) \neq 0\}$

First we recognize that *T4* can be established efficiently together with *T3*; the code is given in Figure 6.

Next, if we assume that $T1(a) = \varnothing$ for all $a \in A$, we can establish *W*, *WC*, and *wcount2(a)* relative to parameter *E* by the code in Figure 7.

Next, let us adjust *wcount*, *W*, and *WC* to take parameter *T1* into account; this code is given in Figure 8.

Observe finally, that *WC*, E_1, and *E* can be equivalenced into the single variable *WC*, that $T1(a)^{-1} = \text{domain}(T2count(a))$ (so that *T1(a)* can be eliminated), that *W* and domain *wcount* can be equivalenced, and that the first loop above that is used to partially establish *W*, *WC*, *wcount2(a)*, and *wcount* can be jammed into the same loop that establishes *T1(a)* and *T2count*. Consequently, we can establish all the invariants needed within the while-loop by executing the code in Figure 9.

Under the same assumptions as was used to analyze the while-loop, we see

```
for a ∈ A loop
    wcount2(a) := ∅
end loop
wcount := ∅
W := ∅
WC := E1
for ⟨p,q⟩ ∈ E1 loop
    if T4{p} ≠ ∅ then
        WC := WC - {⟨p,q⟩}
        W := W ∪ {⟨p,q⟩}
        for a ∈ T4{p} loop
            wcount(p,q) := wcount(p,q) + |Ra{p}|
            wcount2(a)(p,q) := |Ra{p}|
            % assume wcount2(a)(p,q) = 0 if undefined
        end loop
    end if
end loop
```

Figure 7: Establishing W, WC, and *wcount2(a)* ignoring *T1*

that the preceding code takes $O(mn)$ time.

5 Derivation of a RAM Implementation

The critical task of an efficient implementation is to provide data structures supporting unit-time associative access to domain($wcount2(a)$), domain($T2count$), domain$\left(R_a{}^{-1}\right)$, WC, domain($wcount$), domain($T4$), and domain($T3$). The method applied here is based on a real-time simulation of a set machine on a RAM [4].

Unit-time associative access is achieved by array access using integer encodings for the symbols in A and the states in Σ_P and Σ_Q. That is, we assume unique integer identifiers from 1 to $|\Sigma_P|$ for the states in Σ_P, and unique integer identifiers from 1 to $|\Sigma_Q|$ for the states in Σ_Q. Also, assume integer identifiers from 1 to $|A|$ for the symbols in A. For each symbol a belonging to alphabet A, we store three arrays – one dimensional arrays P_a and Q_a of size $|\Sigma_P|$ and $|\Sigma_Q|$ respectively, and two-dimensional array PQ_a of size $|\Sigma_P| \times |\Sigma_Q|$.

For each Σ_P state i, the i-th component of P_a is a record with two fields – the first storing the count $|R_a\{i\}|$ and the second storing a pointer to $R_a{}^{-1}\{i\}$, where set $R_a{}^{-1}\{i\}$ is a one-way list of integers designating Σ_Q-states. For each Σ_Q state i, the i-th component of Q_a is record with only one field – a pointer to $R_a{}^{-1}\{i\}$, where set $R_a{}^{-1}\{i\}$ is a one-way list of integers designating Σ_P-states. Array PQ_a is used to store *wcount2(a)* and *T2count*. For each Σ_P-state i and Σ_Q-state j we have a PQ record with 3 fields. The first field stores $T2count(a)(i,j)$; if $T2count(a)(i,j) \neq 0$, then the second field stores forward and backward pointers connecting this record to a doubly linked list of PQ_a-

```
for a ∈ A loop
    for ⟨qq,p⟩ ∈ T1(a) loop
        for pp ∈ R_a^{-1}{p} loop
            if wcount2(a)(pp,qq) ≥ 1 then
                if wcount(pp,qq) = 1 then
                    W := W - {⟨pp,qq⟩}
                    WC:= WC ∪ {⟨pp,qq⟩}
                end if
                wcount(pp,qq) := wcount(pp,qq) - 1
                wcount2(a)(pp,qq) := wcount2(a)(pp,qq) - 1
            end if
        end loop
    end loop
end loop
```

Figure 8: Establishing *W*, *WC*, and *wcount2(a)* considering *T1*

array components storing records with nonempty *T2count* fields; the third field stores *wcount2(a)(i,j)*. We also need pointers to one end of the doubly linked list storing *T2count* values.

A one dimensional master symbol array *M* of size $|A|$ is used to access the three arrays P_a, Q_a, and PQ_a with any symbol a belonging to A. For each symbol i, the i-th component of *M* is a record with three fields, a pointer to P_a, a pointer to Q_a, and a pointer to PQ_a.

There are three more arrays – one dimensional arrays *P'* and *Q'* of size $|\Sigma_P|$ and $|\Sigma_Q|$ respectively, and two-dimensional array *PQ'* of size $|\Sigma_P| \times |\Sigma_Q|$. For each Σ_P state i, the i-th component of *P'* stores one field – a pointer to *T4*$\{i\}$, where set *T4*$\{i\}$ is a one-way list of integers designating symbols in A and used to access array *M*. Similarly, for each Σ_Q state i, the i-th component of *Q'* stores one field – a pointer to *T3*$\{i\}$, where set *T3*$\{i\}$ is a one-way list of integers designating symbols in A and used to access array *M*. For each Σ_P-state i and Σ_Q-state j we have a *PQ'* record with two fields - one field stores count *wcount(i,j)* and another field stores stores forward and backward pointers connecting to doubly linked lists of *PQ'*-array components with nonempty *W* or *WC* fields (such overlaying being possible because *W* is disjoint from *WC*). We will also need pointers to one end of the doubly linked lists storing *W* and *WC*.

The total space for all of the preceding data structures is $O(|A| \times (|\Sigma_P| + |\Sigma_Q| + |\Sigma_P| \times |\Sigma_Q|))$. In case these three arrays are sparsely occupied, then each array can be initialized in unit time using the solution to [1, exercise 18, p.100].

6 Conclusion

Our Common Lisp implementation was reasonably straightforward, and performed acceptably even using general-purpose Common Lisp primitives rather

```
T3 := ∅ ; T4 := ∅ ; wcount := ∅
for a ∈ A loop
    wcount2(a) := ∅
    T2count := ∅
    for q ∈ rangeR_a do T3{q} := T3{q} ∪ {a}
    for q ∈ domainR_a do T4{q} := T4{q} ∪ {a}
end loop
WC := E_1% pointer assignment is possible
for ⟨p,q⟩ ∈ WC loop
    for a ∈ T3{q} loop
        for qq ∈ R_a^-1{q} do
            T2count(a)(p, qq) := T2count(a)(p, qq) + 1
    end loop
    if T4{p} ≠ ∅ then
        WC := WC - {⟨p,q⟩}
        for a ∈ T4{p} loop
            wcount(p,q) := wcount(p,q) + |(R_a{p})|
            %assume wcount(p,q) = 0 if undefined
            wcount2(a)(p,q) := |(R_a{p})|
        end loop
    end if
end loop
for a ∈ A loop
    for ⟨p, qq⟩ ∈ domainT2count loop
        for pp ∈ R_a^-1{p} loop
            if wcount2(a)(pp, qq) ≥ 1 then
                if wcount(pp,qq) = 1 do WC:= WC ∪ {⟨pp, qq⟩}
                wcount(pp,qq) := wcount(pp,qq) - 1
                wcount2(a)(pp,qq) := wcount2(a)(pp,qq) - 1
            end if
        end loop
    end loop
end loop
W := domain(wcount) % pointer assignment is possible
```

Figure 9: Full Initialization

than the more efficient RAM data structures. We thus have achieved our goal, of giving a good algorithm for computing ready simulations of finite-state processes.

References

[1] A. Aho, J. Hopcroft, and J. Ullman. *Design and Analysis of Computer Algorithms*. Addison-Wesley, 1974.

[2] D. Austry and G. Boudol. Algèbre de processus et synchronisation. *Theoretical Computer Sci.*, 30(1):91–131, 1984.

[3] J. A. Bergstra and J. W. Klop. Process algebra for synchronous communication. *Information and Computation*, 60(1–3):109–137, 1984.

[4] B. Bloom. *Ready Simulation, Bisimulation, and the Semantics of CCS-Like Languages*. PhD thesis, Massachusetts Institute of Technology, Aug. 1989.

[5] B. Bloom, S. Istrail, and A. R. Meyer. Bisimulation can't be traced (preliminary report). In *Conference Record of the Fifteenth Annual ACM Symposium on Principles of Programming Languages*, pages 229–239, 1988. Also appears as MIT Technical Memo MIT/LCS/TM-345.

[6] B. Bloom, S. Istrail, and A. R. Meyer. Bisimulation can't be traced. Technical Report TR 90-1150, Cornell, August 1990. (To appear in JACM).

[7] R. Burstall and J. Darlington. A transformation system for developing recursive programs,. *J. ACM*, 24(1):44–67, Jan 1977.

[8] J. Cai and R. Paige. Program derivation by fixed point computation,. *Science of Computer Programming*, 11(3):197–261, 1988/1989.

[9] J. Cai and R. Paige. Multiset discrimination - a method for implementing programming language systems without hashing,. In *Proceedings of POPL '91*. 1991.

[10] R. Cleaveland, J. Parrow, and B. Steffen. The Concurrency Workbench. In *Proceedings of the Workshop on Automated Verification Methods for Finite State Systems*. Springer-Verlag, 1989. LNCS 407; to appear in ACM TOPLAS.

[11] R. Cleaveland and B. Steffen. Computing behavioural relations, logically. In J. L. Albert, B. Monien, and M. R. Artalejo, editors, *Automata, Languages and Programming: 18th International Colloquium*, volume 510 of *Lect. Notes in Computer Sci.*, pages 127–138. Springer-Verlag, July 1991.

[12] R. de Nicola and M. C. B. Hennessy. Testing equivalences for processes. *Theoretical Computer Sci.*, 34(2/3):83–133, 1984.

[13] R. de Simone. Higher-level synchronising devices in MEIJE-SCCS. *Theoretical Computer Sci.*, 37(3):245–267, 1985.

[14] J. F. Groote and F. Vaandrager. Structured operational semantics and bisimulation as a congruence (extended abstract). In G. Ausiello, M. Dezani-Ciancaglini, and S. R. D. Rocca, editors, *Automata, Languages and Programming: 16th International Colloquium*, volume 372 of *Lect. Notes in Computer Sci.* Springer-Verlag, 1989.

[15] C. A. R. Hoare. Communicating sequential processes. *Comm. ACM*, 21(8):666–677, 1978.

[16] K. Larsen and A. Skou. Bisimulation through probabilistic testing (preliminary report). Technical Report R 88-16, Institut for Elektroniske Systemer, Aalborg Universitetscenter, Aalborg, Danmark, June 1988.

[17] R. Milner. *A Calculus of Communicating Systems*, volume 92 of *Lect. Notes in Computer Sci.* Springer-Verlag, 1980.

[18] R. Milner. Calculi for synchrony and asynchrony. *Theoretical Computer Sci.*, 25(3):267–310, 1983.

[19] R. Milner. *Communication and Concurrency.* Prentice Hall International Series in Computer Science. Prentice Hall, New York, 1989.

[20] L. Ong. *The lazy lambda calculus: an investigation into the foundations of functional programming.* PhD thesis, Imperial College, University of London, 1988.

[21] R. Paige. Real-time simulation of a set machine on a ram. In N. Janicki and W. Koczkodaj, editors, *Computing and Information, Vol II*, pages 69–73. Canadian Scholars' Press, May 1989.

[22] R. Paige and S. Koenig. Finite differencing of computable expressions. *ACM Trans. on Programming Languages and Systems*, 4(3):401–454, 1982.

[23] R. Paige and R. Tarjan. Three partition refinement algorithms. *SIAM J. Computing*, 16(6):973–989, 1987.

[24] I. Phillips. Refusal testing. *Theoretical Computer Sci.*, 50:241–284, 1987.

[25] J. Schwartz. *Programming with Sets: An Introduction to SETL.* Springer-Verlag, 1986.

[26] K. Snyder. The SETL2 programming language. Technical Report 490, Courant Insititute/ New York University, 1990.

Verification of Value-Passing Systems

Zvi Schreiber
Data Connection Limited
100 Church Street, Enfield, Middlesex EN2 6BQ, England
tel +44 81 366 1177

Department of Computing, Imperial College
180 Queen's Gate, London SW7 2BZ, England
e-mail mzs@doc.ic.ac.uk

October 1992

Abstract

There are well known algorithms which automatically verify finite-state concurrent systems. This paper introduces techniques which can automatically verify systems with values of large or infinite sort which are not finite-state.

We show how to represent systems with values, specified in a process algebra, by a graph with parameters. This representation resembles a flow-chart. The size of the graph is independent of the size of the sorts. We show that by using this representation we can, in some cases, automatically check for bisimulation despite large or infinite sorts.

It is often beneficial to treat some parameters of a system combinatorially while others are treated directly. We present a static analysis to decide which parameters should be treated in each of these two ways. We combine this with a liveness analysis. This analysis leads to an automatic verification of the alternating bit protocol, for example, with the message values included in the specification.

Introduction

0.1 Using CCS to verify concurrent systems

One important use of process algebra such as CCS ([Mil89]) is the verification of concurrent systems. Milner ([Mil89, §6.3]), for example, uses CCS to define the behaviour of all the components of the *alternating bit protocol* (Figure 1). This simple protocol consists of a sending and receiving program (*Send* & *Rcv*) with unreliable buffers (*Trans* & *Ack*) providing communication between them. The programs detect the loss of any messages and arrange retransmission.

The protocol is verified by establishing that the composition of all the components (the unreliable buffers and the two programs) behaves in the same way as a reliable buffer. In CCS terms, we establish a weak bisimulation between the composite agent (with internal communications hidden) and a specification:

$$Spec \approx (Send|Trans|Rcv|Ack) \setminus \{send, trans, reply, ack\}.$$

We can test if this equivalence holds by denoting each side by a *labelled transition system* (*l.t.s.*) and going through all the states, provided there are a

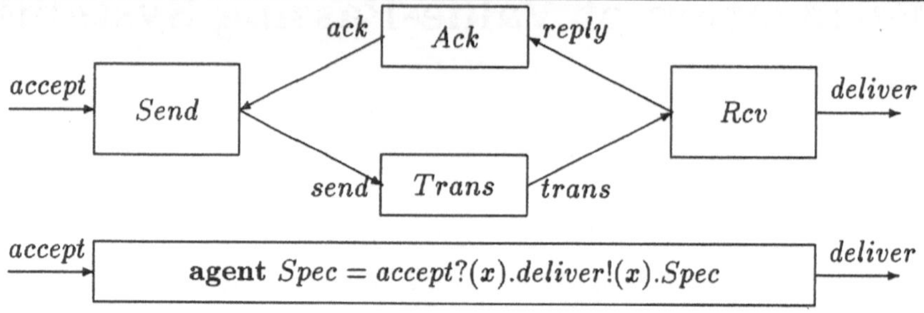

Figure 1: The Alternating Bit Protocol

finite number of states. (Tools such as the Concurrency Workbench ([Mol91]) can do this automatically.)

This approach treats value-passing CCS as a shorthand for basic CCS. For example, the definition of *Spec*[1] in Figure 1 is viewed as a shorthand for the following basic CCS definition:

$$Spec := \sum_{v} accept_v?.deliver_v!.Spec. \tag{1}$$

We call this approach to value-passing *combinatorial*.

0.2 State explosion

The complexity of the combinatorial analysis is polynomial in the number of states of the system, which is the number of nodes in the labelled transition system (see [KS83]). Unfortunately, when considering realistic systems, huge state spaces tend to result from

- large numbers of components ($P_1|P_2|P_3\cdots$); the number of states is exponential in the number of components
- the size of sorts; when representing data values these may be large (typically $2^{\text{a big number}}$) or actually infinite.

In performing a purely combinatorial analysis of the alternating bit protocol, for example, the size of the sorts of the messages being passed through the system will influence the time taken for the analysis[2]. If we assume that the messages are integers, automatic verification becomes impossible.

[1] Our value-passing CCS examples borrow some notation from the concrete value-passing language defined in [Bru91]. Agent definitions are introduced by **agent** and their parameters given by *param:sort* pairs. Declarations of the sorts of channels may be introduced by **label**. One non-standard feature is that here we use the CSP ([Hoa85]) style ! and ? and allow channels to pass tuples of values (so $a?(x,y)$ is valid).

[2] Several authors ([Mil89],[LM86], [Par87] & [BA91]) have taken the approach of ignoring the message values and specifying the protocol — as if it were just passing signals around — in basic CCS. This abstraction is not valid in general — *a protocol which transposes the order of messages will appear to be correct.* For example, suppose C and D are buffers which input two values; C always outputs just the first value, D outputs the second value. If we were to "ignore the message value" we would define $C := in?.in?.out!.C$ and $D := in?.in?.out!.D$ and conclude that $C \sim D$!

0.3 Restrictions

Tractable infinite-state programs This paper only addresses the state explosion caused by infinite sorts because this is a more tractable problem. For example, suppose we define a buffer *Buff* which accepts integers by

$$\begin{aligned}&\textbf{label } in(\mathbb{N}), out(\mathbb{N})\\&\textbf{agent } Buff = in?(x).Buff'(x)\\&\textbf{agent } Buff'(x{:}\mathbb{N}) = out!(x).Buff.\end{aligned} \quad (2)$$

The receiving action on a channel of infinite sort introduces infinite branching for the state *Buff* (via actions $in?(1), in?(2), \ldots$) and an infinite number of states $Buff'(1)$, $Buff'(2)$,....

But we can consider the agent to have just two states, *Buff* and $Buff'(x)$ with the latter parameterised by x. We might be able to analyse such agents without considering the infinite *l.t.s.* because the different instantiations of $Buff(x)$ are, after all, defined together and closely related. We can similarly consider there to be just one transition $in?(x)$, parameterised by the input value, between these two states. In a sense then, our approach is to treat value-passing as fundamental instead of as a shorthand for basic CCS.

Intractable infinite-state programs The less tractable form of infinite-state problems occurs when the agent has an evolving structure. For example if we define $A := a.(A|A)$ then A may evolve to $A|A$ or $A|A|A$ or .... Each of these is substantially different. This behaviour occurs when a static operator $(|, \backslash, [])$ occurs within a recursion.

A related intractable problem is that of infinite-branching caused by either an infinite $\sum$ or unguarded recursion[3].

We do not consider programs with such behaviour. We assume that all programs we analyse have the following properties[4]:

Definition 0.1 (vp-finite-state) *Let P be a value-passing CCS program. P is a* **vp-finite-state** *program if there is no sequence of agent names* $C = C_0, C_1, C_2, \ldots, C_n = C$ *such that for each* i, C_i appears *in the declaration of* C_{i-1}, and, *for some* j, C_j *appears within a static operator in the declaration of* C_{j-1}.

Definition 0.2 (vp-finite-branching) *Let P be a value-passing CCS program with no infinite* $\sum$*s. P is a* **vp-finite-branching** *program if there is no sequence of agent names* $C = C_0, C_1, \ldots, C_n = C$ *such that for each* i, C_i *appears* unguarded *in the declaration of* C_{i-1}.

0.4 Representation

We wish to have an abstract structure on which to perform our analysis, to avoid algorithms having to interpret the syntax and to make our work independent of the underlying process algebra.

---

[3]For example if we define **agent** $A(x{:}\mathbb{N}) = out!(x^2) + A(x{+}1)$ then $A(1)$ has the transitions $out!(1^2), out!(2^2), \ldots$.

[4]These definitions give natural syntactic restrictions on a program to ensure a finite number of states and actions, when value-passing (vp) agents and transitions are considered atomic.

We introduce the notion of a *parameterised graph* (*p.g.*). These are similar to flow charts, with extensions for non-determinism and external communications (and with no global variables).

Assume that the values in the CCS language are all taken from a set V (ranged over by v, w) and that sorts are taken from U. Each element $u \in U$ is a subset of V. Whenever S is a set we will write $\underline{S}$ for $\bigcup_{n\in\mathbb{N}} S^n$.

Description of parameterised graphs Firstly every graph has a set *Act* of channels (consisting of the names of all channels declared in the underlying program) and a function Γ_A with domain *Act* associating each channel with a sort (or tuple of sorts). Let *Act* be ranged over by a.

The set of actions possible in a *p.g.* $\mathcal{P}$ (ranged over by α) is given by:

$$act(\mathcal{P}) = \{\tau\} \cup \{a?(\underline{v}) : a \in Act, \underline{v} \in \Gamma_A(a)\} \cup \{a!(\underline{v}) : a \in Act, \underline{v} \in \Gamma_A(a)\}.$$

A *p.g.* has a set of nodes $\wp$ ranged over by p, q and a map Γ_N associating each node with a tuple of sorts. A node p may represent a simple agent such as *Buff* (with $\Gamma_N(p) = ()$) or an agent with parameters such as $Buff'(x)$ (for which $\Gamma_N(p) = (\mathbb{N})$, for example). Every parameterised graph will have an initial node $\iota \in \wp$.

The set of states of a parameterised graph $\mathcal{P}$ is given by:

$$state(\mathcal{P}) = \{p(\underline{v}) : p \in \wp, \underline{v} \in \Gamma_N(p)\}.$$

We now need to describe the transitions. Suppose p has an internal (τ) transition. We need to know which node q the transition leads to. We also need to know for what values of p's parameters this transition is possible (since this could be restricted by an **if**). This is given by a function γ from $\Gamma_N(p)$ to $\{\mathbf{true}, \mathbf{false}\}$. Finally, we need a function f which, given the value of the parameters of p, gives the values of the parameters of q after the transition. So we let any internal transition be marked by a tuple $tau(q, \gamma, f)$.

A send transition is marked by a tuple $send(a, q, \gamma, f, g)$ where a is the channel name, q, γ, f are as before, and g is a function mapping the values of the parameters of p to the value(s) of the output. A receive transition is marked by $rcv(a, q, \gamma, f)$ where f's domain is now the product of the sort of p and the sort of a. Given that value of p's parameters *and* the value(s) of the input, f gives the value of the parameters of q following the transition.

In each case, a transition may be visualised as an edge from p to q.

Definition 0.3 *A parameterised graph is a tuple* $(Act, \Gamma_A, \wp, \iota, \Gamma_N, T)$ *where*

- *Act and* $\wp$ *are finite sets*
- $\Gamma_A : Act \to \underline{U}$, $\Gamma_N : \wp \to \underline{U}$
- $\iota \in \wp$
- *for each* $p \in \wp$, $T(p)$ *is a finite set of elements of type*[5] $tau(q, \gamma, f)$, $rcv(a, q, \gamma, f)$ *or* $send(a, q, \gamma, f, g)$.

We call the domain of all parameterised graphs[6] *PG, ranged over by* $\mathcal{P}$.

[5] Where $q \in \wp$; $a \in Act$; $\gamma : \Gamma_N(p) \to bool$; $f : \Gamma_N(p) \to \Gamma_N(q)$, or, in a *rcv*, $f : \Gamma_N(p) \times \Gamma_A(a) \to \Gamma_N(q)$; and finally, $g : \Gamma_N(p) \to \Gamma_A(a)$.

[6] Strictly we define two graphs to be isomorphic if one is obtained from the other by relabelling $\wp$ and take *PG* to be the set of equivalence classes.

Example The parameterised graph $\mathcal{P} = (Act, \Gamma_A, \wp, \iota, \Gamma_N, T)$ represents *Buff* of (2) if we let

$$\begin{array}{rcl} Act &=& \{in, out\} \qquad \Gamma_A = \{(in, (\mathbb{N})), (out, (\mathbb{N}))\} \\ \wp &=& \{p_1, p_2\} \qquad\qquad \iota = p_1 \\ \Gamma_N &=& \{(p_1, ()), (p_2, (\mathbb{N}))\} \\ T &=& \{\, (p_1, \{\, rcv(in, p_2, \lambda v \in ().\mathbf{true}, \lambda v \in \mathbb{N}.v)\,\}), \\ && (p_2, \{\, send(out, p_1, \lambda v \in \mathbb{N}.\mathbf{true}, \lambda v \in \mathbb{N}.(), \lambda v \in \mathbb{N}.v)\,\})\,\} \end{array}$$

where we have represented functions by their graphs. □

Operational Semantics We give a *p.g.* $\mathcal{P} = (Act, \Gamma_A, \wp, \iota, \Gamma_N, T)$ an operational semantics in the form of an *l.t.s.* $(state(\mathcal{P}), act(\mathcal{P}), \twoheadrightarrow)$. This is for theoretical purposes only — none of the algorithms compute this *l.t.s.*. The transition relation $\twoheadrightarrow \subseteq state(\mathcal{P}) \times act(\mathcal{P}) \times state(\mathcal{P})$ is defined to be the least relation such that

$$\begin{array}{r} send(a, q, \gamma, f, g) \in T(p) \wedge \underline{v} \in \Gamma_N(p) \wedge \gamma(\underline{v}) \Rightarrow p(\underline{v}) \overset{a!(g(\underline{v}))}{\twoheadrightarrow} q(f(\underline{v})) \\ tau(q, \gamma, f) \in T(p) \wedge \underline{v} \in \Gamma_N(p) \wedge \gamma(\underline{v}) \Rightarrow p(\underline{v}) \overset{\tau}{\twoheadrightarrow} q(f(\underline{v})) \\ rcv(a, q, \gamma, f) \in T(p) \wedge \underline{v} \in \Gamma_N(p) \wedge \underline{w} \in \Gamma_A(a) \wedge \gamma(\underline{v}) \Rightarrow p(\underline{v}) \overset{a?(\underline{w})}{\twoheadrightarrow} q(f(\underline{v}, \underline{w})). \end{array} \tag{3}$$

In the above example we would derive the transitions:

$$\begin{array}{l} p_1() \overset{in?(1)}{\twoheadrightarrow} p_2(1), \quad p_1() \overset{in?(2)}{\twoheadrightarrow} p_2(2), \quad \cdots \\ p_2(1) \overset{out!(1)}{\twoheadrightarrow} p_1(), \quad p_2(2) \overset{out!(2)}{\twoheadrightarrow} p_1(), \quad \cdots \end{array}$$

Obtaining the parameterised graph In [Sch92, §1] the author defines compositional semantic operators which transform a (vp-finite-state vp-finite-branching) CCS program into a *p.g.*. The author introduces a proof that the operational semantics ($\twoheadrightarrow$) on the resulting *p.g.* gives the same transitions as the standard CCS operational semantics ($\rightarrow$) for the original program.

0.5 Comparison with other work

A somewhat similar construct is found in [Kar88] as a form of implementation of a subset of LOTOS; one difference is that parameters are global (there) and that no proof of correctness is offered. In [HL92], *symbolic transition graphs* are developed for a purpose similar to ours. They use symbolic labellings for nodes and transitions while we use abstract functions. There transitions have 'f' functions which are limited to projections. The idea of expanding the parallel operators to achieve one sequential process (a key part of obtaining the *p.g.*) is also found in [CP85] & [Hol89].

0.6 Outline of paper

In Section 1 we show how to check for bisimulation between two *p.g.*s. Section 2 shows how to perform static analysis on *p.g.*s and how to expand parameters with bad behaviour. This may be used to obtain a *p.g.* in a more suitable form for the analysis of Section 1.

1 Verification

1.1 Overview

We review the idea of *bisimulation equivalence* between labelled transition systems. We can apply this to the *l.t.s.* of the states of a *p.g.* as defined in (3). This gives us a notion of bisimulation equivalence[7] between states of two *p.g.*s. It is, of course, not computable in general, since the *l.t.s.* is infinite.

Consider the buffer:

$$\begin{array}{ll} \textbf{label } in(\mathbb{N}), out(\mathbb{N}) & \\ \textbf{agent } \mathit{Buff*} = in?(x).\mathit{Buff*}'(h(x)) & (4) \\ \textbf{agent } \mathit{Buff*}'(x{:}\mathbb{N}) = out!(\overline{h}(x)).\mathit{Buff*} & \end{array}$$

where h and $\overline{h}$ are unary function symbols in the CCS language denoting some encryption and decryption. Infinite state spaces mean that we can't check whether $\mathit{Buff} \sim \mathit{Buff*}$ combinatorially.

We are interested in the following two question:

- Is $\mathit{Buff} \sim \mathit{Buff*}$? We would like an algorithm which will leave us with a proof obligation "$h = (\overline{h})^{-1}$".
- What are the most general conditions on v and w under which $\mathit{Buff}'(v) \sim \mathit{Buff*}'(w)$?

There are many cases where an implementation has some configuration parameter (say the size of its buffers) and we want to know for what values of that parameter it satisfies its specification. We therefore want our algorithm to cope with systems with initial parameters, giving us the most general condition on these parameters under which bisimulation holds. In the second question above, for fixed v, $\mathit{Buff}'(v)$ may be seen as a specification of a buffer with an initial value and $\mathit{Buff*}'(w)$ may be seen as an implementation.

1.2 Strong bisimulation

A possible definition given by Milner ([Mil89, §10.4]) for strong bisimulation equivalence (which he calls *strong equivalence*) involves a decreasing sequence of equivalences on the set of all agents.

Suppose we have a set of agents Pr and a set of actions Act, forming a *labelled transition system* $(Pr, Act, \rightarrow), \rightarrow \subseteq Pr \times Act \times Pr$.

- For any agents A and B, let $A \sim_0 B$.
- For $n = 1, 2, \ldots$ Let $A \sim_n B$ if, for all $a \in Act$,
  - whenever $A \xrightarrow{a} A'$ then, for some state B', $B \xrightarrow{a} B'$ and $A' \sim_{n-1} B'$
  - whenever $B \xrightarrow{a} B'$ then, for some state A', $A \xrightarrow{a} A'$ and $A' \sim_{n-1} B'$.

[7] We only consider what some authors call *early bisimulation*. Late bisimulation distinguishes between agents with identical *l.t.s.*s but different value-passing declarations. This seems too restrictive for many verifications.

Define **bisimulation equivalence**, $\sim$, to be $\bigcap_{n\in\mathbb{N}} \sim_n$. Milner defines and intersects the $\sim_i$ over all ordinals. But the *l.t.s.* of the states of a *p.g.* is *image finite*[8] — it has the property that any state p has only finitely many transitions labelled α for given α; in this case the intersection over the natural numbers will do (see [Sch92, §B]). (Milner also insists that the actions match for any sequences of actions. However, $\sim$ is the same for either definition of the $\sim_i$.)

It is easy to show that $\sim_0 \supseteq \sim_1 \supseteq \cdots$ and that if $\sim_n = \sim_{n+1}$ then $\sim_n = \sim$.

1.3 Alternative characterisation of the $\sim_i$

Let $(Pr, Act, \rightarrow)$ be a labelled transition system. Let θ^0_{pq}=**true** for all $p, q \in Pr$. For $n = 1, 2, \ldots$ and for all such p and q, let

$$\theta^n_{pq} = \left[\bigwedge_{(p,a,p')\in\rightarrow} \bigvee_{(q,a,q')\in\rightarrow} (\theta^{n-1}_{p'q'}) \right] \wedge \left[\bigwedge_{(q,a,q')\in\rightarrow} \bigvee_{(p,a,p')\in\rightarrow} (\theta^{n-1}_{p'q'}) \right]. \tag{5}$$

It is easy to see that $p \sim_n q \iff \theta^n_{pq}$.

1.4 Bisimulation defined on *p.g.*s

We will say states $p(\underline{v}) \in state(\mathcal{P}_1)$ and $q(\underline{w}) \in state(\mathcal{P}_2)$ of two *p.g.*s $\mathcal{P}_i = (Act, \Gamma_A, \wp^i, \iota^i, \Gamma^i_N, T^i)$ $(i = 1, 2)$ are related by $\sim_n$ *iff* they are so related in

$$(state(\mathcal{P}_1) \cup state(\mathcal{P}_2), act(\mathcal{P}_1), \longrightarrow\!\!\!\!\rightarrow) \tag{6}$$

as defined in (3). (Note that this is only defined[9] if $act(\mathcal{P}_1) = act(\mathcal{P}_2)$.)

Theorem 1.1 *For* $i = 1, 2$ *let* $\mathcal{P}_i = (Act, \Gamma_A, \wp^i, \iota^i, \Gamma^i_N, T^i)$ *be p.g.s. For all* $p \in \wp^1, q \in \wp^2$ *and* $n \in \mathbb{N}$ *let* θ^n_{pq} *be a predicate*[10] *over* $\Gamma^1_N(p) \times \Gamma^2_N(q)$ *with* θ^0_{pq} = **true** $(\forall p, q)$ *and for* $n = 1, 2, \ldots$, θ^n_{pq} *as below*[11]. *Then* $p(\underline{v}) \sim_n q(\underline{w}) \iff \theta^n_{pq}(\underline{v}, \underline{w})$.

[8] To see this, consider a state $p(\underline{v})$. Now $p(\underline{v})$ has at most one τ transition for every *tau* element in $T(p)$. It has at most one $a?(\underline{w})$ transition, for fixed $\underline{w}$, for each *rcv* element of $T(p)$. And it has at most one $a!(\underline{w})$ transition for each *send* element of $T(p)$. Since $T(p)$ is finite, it has finitely many of each of these.

[9] In general it is undecideable whether two *p.g.*s have the same channel sorts Γ_A. However, the implementation and specification of a system will generally have syntactically identical label declarations.

[10] Formally, θ is a well-formed formula in a language $\mathfrak{L}$ including: variables v_i, w_i; a constant for every value allowed in the CCS language as well as **true** and **false**; a function symbol for every function in the CCS language (with appropriate arity); a binary predicate '='; and the symbols $\neg, \Rightarrow, \forall, (,)$ and ','. We use $\wedge, \vee$ as shorthands in the normal way. In what follows, f, g are meta-notation for the appropriate function symbols derived from the original representation of the function in the CCS program. $\bigwedge$ and $\bigvee$ are also meta-notation meaning write down the following formula once for each instance of the subscript, with a $\wedge$ or $\vee$ between each such formula. We give $\mathfrak{L}$ the obvious interpretation.

$$
\theta^n_{pq}(\underline{v},\underline{w}) =
\begin{array}{l}
\left[\begin{array}{l}
\bigwedge_{send(a,p',\gamma,f,g)\in T^1(p)} \\
\quad \bigvee_{send(a,q',\gamma',f',g')\in T^2(q)} \\
\qquad \gamma(\underline{v}) \Rightarrow (\gamma'(\underline{w}) \wedge g(\underline{v}) = g'(\underline{w}) \wedge \theta^{n-1}_{p'q'}(f(\underline{v}), f'(\underline{w})))
\end{array}\right] \wedge \\
\left[\begin{array}{l}
\bigwedge_{send(a,q',\gamma,f',g')\in T^2(q)} \\
\quad \bigvee_{send(a,p',\gamma,f,g)\in T^1(p)} \\
\qquad \gamma'(\underline{w}) \Rightarrow (\gamma(\underline{v}) \wedge g(\underline{v}) = g'(\underline{w}) \wedge \theta^{n-1}_{p'q'}(f(\underline{v}), f'(\underline{w})))
\end{array}\right] \wedge \\
\left[\begin{array}{l}
\bigwedge_{rcv(a,p',\gamma,f)\in T^1(p)} \\
\quad \bigvee_{rcv(a,q',\gamma',f')\in T^2(q)} \\
\qquad \gamma(\underline{v}) \Rightarrow (\gamma'(\underline{w}) \wedge \forall \underline{z} \in sort(a) : \theta^{n-1}_{p'q'}(f(\underline{v},\underline{z}), f'(\underline{w},\underline{z})))
\end{array}\right] \wedge \\
\left[\begin{array}{l}
\bigwedge_{rcv(a,q',\gamma,f')\in T^2(q)} \\
\quad \bigvee_{rcv(a,p',\gamma,f)\in T^1(p)} \\
\qquad \gamma'(\underline{w}) \Rightarrow (\gamma(\underline{v}) \wedge \forall \underline{z} \in sort(a) : \theta^{n-1}_{p'q'}(f(\underline{v},\underline{z}), f'(\underline{w},\underline{z})))
\end{array}\right] \wedge \\
\left[\begin{array}{l}
\bigwedge_{tau(p',\gamma,f)\in T^1(p)} \\
\quad \bigvee_{tau(q',\gamma',f')\in T^2(q)} \\
\qquad \gamma(\underline{v}) \Rightarrow (\gamma'(\underline{w}) \wedge \theta^{n-1}_{p'q'}(f(\underline{v}), f'(\underline{w})))
\end{array}\right] \wedge \\
\left[\begin{array}{l}
\bigwedge_{tau(q',\gamma,f')\in T^2(q)} \\
\quad \bigvee_{tau(p',\gamma,f)\in T^1(p)} \\
\qquad \gamma'(\underline{w}) \Rightarrow (\gamma(\underline{v}) \wedge \theta^{n-1}_{p'q'}(f(\underline{v}), f'(\underline{w})))
\end{array}\right]
\end{array}
$$

Proof Start from the alternative characterisation (5) applied to the *l.t.s.* (6). We need to use the semantic rules (3) to change a predicate about the *l.t.s.* for the *p.g.* to a predicate about the *p.g.*s transitions.

The alternative characterisation of $\sim_n$ applied to

$$(state(\mathcal{P}_1) \cup state(\mathcal{P}_2), act(\mathcal{P}_1), \longrightarrow)$$

gives:

$$p(\underline{v}) \sim_n q(\underline{w}) \iff \theta^n_{pq}(\underline{v},\underline{w})$$

where we let θ^n_{pq} be a predicate parameterised by values in $\Gamma^1_N(p) \times \Gamma^2_N(q)$ rather than each $\theta^n_{p(\underline{v})q(\underline{w})}$ being a separate predicate; and where, for $n = 1, 2, \ldots$ and for all p and q,

$$\theta^n_{pq}(\underline{v},\underline{w}) = \left[\begin{array}{l} \bigwedge_{(p(\underline{v}),\alpha,p'(\underline{v}'))\in\longrightarrow} \\ \quad \bigvee_{(q(\underline{w}),\alpha,q'(\underline{w}'))\in\longrightarrow} \\ \qquad \theta^{n-1}_{p'q'}(\underline{v}',\underline{w}') \end{array}\right] \wedge \left[\begin{array}{l} \bigwedge_{(q(\underline{w}),\alpha,q'(\underline{w}'))\in\longrightarrow} \\ \quad \bigvee_{(p(\underline{v}),\alpha,p'(\underline{v}'))\in\longrightarrow} \\ \qquad \theta^{n-1}_{p'q'}(\underline{v}',\underline{w}') \end{array}\right].$$

[11] This predicate is simply saying that for every *send* transition of p there is a *send* transition of q on the same channel such that if p has this transition for parameter values $\underline{v}$ then so does q, for $\underline{w}$. Further, the values output are the same and the transitions lead to states related at the previous iteration. Similarly for every *send* transition of q there is.... The same must be true for receiving and internal transitions. In the receiving case, the resulting states must be related for any input value which is given to both systems.

This predicate may seem big and clumsy. But bisimulation between value-passing agents is complicated by the fact that sending, receiving and internal transitions must be treated differently. This predicate tells the whole story about what we must establish for each of these classes of transitions.

Here, α ranges over $act(\mathcal{P}_1)$, p' over $\wp^1$, q' over $\wp^2$, $\underline{v}'$ over $\Gamma^1_N(p')$ and $\underline{w}'$ over $\Gamma^2_N(q')$.

We split the conjunctions to range over sending, receiving and internal actions separately (replacing α by, respectively, $a!(\underline{v}'')$, $a?(\underline{v}'')$ and τ, with $\underline{v}''$ ranging over $\Gamma_A(a)$), turning these two square brackets into a conjunction of six square brackets. Consider just one of the six square brackets, the *send* subcase of the first bracket above. We will show that it is equal to the first square bracket in the theorem. (The other five are treated similarly.)

Note that, by the rules in (3), $(p(\underline{v}), a!(\underline{v}''), p'(\underline{v}')) \in\twoheadrightarrow$ *iff* there is some $send(a, p', \gamma, f, g) \in T(p)$ such that $\gamma(\underline{v}) = \mathbf{true}, f(\underline{v}) = \underline{v}'$ and $g(\underline{v}) = \underline{v}''$. Thus

$$\left[\begin{array}{c}\bigwedge_{(p(\underline{v}),a!(\underline{v}''),p'(\underline{v}'))\in\twoheadrightarrow} \\ \bigvee_{(q(\underline{w}),a!(\underline{v}''),q'(\underline{w}'))\in\twoheadrightarrow} \\ \theta^{n-1}_{p'q'}(\underline{v}',\underline{w}')\end{array}\right] = \left[\begin{array}{c}\bigwedge_{\{send(a,p',\gamma,f,g)\in T^1(p)\mid\gamma(\underline{v})\}} \\ \bigvee_{\{(q(\underline{w}),a!(\underline{w}''),q'(\underline{w}'))\in\twoheadrightarrow\mid\underline{w}''=g(\underline{v})\}} \\ \theta^{n-1}_{p'q'}(f(\underline{v}),\underline{w}')\end{array}\right].$$

We need only do the same for the $\bigvee$ and then push the side conditions ($\gamma(\underline{v})$ and $\gamma'(\underline{w})$) through to get the format in the theorem. □

Example Apllying this to *Buff* and *Buff** we get, after some reductions[12]:

$$\begin{aligned}\theta^n_{Buff\ Buff*} &= \forall z\in\mathbb{N} : \theta^{n-1}_{Buff'\ Buff*'}(z, h(z)) \\ \theta^n_{Buff'\ Buff*'}(v, w) &= v = \overline{h}(w)_\wedge\theta^{n-1}_{Buff\ Buff*}\end{aligned}$$

Iterating gives:

$$\begin{aligned}\theta^3_{Buff\ Buff*} &= \forall z\in\mathbb{N} : z = \overline{h}(h(z)) \\ \theta^3_{Buff'\ Buff*'}(v, w) &= v = \overline{h}(w)\ _\wedge\ \forall z\in\mathbb{N} : z = \overline{h}(h(z)).\end{aligned}$$

The next iteration introduces no change so $\sim_3 = \sim$ in this case. So the algorithm has automatically provided answers to both the questions posed at the start of this section! □

We have succeeded in stripping the concurrent elements from the problem of establishing $\sim_n$ (any n), leaving just a predicate as a proof obligation. But we still have not succeeded in doing this for $\sim$, in general, because there is no guarantee that the θ^n_{pq} will settle down as $n \to \infty$. The rest of this section concentrates on a special case for which our solution always works.

1.5 Data-independence

Intuitively, a *data-independent* program or process has behaviour which is independent of its input. All it does with input data is to store it and output it unchanged. It neither changes the values by applying functions nor performs any tests on the data's values. Buffers are the most obvious example.

The term *data-independence* was introduced by Wolper in [Wol86] in the context of temporal logic specification of "reactive" programs. The same term

[12] The γs are just **true** and we use $\mathbf{true}_\wedge\phi = \phi$. Later we use $\phi_\wedge\phi = \phi$ and $\forall z{:}\mathbf{true} = \mathbf{true}$.

was used by Jonsson and Parrow ([JP89]) as a syntactic property of CCS agents (no **ifs** and no functions — expressions are constants or parameters).

We define data-independence on *p.g.*s so that a data-independent CCS program (as defined by Jonsson & Parrow) gives a data-independent *p.g.*[13].

Definition 1.2 *A p.g., $(Act, \Gamma_A, \wp, \iota, \Gamma_N, T)$, is* **data-independent** *if whenever $send(a, p', \gamma, f, g) \in T(p)$, $rcv(a, p', \gamma, f) \in T(p)$ or $tau(p', \gamma, f) \in T(p)$, $(p \in \wp,)$ then $\gamma = \lambda \underline{v}.\mathbf{true}$ and, for all i, $(f)_i$ and $(g)_i$ are projection functions or constant functions.*

Theorem 1.3 *Let $\mathcal{P}_i = (Act, \Gamma_A, \wp^i, \iota^i, \Gamma_N^i, T^i)$ $(i = 1, 2)$ be two p.g.s which are data-independent. For all $n \in \mathbb{N}, p \in \wp^1$ and $q \in \wp^2$, let θ_{pq}^n be the predicate over $\Gamma_N^1(p) \times \Gamma_N^2(q)$ defined in Theorem 1.1. Then:*

1. *for all n, p, q, $\theta_{pq}^n(\underline{v}, \underline{w})$ is equivalent to some predicate built out of conjunctions, disjunctions and predicates of the forms*

   - $v_i = w_j$
   - $v_i = c$
   - $w_i = c$,

   c the value of some constant function appearing in $\mathcal{P}_1$ or $\mathcal{P}_2$.

2. *for $\forall p, q$ $\exists n$ such that $\theta_{pq}^n = \theta_{pq}^\infty$ (here '=' means equal for all values).*

Proof The first part is by induction on n. For $n = 0$ and any p, q, θ_{pq}^0 is equal to an empty conjunction. Now assume we have found a canonical form for θ_{pq}^{k-1}, for all p, q. Consider the definition of θ_{pq}^k.

- We can replace all γ expressions with **true**, the empty conjunction.
- $g(\underline{v}) = g'(\underline{w})$ may be replaced by $\bigwedge_i g_i(\underline{v}) = g_i'(\underline{w})$ and each $g_i(\underline{v}) = g_i'(\underline{w})$ is the same as a basic predicate in one of the three forms in the theorem or is of the form $c_1 = c_2$ which may be replaced by **true** or **false**.
- $f(\underline{v})$ is just a tuple of values v_i and/or constants. By the inductive hypothesis, $\theta_{p'q'}^{k-1}(f(\underline{v}, \underline{w}))$ has a canonical form, and now we are just substituting v_is w_is or constants for the v_i and w_is in this predicate. Any resulting expressions $c_1 = c_2$ may be replaced by **true** or **false** to leave a predicate in the required form.
- In the *rcv* case we have the expression $\forall \underline{z} : \theta_{p'q'}^{k-1}(f(\underline{v}, \underline{z}), f'(\underline{w}, \underline{z}))$. Now we may substitute in the canonical form for $\theta_{p'q'}^{k-1}$ to get a predicate built out of the basic predicates in the theorem, and, in addition, basic predicates of the form $z_i = v_j$, $z_i = w_j$, $z_i = z_j$, $z_i = c$ and, as above, $c_1 = c_2$. Because of the universal qualifier, these may respectively be replaced by[14] **false**; **false**; **false** unless $i = j$ in which case **true**; **false**; **false** or **true**, as above.

[13] We do this for consistency. In fact, we could just as well adopt a more general definition allowing **ifs** with conditions built out of constants, parameters, '=' and conjunctions and disjunctions. This shows the power of our approach since the symbolic technique of [JP89] would not be applicable to this more general problem.

[14] If any z_i has a sort of one value, the replacements may be different. We must assume that we can tell if a program has a singleton sort, and, if so, determine its value.

This establishes the first part. Now if we impose some sorting on the canonical predicates, there are only finitely many of them. We also know that for any p, q, $\theta^0_{pq} \Leftarrow \theta^1_{pq} \Leftarrow \cdots$. So for some n, $\forall m > n : \theta^n_{pq} = \theta^m_{pq}$. The second result follows. □

So in the data-independent case, a simple algorithm will produce a predicate saying how the initial parameters of the two systems must be set up for the systems to be strongly equivalent. As it happens, this predicate will have no $\forall$s, and can easily be evaluated for specific values of $\underline{v}$ and $\underline{w}$.

1.6 Comparison with [HL92]

In [HL92], developed independently of the present work, Hennessy and Lin aim to achieve the same things that we set out to achieve in this section: finding a predicate on the parameters of two value-passing agents giving the conditions for the agents to be bisimulation equivalent. Where we use *parameterised graphs*, they use *symbolic transition graphs*; the transitions in this construct may only have 'f' functions with each $(f)_i$ a projection function, meaning that only a narrower set of programs may be represented by a finite graph[15]. Only such programs can be verified automatically by their algorithms. In fact our solution is more general because if a program has a finite symbolic transition graph then it can be represented by a *p.g.* in which each $(f)_i$ is a projection function whence we can again find a finite number of canonical predicates, with every θ^n_{pq} equivalent to one of these[16]. On the other hand, our techniques are not limited to this case.

Note that [HL92] defines a notion of symbolic bisimulation for value-passing agents, rather than directly applying the definition of bisimulation equivalence.

2 Static analysis

2.1 Overview

We show that static analysis may be used to get a value-passing program into a form in which it lends itself to the validation techniques of Section 1.

Consider the following definition of the *Send* component of the alternating bit protocol (Figure 1):

agent *Send* $= \mathit{Accept}(0)$
agent *Accept* $(b{:}\mathit{bit})$ $= \mathit{accept?}(x).\mathit{Send\_data}(b, x)$
agent *Send_data* $(b{:}\mathit{bit}, x{:}\mathbb{N})$ $= \mathit{send!}(b, x).\mathit{Sending}(b, x)$
agent *Sending* $(b{:}\mathit{bit}, x{:}\mathbb{N})$ $= \mathit{send!}(b, x).\mathit{Sending}(b, x) + \mathit{ack?}(b').S(b, b', x)$
agent *S* $(b{:}\mathit{bit}, b'{:}\mathit{bit}, x{:}\mathbb{N})$ $=$ **if** $b = b'$ **then** $\mathit{Accept}(1-b)$ **else** $\mathit{Sending}(b, x)$

[15] A sufficient condition for a sequential vp-finite-branching program to be representable by a finite symbolic transition graph is that every constant expression $C(\underline{e})$ have each e_i just a variable x_i or a constant.

[16] The critical point is that no value ever gets put through more than one function so that compositions of functions do not appear in the θ^n_{pq} predicates. The canonical predicates may be made out of conjunction, disjunction, negation and predicates of the form $\forall \underline{z} : g(\underline{x}) = g'(\underline{y})$ or $\forall \underline{z} : \gamma(\underline{x})$ where g, g', γ are functions from one of the *p.g.*s and the x_i and y_i range over the v_i, w_i and z_i.

The parameters labelled x have infinite sort and are never tested or passed into a function. This agent is nearly data-independent, but the b and b' parameters are tested and changed in the final definition. The following four definitions are equivalent to the above definition of S but don't have this bad behaviour:

$$\begin{array}{llll} \textbf{agent } S_{b,b} & (x{:}\mathbb{N}) & = & Accept(1-b) & (b \in \{0,1\}) \\ \textbf{agent } S_{b,1-b} & (x{:}\mathbb{N}) & = & Sending(b,x) & (b \in \{0,1\}). \end{array}$$

Unfortunately, the definition of *Sending* no longer makes any sense as $S(b, b', x)$ has been split into four distinct nodes. We must therefore expand the b parameter of *Sending* in the same way.

We see that we must always expand any parameter whose value gets passed on to a parameter which we are already expanding. So once we have decided which parameters to expand (to get rid of complicated functions) we need a backwards static analysis to decide what other parameters need be expanded. Of course, if any of these have an infinite (or unmanageably large) sort, we cannot use this technique.

Note that each expansion increases the number of nodes in the *p.g.* and hence the complexity of subsequent analysis.

2.2 Defining the allowed functions

One application of this static analysis is to make a program data-independent. In this case, if a parameter x appears in an expression e then we must expand x unless e equals x for all values (since no non-trivial value expression are allowed in a data-independent program). If we identify a more general class of agents for which the techniques of Section 1 can be applied, we may have some other criterion of "bad behaviour".

All we need to assume is that given some expression e in the definition of $C(x_1, x_2, \ldots)$, the parameters x_i fall into three classes:

- those which don't affect the value of e at all (approximated by 'those which don't appear syntactically in e')
- those that affect the value of e in a way we don't allow (approximated by 'those that appear in e within an argument to a function we don't allow')
- the rest.

Applying this to parameterised graphs We now describe this more formally. Suppose we are given a *p.g.* $(Act, \Gamma_A, \wp, \iota, \Gamma_N, T)$, which we need to expand. Suppose for any sets $\underline{u} \in \underline{U}, u' \in U$ we only want our *p.g.* to involve functions $f : \underline{u} \to u'$ if $f \in \mathcal{F}$ where $\mathcal{F}$ is a class of functions closed under composition[17].

Now suppose we have an approximate but safe algorithm for checking if a function is in $\mathcal{F}$. We expect to be provided with functions $\mathcal{C}, \mathcal{X}, \mathcal{I}$ which tell us how the arguments of an n-ary function f affect its value (compare the three syntactic groups above).

[17] The syntactic restrictions for data-independence, for example, lead to the following class $\mathcal{F}$: for any $\underline{u}, u' \in U$ we let all constant functions from $\underline{u}$ to u' be in $\mathcal{F}$ as well as the projection function onto the i-th argument (for any i such that $u_i = u'$).

- $\mathcal{C}(f)$ is a subset of $\{1,\ldots,n\}$ giving the (positions of) the arguments of f whose value never affect the value of f.

  Formally, $i \in \mathcal{C}(f) \Rightarrow \forall \underline{v}, u : f(\underline{v}) = f(\underline{v}[u/v_i])$. We do not demand the opposite implication — $\mathcal{C}$ is approximate but safe.

- $\mathcal{X}(f)$ gives a subset of $\{1,\ldots,n\}$ giving a set of arguments which may be expanded to leave a function which is in $\mathcal{F}$. $\mathcal{X}(f)$ must be disjoint from $\mathcal{C}(f)$.

  Formally, if $j_1 \leq j_2 \leq \cdots$ are the element of $\mathcal{X}(f)$ and $i_1 \leq i_2 \leq \cdots$ are the remaining elements of $\{1,\ldots,n\}$ then the function

  $$f' = \lambda(v_{i_1}, v_{i_2}, \ldots).f(\underline{v}) \tag{7}$$

  is in $\mathcal{F}$ for any constant values $v_{j_1}, v_{j_2}, \ldots$.

- $\mathcal{I}(f)$ giving the remaining arguments. Intuitively, these arguments influence the value of f but only through functions in $\mathcal{F}$.

The best results will be obtained if $\mathcal{C}(f)$ is maximal and $\mathcal{X}(f)$ is, in some sense, minimal. In practice we would probably obtain $\mathcal{C}$, $\mathcal{X}$ and $\mathcal{I}$ at the time of calculating the *p.g.*, by the syntactic criterion above.

We insist that $\mathcal{C}, \mathcal{X}$ and $\mathcal{I}$ give **consistent** information. By this we mean that if f' is defined as in (7) then $i_k \in \mathcal{S}(f) \Leftrightarrow k \in \mathcal{S}(f')$ ($\mathcal{S}$ any of $\mathcal{C}, \mathcal{X}, \mathcal{I}$). Note that if we obtain these from the syntactic criterion then consistency is satisfied.

We now define the static analysis. The analysis is performed on an abstract version of the *p.g.*. We combine it with a liveness analysis ([ASU86, §10.6]) — there is no need to expand a parameter whose value is passed through a nasty function but then never used! Such a function would anyway not appear in the θ^n_{pq} predicates.

2.3 The abstract parameterised graph

Abstract domain At each iteration we mark each parameter of each node and channel by an element of the following lattice, L (ranged over by $\hat{v}$):

$\top$	...	parameter may have its value modified by a disallowed function *and* the result may influence an output/test
$\vert$		
l	...	parameter may influence an output/test ("live")
$\vert$		
$\bot$	...	nothing known about parameter (once we have reached a fixed point, this indicates a "dead" parameter)

We will expand parameters marked $\top$. We use l to mark parameters which are live, as a precondition of marking them with $\top$.

Abstract function In our abstract graph the transitions go back-to-front and are labelled with abstract functions. We now show how to find the abstract functions. The idea is to abstract away from the details of the function and only take account of which parameters affect each other and whether in a way describable by functions in $\mathcal{F}$.

Let $(Act, \Gamma_A, \wp, \iota, \Gamma_N, T)$ be a *p.g.*. Consider a typical function involved in a transition from node p_1 to node p_2. In fact, start by considering f to be a single component of such a function. Let $f : \Gamma_N(p_1) \to (\Gamma_N(p_2))_j$ where say p_1 has m parameters and p_2 has n.

Assume that we have already assigned an element of L to each of p_2's parameters and, in particular, $\hat{v}$ to the j-th. Now we must decide which parameters of p_1 should be marked by l or $\perp$ by virtue of being passed on to p_2 via function f.

We will mark p_1's i-th parameter by at least $(\tilde{f})_i(\hat{v})$ where $\tilde{f}$, the *inverse dependence function*, is defined by:

$$\tilde{f} : L \longrightarrow L^m \qquad (\tilde{f})_i(\hat{v}) = \begin{cases} \perp & i \in \mathcal{C}(f) \\ t(\hat{v}) & i \in \mathcal{X}(f) \\ \hat{v} & i \in \mathcal{I}(f) \end{cases}$$

where $t(\perp) = \perp, t(l) = t(\top) = \top$. The idea is that if p_1's i-th parameter has no affect on f ($i \in \mathcal{C}(f)$) then we don't need to impose any minimum marking on it. If it does affect f and in a complicated way ($i \in \mathcal{X}(f)$) then, if at least p_2's j-th parameter is live, we must mark this parameter by $\top$. Finally, if p_1's i-th parameter affects p_2's j-th in a way which is describable by functions in $\mathcal{F}$ ($i \in \mathcal{I}(f)$) then it is considered live if p_2's j-th parameter is live ($\hat{v} = l$) but only need be expanded if the latter needs expanding ($\hat{v} = \top$).

By extension if $f : \Gamma_N(p_1) \to \Gamma_N(p_2)$ then we may define

$$\tilde{f} : L^n \longrightarrow L^m \qquad \tilde{f}(\hat{v}_1, \ldots, \hat{v}_n) = \bigsqcup_{j=1,\ldots,n} (\tilde{f}_j)(\hat{v}_j).$$

Now we can say that if all of p_2's parameters are marked by $\underline{\hat{v}}$ then all of p_1's must be marked by at least $\tilde{f}(\underline{\hat{v}})$.

2.4 The static analysis

At the i-th iteration we mark each parameter of each node (by a function ϑ^i) with an element of L and each of the channels' parameters (by φ^i) similarly. We will use α as a function giving the number of parameters of a node or channel (so $\alpha(a)$ is a shorthand for $|\Gamma_A(a)|$ and $\alpha(p)$ for $|\Gamma_N(p)|$).

Formally, let $\vartheta^0, \vartheta^1, \ldots$ be mappings such that for each $i \in \mathbb{N}$ and $p \in \wp$, $\vartheta^i(p) \in L^{\alpha(p)}$. Let $\varphi^0, \varphi^1, \ldots$ be mappings such that for each $i \in \mathbb{N}$ and $a \in Act$, $\varphi^i(a) \in L^{\alpha(a)}$. These are defined in Figure 2.

Whenever there is a transition from p to q, this involves putting the values of the parameters of node p through a function f to get the values of the parameters of q. If we have already marked (the parameters of) q with $\vartheta^n(q)$, we must mark p with at least $\tilde{f}(\vartheta^n(q))$.

[18]Note that in the case of receiving actions, f is of type $\Gamma_N(p) \times \Gamma_A(a) \to \Gamma_N(q)$ so $\tilde{f}$ will be of type $L^{\alpha(q)} \longrightarrow L^{\alpha(p)+\alpha(a)}$. The first $\alpha(p)$ values indicate which parameters of p influence behaviour, while the next $\alpha(a)$ values indicate which parameters of a influence behaviour. Thus, these are considered separately (one in calculating ϑ^i, the other in calculating φ^i).

Let $\vartheta^0(p) = \{\bot, \bot, \ldots\}, \forall p$; $\varphi^0(a) = \{\bot, \bot, \ldots\}, \forall a$.
For $i = 1, 2, \ldots, p \in \wp, a \in Act$ let

$$\vartheta^{i+1}(p) = \left[\bigsqcup_{send(a,q,\gamma,f,g)\in T(p)} \tilde{f}(\vartheta^i(q)) \sqcup \tilde{g}(\varphi^i(a)) \sqcup \tilde{\gamma}(l) \right] \sqcup \left[\bigsqcup_{rcv(a,q,\gamma,f)\in T(p)} \tilde{f}_{1,\ldots,\alpha(p)}(\vartheta^i(q)) \sqcup \tilde{\gamma}(l) \right] \sqcup \left[\bigsqcup_{tau(q,\gamma,f)\in T(p)} \tilde{f}(\vartheta^i(q)) \sqcup \tilde{\gamma}(l) \right] \quad (8)$$

$$\varphi^{i+1}(a) = \left[\bigsqcup_{p\in\wp, rcv(a,q,\gamma,f)\in T(p)} \tilde{f}_{\alpha(p)+1,\ldots,\alpha(p)+\alpha(a)}(\vartheta^i(q)) \right] \sqcup (l, \ldots, l) \quad (9)$$

Figure 2: The static analysis[18]

Note that if we expand $p(x)$ into distinct nodes p_x then $a?(\ldots, x, \ldots).p(x)$ no longer makes sense and we must expand channel a into channels $a_1, a_2, \cdots$ (and use $\sum a_v?(\ldots).p_v$). The φs keep track of which channels must be expanded. We must then expand parameters x which are involved in an output action $a!(\ldots, f(x), \ldots)$ in parameters of a marked for expansion; this is ensured by the expression $\tilde{g}(\varphi^i(a))$. In this case liveness is certain — the value is being output here and now. Equation (9) therefore makes sure all the φs have values of at least l.

Finally, we must expand parameters which are passed through tests (γs) involving complicated functions. Here again, liveness is certain — the result of the test will affect all subsequent behaviour. Hence the expression $\tilde{\gamma}(l)$.

Lemma 2.1 *For each $p \in \wp, a \in Act$: $\vartheta^0(p) \sqsubseteq \vartheta^1(p) \sqsubseteq \cdots$ and $\varphi^0(a) \sqsubseteq \varphi^1(a) \sqsubseteq \cdots$ (pointwise $\sqsubseteq$).*

Proof The *inverse dependence functions* are monotonic. So if, for some i, $\vartheta^i(p) \sqsubseteq \vartheta^{i+1}(p)$ for all p and $\varphi^i(a) \sqsubseteq \varphi^{i+1}(a)$ for all a then it follows that $\vartheta^{i+1}(q) \sqsubseteq \vartheta^{i+2}(q)$ for each q and $\varphi^{i+1}(b) \sqsubseteq \varphi^{i+2}(b)$ for each b.

But trivially $\forall p, a : \vartheta^0(p) \sqsubseteq \vartheta^1(p), \varphi^0(a) \sqsubseteq \varphi^1(a)$. The result follows by induction. □

Corollary 2.2 *If $|\wp| + |Act| = m$ and $\forall p \in \wp, a \in Act : \alpha(p) \leq k, \alpha(a) \leq k$ then the static analysis reaches a fixed point in at most $2mk$ iterations.*

Proof If for all p and a, $\vartheta^n(p) = \vartheta^{n+1}(p)$ and $\varphi^n(a) = \varphi^{n+1}(a)$, it follows that for all a and p, $\vartheta^{n+1}(p) = \vartheta^{n+2}(p)$ and $\varphi^{n+1}(a) = \varphi^{n+2}(a)$ (etc.). So some value $(\vartheta^i(p))_j$ or $(\varphi^i(a))_j$ must increase at each iteration up to the fixed point, by Lemma 2.1, and any value can only increase twice since $|L| = 3$. Hence we must reach a fixed point within $2mk$ iterations. □

2.5 Expanding the *p.g.*

Suppose we have a *p.g.* $(Act, \Gamma_A, \wp, \iota, \Gamma_N, T)$ to which we have applied our static analysis reaching a fixed point (ϑ, φ) and suppose that all the parameters marked by $\top$ have finite sort. We derive an expanded parameterised graph $(Act', \Gamma'_A, \wp', \iota', \Gamma'_N, T')$, by simultaneously performing the following expansions. This is equivalent to the syntactic expansion described at the beginning of this section.

[Nodes] For each node p, suppose the j_1-th, j_2-th,...elements of $\vartheta(p)$ are $\top$ and the i_1-th, i_2-th,...are not ($i_1 \leq i_2 \leq \cdots, j_1 \leq j_2 \leq \cdots$), we will have, in our expanded *p.g*, nodes[19] $p_{\underline{w}}$ for every $\underline{w} \in (\Gamma_N(p))_{j_1} \times (\Gamma_N(p))_{j_2} \times \cdots$. We let $\Gamma'_N(p_{\underline{w}}) = (\Gamma_N(p))_{i_1} \times (\Gamma_N(p))_2 \times \cdots$ for each $\underline{w}$.

For $p \in \wp$ define π^1_p by $\pi^1_p(\underline{v}) = (v_{i_1}, v_{i_2}, \ldots)$ and π^2_p by $\pi^2_p(\underline{v}) = (v_{j_1}, v_{j_2}, \ldots)$. Thus, for a set of parameter values $\underline{v}$ for node p, π^2_p selects the values which need expanding, π^1_p the ones which don't.

We have a correspondence for each $p \in \wp$ and $\underline{v} \in \Gamma_N(p)$ between state $p(\underline{v})$ in the old system, and $p_{\pi^2_p(\underline{v})}(\pi^1_p(\underline{v}))$ in the new.

[Channels] For each channel a, suppose the j_1-th, j_2-th,...elements of $\varphi(a)$ are $\top$ and the i_1-th, i_2-th,...are not ($i_1 \leq i_2 \leq \cdots, j_1 \leq j_2 \leq \cdots$), we will have, in our expanded *p.g*, channels $a_{\underline{w}}$ for every $\underline{w} \in (\Gamma_N(a))_{j_1} \times (\Gamma_N(a))_{j_2} \times \cdots$. We let $\Gamma'_A(a_{\underline{w}}) = (\Gamma_N(a))_{i_1} \times (\Gamma_N(a))_{i_2} \times \cdots$ for every $\underline{w}$.

Again, for each such channel define π^1_a by $\pi^1_a(\underline{v}) = (v_{i_1}, v_{i_2}, \ldots)$ and π^2_a by $\pi^2_a(\underline{v}) = (v_{j_1}, v_{j_2}, \ldots)$.

Here too we get a simple bijection between the actions in the old and new *p.g.*s.

[Transitions] We must now expand the sets of transitions given by T, so that they are expressed in terms of our new, bigger sets of nodes and channels.

Let $tau(q, \gamma, f) \in T(p)$. We are going to have to expand this to several elements since p is split into several nodes (its 'children'!) $p_{\underline{w}}$. For each such child we will need to calculate which of q's children is the result. The static analysis should guarantee that given the particular 'child' of p we always get to the same child of q, no matter what the parameters[20].

We will also need to restrict γ and f to deal with the more restricted set of parameters which the children have.

Formally, for each $\underline{w} \in (\Gamma_N(p))_{j_1} \times (\Gamma_N(p))_{j_2} \times \cdots$ (the j_i defined as above for p), let $tau(q_{h(\underline{w})}, \gamma_{\underline{w}}, f_{\underline{w}}) \in T'(p_{\underline{w}})$ where

- $\gamma_{\underline{w}}(\underline{z}) = \gamma(\underline{v})$ where $\underline{v}$ is picked so that $\pi^1_p(\underline{v}) = \underline{z}$, $\pi^2_p(\underline{v}) = \underline{w}$.

[19] Here we assume that we have a partial injective function $\hat{\mathcal{K}} \times \underline{V} \to \hat{\mathcal{K}}$ for selecting new node names. We make the same assumption about channel names. In fact, since all the parameters we expand must have finite sort, we only need such a function for finitely many tuples in any given expansion, so it must be possible to construct such a function (as $\hat{\mathcal{K}}$ is infinite).

[20] That is, if $\pi^2_p(\underline{v}) = \pi^2_p(\underline{w})$ then $\pi^2_q(f(\underline{v})) = \pi^2_q(f(\underline{w}))$.

- $f_{\underline{w}}(\underline{z}) = \pi_q^1(f(\underline{v}))$ where $\underline{v}$ is picked so that $\pi_p^1(\underline{v}) = \underline{z}, \pi_p^2(\underline{v}) = \underline{w}$.
- $h(\underline{w}) = \pi_q^2(f(\underline{v}))$ where $\underline{v}$ is picked so that $\pi_p^2(\underline{v}) = \underline{w}$. The values of the other components of $\underline{v}$ can be chosen arbitrarily. These cannot impact the value of $\pi_q^2(f(\underline{v}))$ since if they did, they too would be marked $\top$ (and would therefore be selected by π_p^2).

The receiving and sending actions are treated similarly, paying attention to the actions as well as the nodes.

Definition 2.3 *For $i = 1, 2$, let $\mathcal{P}_i = (Act^i, \Gamma_A^i, \wp^i, \iota^i, \Gamma_N^i, T^i)$ be p.g.s and let $\phi : act(\mathcal{P}_1) \rightarrow act(\mathcal{P}_2)$ be a bijection.*

*Say a bijection $\psi : state(\mathcal{P}_1) \rightarrow state(\mathcal{P}_2)$ is a p.g.-**isomorphism** with respect to ϕ if*

$$p \stackrel{\alpha}{\twoheadrightarrow} q \iff \psi(p) \stackrel{\phi(\alpha)}{\twoheadrightarrow} \psi(q).$$

We now show that the above expansion only changes the *p.g.* by an isomorphism. In the next subsection we prove that the expanded *p.g.* has the desirable properties. For full proofs see [Sch92].

Theorem 2.4 *The mapping $p(\underline{v}) \longmapsto p_{\pi_p^2(\underline{v})}(\pi_p^1(\underline{v}))$. is a p.g.-isomorphism to the original p.g. with respect to*

$$a!(\underline{v}) \mapsto a_{\pi_a^2(\underline{v})}!(\pi_a^1(\underline{v})), \quad a?(\underline{v}) \mapsto a_{\pi_a^2(\underline{v})}?(\pi_a^1(\underline{v})), \quad \tau \mapsto \tau.$$

This means that if we expand two *p.g.*s and prove the expansions are bisimilar then the original *p.g.*s must be bisimilar.

2.6 Correctness

We must now prove that the expansion achieves what we want it to by first proving that it gets rid of all parameters marked $\top$ and then proving that if no parameters are marked $\top$, no disallowed functions will appear in the θ_{pq}^n predicates. We do not give full proofs here and refer the interested reader to [Sch92].

Lemma 2.5 *Suppose we repeat our analysis on $(Act', \Gamma'_A, \wp', \iota', \Gamma'_N, T')$, the expanded p.g., with result (ϑ', φ'). Then for all $p \in \wp, a \in Act$ and for every $\underline{w}$ we have*[21]*:*

$$\vartheta'(p_{\underline{w}}) = \pi_p^1(\vartheta(p)) \qquad \varphi'(a_{\underline{w}}) = \pi_a^1(\varphi(a)).$$

In words, the remaining parameters in the expanded graph are marked by the same element of L as they were in the unexpanded graph.

Proof The proof depends on the consistency of $\mathcal{C}, \mathcal{X}, \mathcal{I}$ (Subsection 2.2). □

Corollary 2.6 *ϑ' and φ' mark nothing with $\top$.* □

[21] Here the πs give the same projections as defined above but the sort of each parameter is L. This is fine if we take the πs to be polymorphic projection functions.

If the static analysis marks nothing with $\top$ then all functions h appearing in the *p.g.* (either as a γ or as a component of an f or g) are in $\mathcal{F}$, with the possible exception of functions whose value is only passed to dead parameters[22].

We further claim that since only dead parameters are passed to unallowed functions, these would never occur in the θ predicates of Theorem 1.1. We now prove this.

However, we can only prove this if we assume that the way we choose to translate the functions into syntactic predicates is reasonable (otherwise we may just slip in a disallowed symbol unnecessarily). So assume that[23]

- if $i \in \mathcal{C}(f)$ then the syntax we put in a predicate for the meta-notation $f(\underline{u})$ will not contain the symbol u_i
- if $f \in \mathcal{F}$ then the syntax we put in a predicate for the meta-notation $f(\underline{u})$ will only contain function symbols whose interpretation is in $\mathcal{F}$.

Theorem 2.7 *Suppose we are given p.g.s* $(Act, \Gamma_A, \wp^i, \iota^i, \Gamma_N^i, T^i)$ $(i = 1,2)$, *which are expanded in the sense that performing the static analysis marks nothing with* $\top$.

Suppose for $n = 0, 1, 2, \ldots$ *and for every* $p \in \wp^1, q \in \wp^2$ *we calculate the predicate* θ^n_{pq} *of Theorem 1.1. Then all function symbols in the predicates have interpretations in* $\mathcal{F}$.

The theorem follows immediately from the following lemma and corollary.

Lemma 2.8 *In the p.g.s from the theorem, if* θ^n_{pq} *involves the symbol* u_i *then* $(\vartheta^n(p))_i \sqsupseteq l$.

Proof By induction on n. □

Corollary 2.9 *If some component* $(f)_i$ *of* f *in* $send(a, p', \gamma, f, g) \in T(p)$ *or* $rcv(a, p', \gamma, f) \in T(p)$ *or* $tau(p', \gamma, f) \in T(p)$ *appears in the meta-notation for some predicate* θ^n_{pq} *then* $(\vartheta(p'))_i \sqsupseteq l$. □

2.7 Using the expansions

Suppose we are given two *p.g.s*, $(Act, \Gamma_A, \wp^i, \iota^i, \Gamma_N^i, T^i)$ $(i = 1,2)$, and we wish to find conditions for the two to be bisimilar. If we expanded both independently we might end up with different (Act, Γ_A) pairs. We can avoid this by adding an intermediate step:

$$\varphi_1^{i\frac{1}{2}}(a) = \varphi_1^i(a) \sqcup \varphi_2^i(a) \qquad \varphi_2^{i\frac{1}{2}}(a) = \varphi_1^i(a) \sqcup \varphi_2^i(a),$$

so that channels marked for expansion in one *p.g.* are also marked for expansion in the other. The ϑs are unchanged in this intermediate step. This extra step is safe since it can only increase the result.

[22] The proof is simple: if $h \notin \mathcal{F}$ then for some i, $i \in \mathcal{X}(h)$ and then if the value of h was passed on to a parameter of q, say, which was live (has been marked by l or $\top$) then the next iteration would mark p's i-th parameter with $\top$, a contradiction.

[23] In practice we would expect functions f, g, γ to be stored syntactically and for this syntax to be copied into the predicate when required. We also expect $\mathcal{C}$ to be calculated syntactically. In such a scenario, these assumptions hold trivially.

Conclusion and Further Research

Automatic verification is likely to become increasingly important because the expertise needed for manual verification is very scarce. Most systems have some values of large sort and cannot be verified by calculating the labelled transition systems. The sections of this paper have respectively:

- provided a representation of value-passing agents (except for agents that are not vp-finite-state/branching) suitable for any type of analysis
- shown how to find a proof obligation for the approximations to bisimulation equivalence, and, in special cases, for bisimulation equivalence
- shown how to combine the above analysis with the well developed combinatorial approach.

Several areas for further research suggest themselves:

- Can we determine the broadest class of agents for which the θ^n_{pq} of Section 1 settle down as $n \to \infty$. Could we then use approximate methods to verify more general agents? Some work is needed on reducing the θ^n_{pq}.
- This work could be adapted to other equivalences and properties.

Thanks to my supervisor, Chris Hankin, for extensive help especially in reviewing drafts of this paper. Also to the following for comments, suggestions and encouragement: Samson Abramsky, Iain Philips and David Sands (at Imperial), Bent Thomsen (ECRC), Faron Moller and Glenn Bruns (Edinburgh) and Bengt Jonsson (SICS).

This research was carried out as part of the formal methods work at Data Connection Limited.

References

[ASU86] A. V. Aho, R. Sethi, and J. D. Ullman. *Compilers: Principles, Techniques and Tools.* Addisson-Wesley, 1986.

[BA91] G. Bruns and S. Anderson. The formalization and analysis of a communications protocol. Technical Report ECS-LFCS-91-137, LFCS, Dept of Computer Science, University of Edinburgh, The King's Buildings, Edinburgh EH9 3JZ, April 1991.

[Bru91] G. Bruns. A language for value-passing CCS. Technical Report ECS-LFCS-91-175, LFCS, Dept of Computer Science, University of Edinburgh, The King's Buildings, Edinburgh EH9 3JZ, August 1991.

[CP85] L. Cardelli and R. Pike. Squeak: a language for communicating with mice. In *ACM SIGRAHP*, volume 19, number 3, pages 199–204, July 1985.

[HL92] M. Hennessy and H. Lin. Symbolic bisimulation. Technical Report 1/92, University of Sussex, Computer Science, School of Cognitive and Computing Sciences, Brighton BN1 9QH, April 1992.

[Hoa85] C. A. R. Hoare. *Communicating Sequential Processes.* Prentice-Hall, 1985.

[Hol89] U. Holmer. Translating static CCS agents into regular form. Technical Report 51, Programming Methodology Group, University of Göteborg and Chalmers University of Technology, S-412 96 Göteborg, Sweden, March 1989.

[JP89] B. Jonsson and J. Parrow. Deciding Bisimulation equivalences for a class of non-finite-state programs. In *Proc. 6th Symposium on Theoretical Aspects of Computer Science*, volume 349 of *LNCS*, pages 421–433. SV, 1989.

[Kar88] G. Karjoth. Implementing process algebra specifications by state machines. In *Protocol Specification, Testing, and Verification, VIII*, pages 47–62. IFIP, 1988.

[KS83] P. Kannellakis and S. Smolka. CCS expressions, finite state processes, and three problems of equivalence. *Proceedings of the ACM Symposium on Principles of Distributed Computing*, pages 228–240, 1983.

[LM86] K. Larsen and R. Milner. A complete protocol verification using Relativized Bisimulation. Technical Report ECS-LFCS-86-13, LFCS, Dept of Computer Science, University of Edinburgh, The King's Buildings, Edinburgh EH9 3JZ, September 1986.

[Mil89] R. Milner. *Communication and Concurrency*. Prentice-Hall, 1989.

[Mol91] F. Moller. *The Edinburgh Concurrency Workbench (Version 6.0)*. Dept of Computer Science, University of Edinburgh, The King's Buildings, Edinburgh EH9 3JZ, 1991.

[Par87] J. Parrow. Verifying a CSMA/CD-protocol with CCS. Technical Report ECS-LFCS-87-18, Dept of Computer Science, University of Edinburgh, The King's Buildings, Edinburgh EH9 3JZ, 1987.

[Sch92] Z. Schreiber. Verification and analysis of value-passing CCS programs with infinite sorts. Technical Report DoC 92/9, Department of Computing, Imperial College of Science, Technology & Medicine, 180 Queen's Gate, London SW7 2BZ, June 1992.

[Wol86] P. Wolper. Expressing interesting properties of programs in propositional temporal logic (extended abstract). In *Proc. 13th ACM Principles of Programming Languages*, pages 184–193, January 1986.

An Extension of the Testing Method for Processes Passing Infinite Values

Shoji Yuen, Toshiki Sakabe and Yasuyoshi Inagaki

School of Engineering, Nagoya University
Furo-cho, Chikusa-ku, Nagoya 464-01, JAPAN

Summary

In this abstract, we present an extended framework for communicating processes capable of passing infinite values. An operational semantics is given by extending the *Testing Method* which is advocated by De Nicola and Hennessy. Value passing in a communication is observed by a set of tests which approximate the value syntactically in order to treat infinite values. For this purpose, we propose an extended testing system to identify a process by the result of finite testing sequences. It is shown that passing infinite values is characterized by the system we propose in the sense of testing.

1 Introduction

Recently, communicating process has been studied as a model of computation with concurrency. A family of formal frameworks like CCS[10][11] CSP[8], ATP[5] etc. to express communicating process have been proposed. In these frameworks, concurrency is treated as nondeterminacy by interleaving and communication by handshake. The semantic treatments for these frameworks are algebraic. Thus, it is suitable for modeling concurrent programs with the modular basis.

From this point of view, the authors of [7] [10, (Chapter 9)][16] have proposed the formal descriptions of the concurrent behaviour of programs by translating program statements into expressions of communicating processes. By this translation, one can formally argue such properties of concurrent programs as the possibilities of deadlock or divergence.

In order to model programs in this manner, it is natural for the communicating process to have the value passing facility. But the semantic treatment of values in the systems mentioned above are implicit and lack of structures.

To treat a value as a formal object, we propose an extended framework for processes based on ATP[5]. In this abstract, we deal with infinite structure of values. The notion of infinite values is useful in specifications of values. It can be treated operationally by the lazy evaluation mechanism[4]. Our purpose here is to introduce a formal treatment of values (including infinite ones) into process calculi.

In order to treat infinite values, we extend the testing method[5][15] by introducing a new notion called "Value Precision" to approximate infinite values passed in communications. A communication passing an infinite value is observed by the most precise testing w.r.t. values.

We present an extended testing system, where a value is defined to be the normal form given by a TRS (term rewriting system) associated with the behaviour expression. This enables us to treat values as formal objects and to have an formal semantics for a value passing calculus like full CCS. Moreover, by interpreting a nonterminating rewriting as an infinite generation of data items, we can treat an infinite value like a stream.

A value expression is not evaluated at the moment of passing in a communication, but unfolded internally by the rewriting system if the evaluation is needed for conditional branches. But when an experimenter tests a communication, a value is externally observed with syntactic approximation by finite partial terms. And we show that this observation is enough to identify a process with *for all* testing sequences.

2 Value Expression and Rewriting System

2.1 Value Expression

A *value expression* is a term constructed from the signature Σ_v and variables. The set of value expressions with variables X is denoted by $T_{\Sigma_v}(X)$ and by T_{Σ_v} if $X = \emptyset$. We here assume that Σ_v is partitioned into a pair of sets: Σ_{vc} and Σ_{vf}. The former is the set of symbols called *constructors* which are intended not to be evaluated any more. And the latter is the set of symbols called *defined functions* which are intended to be evaluated.

We assume that Σ_{vc} has a special constant ω which informally means being "undefined" and is used for approximation. A value expression containing ω is called *partial*, while ω-free one is called *total*. We call a term constructed from Σ_{vc} and variables as a *constructor term*.

2.2 Rewriting System

Value expressions are evaluated by rewriting in a *value rewriting system*, VRS for short, which is a term rewriting system[2] satisfying several conditions. A VRS is a set of rewriting rules:

$$\{t_{l_1} \to t_{r_1}, \ldots, t_{l_n} \to t_{r_n}\}$$

where t_{l_i} and t_{r_i} is total for every i. We consider value expressions have the same meanings as values if they are rewritten to the same expression. For this purpose, a VRS satisfies the following four conditions:

(1) Constructor system w.r.t. Σ_{vc}:
The left hand side of each rewriting rule is in the form of $f(t_1, \ldots t_m)$ where $f \in \Sigma_{vf}$ and every t_i is a total constructor term.

(2) Left Linearity:
Any variable can occur in t_{l_i} at most once.

(3) Non-ambiguity:
For a pair of left hand sides of rules, t_{l_i} and t_{l_j}, a non-variable subterm of t_{l_i}, say t'_l, is not unifiable with t_{l_j}, except the case that $i = j$ and $t_{l_i} = t'_{l_i}$.

(4) Perfectness[14]
Any term in the form of $f(t_1, \ldots, t_m)$ is reducible, where $f \in \Sigma_{vf}$ and every t_i is a total constructor term with no variable.

A total value expression containing defined functions is reducible by a VRS. Note that a VRS has the confluent property, but does not have to be terminating.

2.3 Interpretation of Value Expression

In this section, we give an interpretation to ground value expressions, which may be infinite in its structure. First we define a partial order $\preceq$ over $T_{\Sigma_{vc}}$ to be the least relation satisfying:

(1) $\omega \preceq t$, $t \preceq t$ for any t.

(2) If $t_1 \preceq t_2$ and $t_2 \preceq t_3$, then $t_1 \preceq t_3$.

(3) If $t_i \preceq t'_i$ for $1 \leq i \leq n$, then $c(t_1, \ldots, t_m) \preceq c(t'_1, \ldots, t'_m)$ for $c \in \Sigma_{vc}$.

The ground constructor terms are ordered by $\preceq$ to form a po-set with the least element ω. This domain is uniquely extended to the algebraic-cpo $(T^{\infty}_{\Sigma_{vc}}, \preceq)$ which is isomorphic to $(\mathcal{I}(T_{\Sigma_{vc}}), \subseteq)$ where $\mathcal{I}(T_{\Sigma_{vc}}) = \{I(t)\ ;\ t \in T_{\Sigma_{vc}}\}$ and $I(t) = \{u\ ;\ u \preceq t\}$ [5].

Here, we give an interpretation to a value expression as the limit of the rewriting sequence from it in order to treat an infinite value. If the rewriting from t terminates, one can interpret a value expression t as the normal form. But if the rewriting does not terminate, we collect the intermediate results as *approximate normal forms*[13] and take their limit as its meaning. In our VRS, a subterm whose top symbol is a defined function has no information yet, since it must be rewritten afterwards. Thus, the information that a value expression t carries so far is the top part made up by constructor symbols. This is formally defined as $\omega(t)$:

$$\omega(t) = t[p \leftarrow \omega\ ;\ p \in min(occ_{\Sigma_{vf}}(t))]$$

where $occ_{\Sigma_{vf}}(t)$ is the set of occurrences of Σ_{vf} in t and $min(P)$ is the set of occurrences with no proper ancestors in P. $t[p \leftarrow u]$ stands for the term obtained by replacing the subterm at p with u. For more general TRS's, please refer [13].

For a VRS R, we write $\Rightarrow_R$ for the rewriting relation by the outer-most strategy[1].

[1]The essential factor for the rewriting strategy is fairness, since a defined function *must* be rewritten.

Next, we define an interpretation of a value expression by R. For a value expression t, the collection of the intermediate results is the set $\{\omega(t')\ ;\ t \Rightarrow_R^* t'\}$. t is interpreted as the *symbolic value* in R, denoted by $Val_R(t)$:

$$Val_R(t) = lub\{\omega(t')\ ;\ t \Rightarrow_R^* t'\}$$

For arbitrary t, $\{\omega(t')\ ;\ t \Rightarrow_R^* t'\}$ is directed and $\omega(t') \in T_{\Sigma_{vc}}$. These facts make $Val_R(t)$ well-defined. We take the *value* denoted by a value expression t as $Val_R(t)$ including the infinite case. If we restrict the domain of ω to ground value expressions, we can have the unique interpretation Val_R for $T^{\infty}_{\Sigma_{vc}}$ as a special case of [13]. We consider infinite elements in $T^{\infty}_{\Sigma_{vc}}$ as infinite values. Thus, an infinite value may have a finite representation to be passed in a communication.

Example

A VRS is given by:

$$R = \{nlist(x) \rightarrow cons(x, nlist(s(x)))\}$$

with $\Sigma_{vc} = \{0, s, cons\} \cup \{\omega\}$, then $Val_R(nlist(0))$ gives the interpretation:

$$cons(0, cons(s(0), cons(s^2(0), \cdots$$

which means the infinite list of natural numbers intuitively, since $Val_R(nlist(0)) =$

$$lub\{\omega, cons(0,\omega), cons(0, cons(s(0),\omega)), cons(0, cons(s(0), cons(ss(0),\omega)))\ldots\}$$

3 Behaviour Expression

3.1 Syntax

Given the set of channels C and the signature of value expressions $\Sigma_v (= \Sigma_{vc} + \Sigma_{vf})$, a behaviour expression E is defined in the BNF-like manner:

$$\texttt{NIL} \parallel c?x.E \parallel c!t.E \parallel E+E \parallel E|E \parallel [t == u].E$$

where $c \in C$, $t \in T_{\Sigma_v}(X)$, $var(t) \cap var(u) = \emptyset$, $u \in T_{\Sigma_{vc}}(X)$ and u is total.

$[t == u].E$ is the new syntactic element in our calculus which intuitively means the conditional branch: **if** $t = u$ **then** E. But it differs from the ordinary if–then branches in that when t matches u, the condition holds and the free variables in E are substituted by the matching since the free occurrences $var(u)$ in E are bound. $c?x$ and $c!t$ mean to *import* a value to x through the channel c and to *export* a value t through the channel c, respectively. $+$ and $|$ are *choice* and *composition*, respectively.

A variable occurrence in an expression may be *bound*. The bound variables are defined as follows:

- The occurrences of x in the form of $c?x.E$ are bound.
- The occurrences of $var(u)$ in the form of $[t == u].E$ are bound.

A variable not bound is *free*. A behaviour expression with no free variable occurrence is called a *process expression*. The set of all the process expressions is denoted by $\mathcal{P}$. Process expressions are closed under the derivation. This is easily shown by the induction of applications of the rules and axioms.

A *program* is a pair of a process expression P and a VRS R, denoted by (P, R).

3.2 Semantics by Derivation

A behaviour expression ready to communicate is related to the expression after the communication by the derivation labelled with an *event* e. An event is in the form of $< c?t >$, $< c!t >$, where c is a channel and t is a ground value expression. $< c?t >$ is an event importing a value expressed by t through a channel c, while $< c!t >$ is an event exporting a value expressed by t through a channel c.

The derivation semantics of a behaviour expression is defined by the following SOS formulas:

$$(\mathrm{I})\quad \frac{}{c?x.E \xrightarrow{< c?t >}_R E\{t/x\}} \qquad (\mathrm{O})\quad \frac{}{c!t.E \xrightarrow{< c!t >}_R E}$$

$$(\mathrm{S_{01}})\quad \frac{E_1 \longrightarrow_R E_1'}{E_1 + E_2 \longrightarrow_R E_1' + E_2} \qquad (\mathrm{S_{02}})\quad \frac{E_2 \longrightarrow_R E_2'}{E_1 + E_2 \longrightarrow_R E_1 + E_2'}$$

$$(\mathrm{S_1})\quad \frac{E_1 \xrightarrow{e}_R E_1'}{E_1 + E_2 \xrightarrow{e}_R E_1'} \qquad (\mathrm{S_2})\quad \frac{E_2 \xrightarrow{e}_R E_2'}{E_1 + E_2 \xrightarrow{e}_R E_2'}$$

$$(\mathrm{C_{01}})\quad \frac{E_1 \longrightarrow_R E_1'}{E_1|E_2 \longrightarrow_R E_1'|E_2} \qquad (\mathrm{C_{02}})\quad \frac{E_2 \longrightarrow_R E_2'}{E_1|E_2 \longrightarrow_R E_1|E_2'}$$

$$(\mathrm{C_1})\quad \frac{E_1 \xrightarrow{e}_R E_1'}{E_1|E_2 \xrightarrow{e}_R E_1'|E_2} \qquad (\mathrm{C_2})\quad \frac{E_2 \xrightarrow{e}_R E_2'}{E_1|E_2 \xrightarrow{e}_R E_1|E_2'}$$

$$(\mathrm{C_3})\quad \frac{E_1 \xrightarrow{< c?t >}_R E_1'\ ,\ E_2 \xrightarrow{< c!t >}_R E_2'}{E_1|E_2 \longrightarrow_R E_1'|E_2'}$$

$$(\mathrm{C_4})\quad \frac{E_1 \xrightarrow{< c!t >}_R E_1'\ ,\ E_2 \xrightarrow{< c?t >}_R E_2'}{E_1|E_2 \longrightarrow_R E_1'|E_2'}$$

$$(\mathrm{G_1})\quad \frac{t \Rightarrow_R t', t \neq u\theta \text{ for all } \theta}{[t == u].E \longrightarrow_R [t' == u].E} \qquad (\mathrm{G_2})\quad \frac{t = u\theta \text{ for some } \theta}{[t == u].E \longrightarrow_R E\theta}$$

$$(\mathrm{G_3})\quad \frac{t \not\Rightarrow_R t',\ t \neq u\theta \text{ for all } t' \text{ and } \theta,\ t \text{ is partial}}{[t == u].E \longrightarrow_R [t == u].E}$$

In the derivation, a value term is rewritten, internally on the left hand side of '=='. If the demanded information is obtained by the right hand side of '==', the computation gets through the guard. And the information obtained by pattern matching can be passed through the variables on the right hand side of '==' if exists. However, if pattern matching fails, the process either terminates or diverges. Here, a *partical* term, i.e. a term containing ω's, may make a process diverge if pattern matching is not possible, in the sense that a meaningless value computation is treated as a meaningless process. Thus, ω is also treated as meaningless in viewe of its behaviour.

We now introduce some notations for the following argument.

For process expressions P, Q and an event e, let $P \stackrel{e}{\Longrightarrow}_R Q$ if $P(\longrightarrow_R)^* P' \stackrel{e}{\longrightarrow}_R Q'(\longrightarrow_R)^* Q$ for some P' and Q'. And we write $S(P)$ for the set of channels in which events are ready to happen.

$$S(P) = \{c?; P \stackrel{<c?v>}{\Longrightarrow}_R P' \text{ for some } P'\} \cup \{c!; P \stackrel{<c!v>}{\Longrightarrow}_R P' \text{ for some } P'\}$$

4 Testing System

4.1 Testing Term

We give an experimenter in the testing method as a *testing terms*. A testing term T are defined in the BNF-like manner:

$$\texttt{SUCC} \parallel \texttt{FAIL} \parallel c?v.T \parallel c!v.T \parallel 1.T \parallel T + T$$

where v is a ground constructor term and 1 represents the autonomous state transition of the experimenter. SUCC and FAIL denote the successful and unsuccessful termination of testing respectively. The set of all testing terms is denoted by $\mathcal{T}$.

The testing sequences generated by a testing term is defined by the SOS formulas:

$$(\mathbf{I_T}) \quad \frac{}{c?t.T \xrightarrow{<c?t>}_T T} \qquad (\mathbf{O_T}) \quad \frac{}{c!t.T \xrightarrow{<c!t>}_T T}$$

$$(\mathbf{1_T}) \quad \frac{}{1.T \longrightarrow_T T}$$

$$(\mathbf{S_{T01}}) \quad \frac{T_1 \longrightarrow_T T_1'}{T_1 + T_2 \longrightarrow_T T_1'} \qquad (\mathbf{S_{T02}}) \quad \frac{T_2 \longrightarrow_T T_2'}{T_1 + T_2 \longrightarrow_T T_2'}$$

$$(\mathbf{S_{T1}}) \quad \frac{T_1 \xrightarrow{e}_T T_1'}{T_1 + T_2 \xrightarrow{e}_T T_1'} \qquad (\mathbf{S_{T2}}) \quad \frac{T_2 \xrightarrow{e}_T T_2'}{T_1 + T_2 \xrightarrow{e}_T T_2'}$$

A derivation from a testing term T means a query to a process being tested. Note that a testing term contains no reducible value expression, thus a query made by a testing term is finite. We intend to identify a process by finite testings.

4.2 Testing with Infinite Values

In this section, we present a testing system for testing with infinite values. It is determined by the limit of the finite collection to approximate values in testings. First, we introduce an order $\triangleleft$ over testing terms:

(i) $\mathtt{SUCC} \triangleleft \mathtt{SUCC}$, $\mathtt{FAIL} \triangleleft \mathtt{FAIL}$

(ii) $v_1 \preceq v_2$ and $T_1 \triangleleft T_2$ implies
$c?v_1.T_1 \triangleleft c?v_2.T_2$ and $c!v_1.T_1 \triangleleft c!v_2.T_2$

(iii) $T_1 \triangleleft T_2$ implies $\mathtt{1}.T_1 \triangleleft \mathtt{1}.T_2$

(iv) $T_1 \triangleleft T_1'$ and $T_2 \triangleleft T_2'$ implies $T_1 + T_2 \triangleleft T_1' + T_2'$

Intuitively, $T_1 \triangleleft T_2$ means that T_1 and T_2 have the same capabilities for communications, but T_1 has the less observation for values than that of T_2.

Based on $\triangleleft$, one can recursively construct a test with a value d including infinite one in the following manner:

(i) *Failure and Success*
Success and failure of a test are $\mathtt{SUCC}$ and $\mathtt{FAIL}$ respectively.

(ii) *Value Input*
For $d \in T^{\infty}_{\Sigma_{vc}}$, there always exists the set $D = \{v; v \in T_{\Sigma_{vc}}, v \preceq d\}$ such that *lub* $D = d$. The test which becomes T' by importing d at the channel c is constructed as the limit of the directed set:
$\{c?v.T' \ ; \ v \in T_{\Sigma_{vc}}, v \preceq d\}$.

(iii) *Value Output*
For $d \in T^{\infty}_{\Sigma_{vc}}$, there always exists the set $D = \{v; v \in T_{\Sigma_{vc}}, v \preceq d\}$ such that *lub* $D = d$. The test which becomes T' by exporting d at the channel c is constructed as the limit of the directed set:
$\{c!v.T' \ ; \ v \in T_{\Sigma_{vc}}, v \preceq d\}$.

The construction is characterized as follows. Let T_ω be the testing term obtained by substituting ω for all the value expressions in T. Since T_ω has the least value observations among those with the same communicating capabilities, the least equivalence $[T]$ including $\triangleleft$ for a testing term T will be:

$$[T] = \{U \ ; \ T_\omega \triangleleft U\}$$

For each T, $([T], \triangleleft)$ forms a po-set with the least element T_ω. As done for value expressions, one can construct the unique algebraic cpo $([T]^\infty, \triangleleft)$ isomorphic to $(\mathcal{I}([T]), \subseteq)$, where $\mathcal{I}([T]) = \{I(U) \ ; \ U \in [T]\}$ and $I(U) = \{U' \ ; \ U' \triangleleft U\}$. The set of testing terms obtained by completing $[T]$ for every $T \in \mathcal{T}$ is denoted by $\mathcal{T}^\infty$. We call $U \in \mathcal{T}^\infty$ *finite* when all the value expressions appearing in U are in $T_{\Sigma_{vc}}$.

The derivation relation $\xrightarrow{e}_{\mathcal{T}}$ is uniquely extended to $\xrightarrow{e'}_{\mathcal{T}^\infty}$ over $\mathcal{T}^\infty$, where an event is extended to have a value in $T^{\infty}_{\Sigma_{vc}}$:

(1) $T_1 \xrightarrow{<c?d>}_{T_\infty} T_2$ if for all $T_1' \lhd T_1$,
there exists T_2' such that $T_1' \xrightarrow{<c?v>}_{T} T_2'$ where $v \preceq d$ and $T_2' \lhd T_2$

(2) $T_1 \xrightarrow{<c!d>}_{T_\infty} T_2$ if for all $T_1' \lhd T_1$,
there exists T_2' such that $T_1' \xrightarrow{<c!v>}_{T} T_2'$ where $v \preceq d$ and $T_2' \lhd T_2$

(3) $T_1 \longrightarrow_{T_\infty} T_2$ if for all $T_1' \lhd T_1$,
there exists T_2' such that $T_1' \longrightarrow_{T} T_2'$ where $v \preceq d$ and $T_2' \lhd T_2$

This definition is well-defined since $\lhd$ is compatible with e w.r.t. $\preceq$ of events to each other.

[Lemma 1]

(i) Let $T \xrightarrow{<c?d>}_{T_\infty} T'$ and $T = \sqcup_\lhd D_1$ where $D_1 = \{U; U \lhd T,\ U \text{ is finite}\}$.
Then $T' = \sqcup_\lhd \{U'; U \xrightarrow{<c?v>}_{T} U', U \in D_1\}$ and
$d = lub\{v; U \xrightarrow{<c?v>}_{T} U', U \in D_1\}$

(ii) Let $T \xrightarrow{<c!d>}_{T_\infty} T'$ and $T = \sqcup_\lhd D_2$ where $D_2 = \{U; U \lhd T,\ U \text{ is finite}\}$.
Then $T' = \sqcup_\lhd \{U'; U \xrightarrow{<c!v>}_{T} U', U \in D_2\}$ and
$d = lub\{v; U \xrightarrow{<c!v>}_{T} U', U \in D_2\}$

(iii) Let $T \longrightarrow_{T_\infty} T'$ and $T = \sqcup_\lhd D_3$ where $D_3 = \{U; U \lhd T,\ U \text{ is finite}\}$.
Then $T' = \sqcup_\lhd \{U'; U \longrightarrow_{T} U', U \in D_3\}$

□

For example, the limit of the sequence by $\lhd$:

$$a!\omega.\texttt{SUCC} \ \lhd\ a!cons(0,\omega).\texttt{SUCC} \ \lhd\ a!cons(0, cons(s(0),\omega)).\texttt{SUCC} \ \ldots$$

is the test for $a!d.\texttt{SUCC}$ where $d = lub\{cons(0, cons(s(0), \cdots cons(s^n(0),\omega)))\}$. This test succeeds after exporting a list of natural numbers from the channel a and the test is successful if all the finite approximation is correct.

Next, we shall define a testing system $\mathcal{ES}_I$, called the *ideal testing system*, which tests a process with infinitely structured values.

[Definition 2] (Ideal Testing System)
An *ideal testing system* $\mathcal{ES}_I$ is a transition system, $(\mathcal{S}_I, \rightarrowtail_I)$. The set of states $\mathcal{S}_I$ is given by:

$$\mathcal{S}_I = \{(P||T)\ ;\ P \in \mathcal{P}, T \in \mathcal{T}^\infty\}$$

The transition relation $\rightarrowtail_I$ is defined by the following SOS-like formulas:

$$(\mathbf{N_I}) \quad \frac{E \ \longrightarrow_R\ E'}{(E||T) \rightarrowtail_I (E'||T)},\ T \neq \texttt{SUCC}, T \neq \texttt{FAIL}$$

$$(\mathbf{1_I}) \quad \frac{T \ \longrightarrow_{T_\infty}\ T'}{(E||T) \rightarrowtail_I (E||T')}$$

$$(\mathbf{I_I}) \quad \frac{E \xrightarrow{<c?t>}_R E' \;,\; T \xrightarrow{<c?d>}_{T_\infty} T'}{(E\|T) \rightarrowtail_I (E'\|T')}, \quad Val_R(t) = d$$

$$(\mathbf{O_{I_1}}) \quad \frac{E \xrightarrow{<c!t>}_R E' \;,\; T \xrightarrow{<c!d>}_{T_\infty} T'}{(E\|T) \rightarrowtail_I (E'\|T')}, \quad d = Val_R(t)$$

$$(\mathbf{O_{I_2}}) \quad \frac{E \xrightarrow{<c!t>}_R E' \;,\; T \xrightarrow{<c!d>}_{T_\infty} T'}{(E\|T) \rightarrowtail_I (E'\|\mathbf{FAIL})}, \quad d \not\leq Val_R(t)$$

□

$\mathcal{ES}_I$ models the behaviours of a process w.r.t. values, including infinite cases. Note that (O_{I_2}) makes a process stop when the communication link matches but the value does not match. This shows that in our model a value is evaluated not before the communication, but after the communication.

The computation sequence is also defined similar to that of $\mathcal{ES}$ as follows:

An *ideal value testing* for P by T is a sequence of states $s_0, s_1, s_2, \ldots$ begining with the $(P\|T)$: where $s_0 = (P\|T)$ and

a. $s_i \rightarrowtail_I s_{i+1}$ if s_i is not the final state of the sequence

b. there is no derivative of s_i by $\rightarrowtail_I$, otherwise

We shall write $Comp_R^I(P,T)$ for the set of ideal value testings for P by T.

The result of an ideal value testing s, written as $Res_I(s)$, is defined as follows:

$Res_I(s) = \top$ if s ends in the state of $(P'\|\mathrm{SUCC})$

$Res_I(s) = \bot$ otherwise

where $\top$ and $\bot$ denote success and failure of the value testing s respectively.

We write $Res_I(P,R,T)$ for $\cup\{Res_I(s) \; ; \; s \in Comp_R^I(P,T)\}$. $Res_I(P,R,T)$ is one of $\{\top\}, \{\bot\}, \{\top,\bot\}$.

4.3 Testing System with Finite Approximation

Although the behaviours of processes capable of passing infinite values are defined by the ideal testing system, the system is just a semantical model since one cannot check a test with infinite values. Thus, we now present a extended testing system with the approximation mechanism. Now, we introduce the mechanisms for approximation called *value precision* and *testing precision.*

[Definition 3] (Value precision and Testing precision)
A *value precision* at channel c is a triple denoted as either $< c?, t \frown v >$ or $< c!, t \frown v >$, where t is a ground value expression and v is a ground constructor

term. A finite sequence of value precisions is called a *testing precision*. We write $\mathcal{Q}$ for the set of all testing precisions. □

Next, we define the testing system.

[Definition 4] (Testing System)
A *testing system* $\mathcal{ES}$ is a transition system, $(\mathcal{S}, \rightarrowtail)$. The set of states $\mathcal{S}$ is given by:

$$\mathcal{S} = \{(P\|T, Q)\ ;\ P \in \mathcal{P}, T \in \mathcal{T}, Q \in \mathcal{Q}\}$$

The transition relation $\rightarrowtail$ is defined by the following SOS-like formulas:

$$\textbf{(N)}\quad \frac{E \longrightarrow_R E'}{(E\|T, Q) \rightarrowtail (E'\|T, Q)},\ T \neq \texttt{SUCC}, T \neq \texttt{FAIL}$$

$$\textbf{(1)}\quad \frac{T \longrightarrow_T T'}{(E\|T, Q) \rightarrowtail (E\|T', Q)}$$

$$\textbf{(I)}\quad \frac{E \xrightarrow{<c?t>}_R E'\ ,\ T \xrightarrow{<c?v>}_T T'}{(E\|T, Q) \rightarrowtail (E'\|T', Q\cdot < c?, t \frown v >)},\ v \preceq Val_R(t)$$

$$\textbf{(O)}\quad \frac{E \xrightarrow{<c!t>}_R E'\ ,\ T \xrightarrow{<c!v>}_T T'}{(E\|T, Q) \rightarrowtail (E'\|T', Q\cdot < c!, t \frown v >)}$$

□

One step of a testing is successful if a communication is possible at the named channel. The value precision to be associated with the communication is either $< c?, t \frown v >$ or $< c!, t \frown v >$. The testing system only tests the communication capability at channels. The correctness of the testing is checked after testing communications. This point is discussed at the end of the abstract.

The difference between $\mathcal{ES}$ and $\mathcal{ES}_I$ is as follows:

- $\mathcal{ES}$ can test only by finite testing terms w.r.t. values. While $\mathcal{ES}_I$ can test with infinite testing terms.
- In $\mathcal{ES}$, a process being tested can take more information if it is compatible with the testing term. While $\mathcal{ES}_I$ have to take the exact information offered by the testing term.

A testing sequence, called a *value testing* is also defined for $\mathcal{ES}$ as follows:

[Definition 5] (Value Testing)
A *value testing* for P by T is a sequence of states, $s_0, s_1, s_2 \ldots$, beginning with the state $s_0 = (P\|T, \epsilon)$ as follows:

a. $s_i \rightarrowtail s_{i+1}$ if s_i is not the final state of the sequence.

b. Otherwise, there is no derivative by $\rightarrowtail$.

The set of value testings for P by T is written by $Comp_R(P, T)$. □

4.4 Result of Value Testings

We shall define which state is successful in value testings in order to distinguish processes. The result of a value testing should be either success, $\top$, or failure, $\bot$. But in our framework, values in testings may be partial i.e. not fully specified for the purpose of approximation in values. Thus, the results obtained by a value testing are relative to the values used in the testing. We observe some extra information to indicate to what extent the result is informative w.r.t. values.

First, we define a predicate $correct_R$ for a value precision w.r.t. a VRS R in the following manner:

$$correct_R(p) = \begin{cases} true & \text{if } \exists t' : t \Rightarrow_R^* t', v \preceq t' \\ false & \text{otherwise} \end{cases}$$

where p is either $< c?, t \frown v >$ or $< c!, t \frown v >$.

If $correct_R(p)$ is true, a value t offered by a process being tested can be *observed* by v. In other words, t is more informative than required by v at the communication channel c.

The following property is obvious from the definition.

[Proposition 6] Let p either $< c?, t \frown v >$ or $< c!, t \frown v >$.
$correct_R(p)$ is true if and only if $v \preceq Val_R(t)$. □

$correct_R(c?, t \frown v)$ or $correct_R(c!, t \frown v)$ is true means that the value t is *correctly observed* by v at the communication channel.

$correct_R$ is extended to a testing precision Q as follows:

$$correct_R(\epsilon) = true$$

$$correct_R(Qp) = \begin{cases} correct_R(Q) & \text{if } correct_R(p) \\ false & \text{otherwise} \end{cases}$$

A testing precision is correct when each communication is correctly observed.

For a value testing s beginning with $(P\|T, \epsilon)$, the result of s written by $Res(s)$ is defined as follows:

- **Case 1:** s is finite

  Let the last state of s be $(P'\|T', Q)$.

$$Res(s) = \begin{cases} (\top, Q', S(P')) & \text{if } T' = \texttt{SUCC} \text{ and } correct_R(Q') \\ (\bot, Q', S(P')) & \text{otherwise} \end{cases}$$

- **Case 2:** s is infinite

  There must be a state that $(P'\|T', Q')$ such that T' doesn't change any more in the sequence, because a testing term has only a finite structure. In this case,

$$Res(s) = (\bot, Q', \Delta)$$

  where Δ represents that the test is incomplete.

We write $Res(P, R, T)$ for $\{Res(s) \ ; \ s \in Comp_R(P, T)\}$.

Example

Let $P_1 = a?x.b!x.\mathtt{NIL}$, $P_2 = a?x.[x == y].b!y.\mathtt{NIL}$,
$T_1 = a?0.b!\omega.\mathtt{SUCC}$ and $T_2 = a?\omega.b!0.\mathtt{SUCC}$
where $R = \emptyset$ with $\Sigma_{vc} = \{0, \omega\} andSigma_{vf} = \emptyset$. Then:

(1) $Comp(P_1, R, T_1) = \{(P_1 || T_1, \epsilon)$
$\rightarrowtail (b!0.\mathtt{NIL} || b!\omega.\mathtt{SUCC}, < a?, 0 \frown 0 >)$
$\rightarrowtail (\mathtt{NIL} || \mathtt{SUCC}, < a?, 0 \frown 0 >< b!, 0 \frown \omega >)\}$

$Res(P_1, R, T_1) = \{(\top, < a?, 0 \frown 0 >< b!, 0 \frown \omega >, \emptyset)\}$

(2) $Comp(P_1, R, T_2) = \{(P_1 || T_2, \epsilon)$
$\rightarrowtail (b!0.\mathtt{NIL} || b!0.\mathtt{SUCC}, < a?, 0 \frown \omega >)$
$\rightarrowtail (\mathtt{NIL} || \mathtt{SUCC}, < a?, 0 \frown \omega >< b!, 0 \frown 0 >)$,
$(P_1 || T_2, \epsilon)$
$\rightarrowtail (b!\omega.\mathtt{NIL} | b!0.\mathtt{SUCC}, < a?, \omega \frown \omega >)$
$\rightarrowtail (\mathtt{NIL} || \mathtt{SUCC}, < a?, \omega \frown \omega >< b!, \omega \frown 0 >)\}$

$$\begin{aligned} Res(P_1, R, T_2) &= \{(\top, < a?, 0 \frown \omega >< b!, 0 \frown 0 >, \emptyset), \\ &\quad (\bot, < a?, \omega \frown \omega >< b!, \omega \frown 0 >, \emptyset)\} \end{aligned}$$

(3) $Comp(P_2, R, T_1) = \{(P_2 || T_1, \epsilon)$
$\rightarrowtail ([0 == y].b!y.\mathtt{NIL} || b!0.\mathtt{SUCC}, < a?, 0 \frown \omega >)$
$\rightarrowtail (b!0.\mathtt{NIL} || b!0.\mathtt{SUCC}, < a?, 0 \frown \omega >)$
$\rightarrowtail (\mathtt{NIL} || \mathtt{SUCC}, < a?, 0 \frown \omega >< b!, 0 \frown 0 >)\}$

$$\begin{aligned} Res(P_2, R, T_1) &= \{(\top, < a?, 0 \frown 0 >< b!, 0 \frown \omega >, \emptyset), \\ &\quad (\bot, < a?, \omega \frown \omega >< b!, \omega \frown 0 >, \emptyset)\} \end{aligned}$$

(4) $Comp(P_2, R, T_2) = \{(P_2 || T_2, \epsilon)$
$\rightarrowtail (P_2' || T_2', < a?, \omega \frown \omega >)$
$\rightarrowtail (P_2' || T_2', < a?, \omega \frown \omega >) \rightarrowtail \ldots$,
$(P_2 || T_2, \epsilon)$
$\rightarrowtail ([0 == y].b!y.\mathtt{NIL} || b!0.\mathtt{SUCC}, < a?, 0 \frown \omega >)$
$\rightarrowtail (b!0.\mathtt{NIL} || b!0.\mathtt{SUCC}, < a?, 0 \frown \omega >)$
$\rightarrowtail (\mathtt{NIL} || \mathtt{SUCC}, < a?, 0 \frown \omega >< b!, 0 \frown 0 >)\}$

$$\begin{aligned} Res(P_2, R, T_2) &= \{(\bot, < a?, \omega \frown \omega >, \Delta), \\ &\quad (\top, < a?, 0 \frown \omega >< b!, 0 \frown 0 >, \emptyset)\} \end{aligned}$$

where $P_2' = [\omega == y].b!y.\mathtt{NIL}$ and $T_2' = b!0.\mathtt{SUCC}$.

4.5 Observation for Value Testing

We are going to identify processes in the finite observations by the result precise enough to determine w.r.t. values whether the value testing is successful or not.

To measure the precision of a value testing, a value precision is preordered by $\sqsubseteq_p$ according to the value from testing terms:

When $v_1 \preceq v_2$,

$$(c?, t_1 \frown v_1) \sqsubseteq_p (c?, t_2 \frown v_2) \text{ and } \qquad (c!, t_1 \frown v_1) \sqsubseteq_p (c!, t_2 \frown v_2)$$

The preorder $\sqsubseteq_q$ between testing precisions is given based on $\sqsubseteq_p$:

(i) $\epsilon \sqsubseteq_q \epsilon$

(ii) $pQ \sqsubseteq_q p'Q'$ if $p \sqsubseteq_p p'$ and $Q \sqsubseteq_q Q'$

Based on this preorder, we classify the results of value testings by $\sqsubseteq_r$ as follows:

$(r_1, Q_1, A_1) \sqsubseteq_r (r_2, Q_2, A_2)$ if $Q_1 \sqsubseteq_q Q_2$ and $A_1 = A_2$
where $|Q|$ means the length of a testing precision.

Intuitively, $Res(s_1) \sqsubseteq_r Res(s_2)$ indicates that s_2 gives a more precise result than that of s_1 w.r.t. values.

For a set of results S, we define the *observation* $Obs(S) \subseteq \{\bot, \top\}$.

[Definition 7] (Observation)
For a set $S \subseteq \mathcal{Q}$, $Obs(S)$ is the least set satisfying the following two conditions:

(i) $\top \in Obs(S)$ if there exists $(\top, Q, A) \in S$ such that for all $(\bot, Q', A') \in S$, $(\bot, Q', A') \sqsubseteq_r (\top, Q, A)$ or $(\top, Q, A) \not\sqsubseteq_r (\bot, Q', A')$

(ii) $\bot \in Obs(S)$ if there exists $(\bot, Q, A) \in S$ such that for all $(\top, Q', A') \in S$, $(\top, Q', A') \sqsubseteq_r (\bot, Q, A)$ or $(\bot, Q, A) \not\sqsubseteq_r (\top, Q', A')$

□

Intuitively, Obs gets a *maximally precise* result. We shall write $Obs(P, R, T)$ for $Obs(Res(P, R, T))$. An observation will be stabilized eventually and thus there is the observation for a process.

[Lemma 8] There exists $(r, Q, A) \in Comp(P, R, T)$ such that for all $(r', Q', A') \in Comp(P, R, T)$, if $(r, Q, A) \sqsubseteq_r (r', Q', A')$ and $(r', Q', A') \not\sqsubseteq_r (r, Q, A)$, then $r = r'$ □

Proof: Let $S = Comp(P, R, T)$. Suppose that for all $(r, Q, A) \in S$ there exists $(r', Q', A') \in S$ such that $(r, Q, A) \sqsubseteq_r (r', Q', A')$, $(r', Q', A') \not\sqsubseteq_r (r, Q, A)$ and $r \neq r'$. Then, for all $(\top, Q, A) \in S$ there exists $(\bot, Q', A') \in S$ such that $(\top, Q, A) \sqsubseteq_r (\bot, Q', A')$ and $(\bot, Q', A') \not\sqsubseteq_r (\top, Q, A)$. While, for $(\bot, Q', A') \in S$ there exists $(\top, Q'', A'') \in S$ such that $(\bot, Q', A') \sqsubseteq_r (\bot, Q'', A'')$, $(\bot, Q'', A'') \not\sqsubseteq_r (\top, Q', A')$ and $A = A' = A''$.

Thus, there is a sequence: $(\top, Q, A) \sqsubseteq_r (\bot, Q', A) \sqsubseteq_r (\top, Q'', A)$ such that $Q \sqsubseteq_q Q' \sqsubseteq_q Q''$, $Q' \not\sqsubseteq_q Q$,$Q'' \not\sqsubseteq_q Q'$ and $Q \not\sqsubseteq_q Q''$. This also holds for $(\bot, Q, A) \in S$.

Since S is not empty, by repeating the above argument there must be an infinite sequence which gets more precise strictly:

$$(r, Q_0, A_0) \sqsubseteq_r (r', Q_1, A_1) \sqsubseteq_r (r, Q_2, A_2) \sqsubseteq_r (r', Q_3, A_3) \ldots$$

where $r \neq r'$ and r is either $\perp$ or $\top$. But this is impossible because a testing term T is in $\mathcal{T}$. ■

[Proposition 9] $Obs(P, R, T)$ is not empty. □

Proof: Suppose $Obs(P, R, T)$ is empty. Then, neither of the conditions for including $\perp$ and $\top$ holds. But, this contradicts lemma8. ■

Thus, $Obs(P, R, T)$ is one of $\{\top\}$, $\{\perp\}$ and $\{\top, \perp\}$.

The results for P by $T \in \mathcal{T}^\infty$ are fully determined by Obs and the set of finite approximating testing terms.

[Theorem 10] For $T \in \mathcal{T}^\infty$,

(i) $\top \in Res_I(P, R, T)$ if and only if
for all $T' \lhd T$ where $T' \in \mathcal{T}$, there exists $T_0 \in \mathcal{T}$ such that
$T' \lhd T_0$ and $\top \in Obs(P, R, T_0)$

(ii) $\perp \in Res_I(P, R, T)$ if and only if
for all $T' \lhd T$ where $T' \in \mathcal{T}$, there exists $T_0 \in \mathcal{T}$ such that
$T' \lhd T_0$ and $\perp \in Obs(P, R, T_0)$

□

The proof is shown in the appendix.

Finally, the property of "must-succeed" is characterized as follows:

[Corollary 11] $Res_I(P, R, T) = \{\top\}$ if and only if
there exists $T_0 \lhd T$ such that $T_0 \lhd T'$ implies $Res_I(P, R, T') = \{\top\}$ □

Proof: $Res_I(P, R, T) = \{\top\}$ iff $\perp \notin Res(P, R, T)$ ■

Example

Consider a program (P, R) where

$$P = \quad a?x.[nlist(x) = cons(y, z)].b!y.c!z.\mathtt{NIL}$$
$$R = \quad \{nlist(x) \rightarrow cons(x, nlist(s(x)))\}$$

(P, R) is successfully tested by the test determined by the set $\mathbf{T}$:

$$\mathbf{T} = \{a?0.b!0.c!cons(s(0) \cdots cons(s^n(0), \omega) \cdots).\mathtt{SUCC} \;;\; n \geq 1\}$$

$\sqcup\mathbf{T}$ means that the test terminates successfully if the output from c which is equal to the cdr part of $nlist(0)$. We have $Res_I(P, R, \sqcup\mathbf{T}) = \{\top\}$. For all $U \lhd \sqcup\mathbf{T}$, $\top \in Obs(P, R, U)$. Some $Obs(P, R, U)$ contains $\perp$, say $U = a?\omega.b!0.c!\omega.\mathtt{SUCC}$, but the more precise tests will override $\perp$. For example, let P_1 and P_2 be:

Example

P_1: $a?x.(b!f(x).\mathtt{NIL} + b!s(f(x)).\mathtt{NIL})$

P_2: $a?x.b!f(x).\mathtt{NIL}$

and a VRS $R = \{f(x) \to s(f(x))\}$ with $\Sigma_{vc} = \{s, 0\}$. Then, one cannot distinguish P_1 from P_2 by any value testing, since $Val_R(f(c)) = Val_R(s(f(c)))$ for all $c \in T_{\Sigma_{vc}}$,

5 Concluding Remarks

We have presented an extension of the testing method for processes with value passing, where a value is allowed to be infinitely structured. The infinite structure is denoted by a VRS which rewrites a value expression to a constructor term. By this, we intend to treat a value as a syntactic object, i.e. a constructor term. If an infinite value is represented by a value expression, it can be passed in a communication and evaluated inside a process.

Our extension for the value precision preordering $\sqsubseteq_r$ is essentially needed when value with infinite structures, namely the terms which may be unwound by VRS, are present, If values are finite, we can have the most precise result directly without approximation. But even if values are finite, our treatment of values passed in communications is slightly different from that of indexed channels by values as the following example shows:

P_1 : $c?x.(d!x.\mathtt{NIL} + d!not(x).\mathtt{NIL})$

P_2 : $c?x.d!x.\mathtt{NIL} + c?x.d!not(x).\mathtt{NIL}$

Let the associated VRS be:

$$R = \{not(true) \to false, not(false) \to true\} \text{ with } \Sigma_{vc} = \{true, false\}$$

One cannot distinguish (P_1, R) and (P_2, R) by any value testing we have presented here, but these can be distinguished if c and d are indexed by $true$ and $false$. Namely, by our testing method, a testing term can choose only a channel externally, but cannot observe the value to be passed before the communication link is established. This assumption is justified by the fact that the value offered by a process cannot be fully evaluated prior to the communication in general. Thus, the power of our observation is less than that of the indexed communications by finite values.

The model of value passing in communication process has been studied recently in various calculi as in [6][11][12]. The syntactic counterpart of our treatment of values is the language such as LOTOS[3], where algebraic specifications described by equations are combined with behaviour expressions of CCS. When implementing a LOTOS interpreter like [9], the system must have some proof procedure for equations which is behaviourally reflected in conditional branches. We have here treated the procedure as rewriting by a restricted form of TRS and included the lazy evaluation mechanism.

We could have introduced the infinite structure of values by the recursive process structure. For example, the process $rec\ b(x).c!x.b(s(x))$ produces an infinite stream of natural numbers from channel c by initializing x to be 0. But in this approach there is no syntactic counterpart for infinite values. And one has to restrict the channel c to ensure passing all the information that the process exports, since evaluation of the value is made externally. In our framework, value expressions are assured to be rewritten in a deterministic manner by the VRS. Some equivalence may be introduced between these approaches, but this is a topic of the future research.

In this abstract, we have merely shown the *operational* extension of the testing semantics for communicating process. To establish the denotational model and the proof system is under research. To investigate the relation between the recursive structure of processes and that of values is a very interesting topic although we have not treated it here. And much work is left to be done.

Acknowledgement

The authors wish to thank Professor Tohru Naoi, Mr. Ken Mano, Mr. Fabio Casablanca and Professor Tomio Hirata for their helpful discussions and comments. They also thank Professor Michio Ohyamaguti and Dr.Yoshito Toyama for their helpful suggestions.

References

[1] S. Abramsky. Observation Equivalence as a Testing Equivalence. *Theoretical Computer Science*, 53:225–241, 1987.

[2] G. Boudul. Computational semantics of term rewritng systems. In M. Nivat and J.C.Reynolds, editors, *Algebraic Methods in Semantics*, pages 169–236. Cambridge University Press, 1985.

[3] E. Brinksma. A Tutorial on LOTOS. *Protocol Specification, testing, and Verification, V*, pages 171–194, 1986.

[4] P. Henderson. *Functional Progamming (Application and Implementation)*. Prentice-Hall, 1980.

[5] M. Hennessy. *Algebraic Theory of Processes.* MIT Press, 1988.

[6] M. Hennessy and A. Ingólfsdóttir. A Theory of Communicating Processes With Value-passing. *Lecture Notes in Computer Science*, 443:209–219, 1990.

[7] M.C.B. Hennessy and W. Li. Translating A Subset of ADA into CCS. In D.Bjørner, editor, *Formal Description of Programming Concepts–II*, pages 227–249. North-Holland, 1982.

[8] C.A.R. Hoare. *Communicating Sequential Processes.* Prentice-Hall, 1985.

[9] L. Logrippo, A. Obaid, J.P. Briand, and M.C. Fehri. An Interpreter for LOTOS, A Specification Language for Distributed Systems. *Software–Practice and Experience*, 18(4):365–385, 1988.

[10] R. Milner. A Calculus of Communicating Systems. *Lecture Note in Computer Science*, 92, 1980.

[11] R. Milner. *Communication and Concurrency*. Prentice-Hall, 1989.

[12] R. Milner, J. Parrow, and D. Walker. A Calculus of Mobile Processes. Technical Report ECS-LFCS-89-85, LFCS Department of Computer Science, Edinburgh University, 1989.

[13] T. Naoi and Y. Inagaki. Algebraic Semantics and Complexity of Term Rewritng Systems. *Lecture Notes in Computer Science*, 355:311–325, 1988.

[14] T. Naoi and Y. Inagaki. Recursive Functions on a Completed Set of Natural Numbers. *Technical Report of IEICE*, COMP90(1), 1990.

[15] R. De Nicola and M. Hennessy. Testing Equivalences For Processes. *Theoretical Computer Science*, 34:83–133, 1984.

[16] S. Smolka and R. Strom. A CCS Semantics for NIL. *Formal Description of Programming Concepts - III*, pages 347–373, 1987.

Appendix: The Proof of Theorem 10

[Theorem 10] For $T \in \mathcal{T}^{\infty}$,

(i) $\top \in Res_I(P, R, T)$ if and only if
for all $T' \lhd T$ where $T' \in \mathcal{T}$, there exists $T_0 \in \mathcal{T}$ such that
$T' \lhd T_0$ and $\top \in Obs(P, R, T_0)$

(ii) $\bot \in Res_I(P, R, T)$ if and only if
for all $T' \lhd T$ where $T' \in \mathcal{T}$, there exists $T_0 \in \mathcal{T}$ such that
$T' \lhd T_0$ and $\bot \in Obs(P, R, T_0)$

□

Proof: By the induction on the forms of T. For (i):

- The base case is when T is either SUCC or FAIL, where the theorem(i) obviously holds.
- $T \equiv c?d.U$

($\Rightarrow$) Since $\top \in Res_I(P, R, T)$, there exists P_1, P_2 and P_3 such that

$$(P||T) \rightarrowtail_I{}^* (P_1||T) \rightarrowtail_I (P_2||U) \rightarrowtail_I{}^* (P_3||\mathtt{SUCC})$$

where $P \longrightarrow_R^* P_1$ and $P_1 \xrightarrow{< c?t >}_R P_2$, $Val_R(t) = d$.
This implies $\top \in Res_I(P_2, R, U)$. Then by the induction hypothesis, for all $U' \lhd U$ there exists U_0 such that $U' \lhd U_0$ and $\top \in Obs(P_2, R, U_0)$. (*)

$T' \lhd T$ follows $T' \equiv c?v'.U'$ where $v' \preceq d$ and $U' \lhd U$. Then $(P||c?v'.U', \epsilon) \rightarrowtail^* (P_1||c?v'.U', \epsilon) \rightarrowtail (P_2||U', < c?, t \frown v' >)$. From the fact (*), $(P_2||U_0, \epsilon) \rightarrowtail^* (P_4||\mathtt{SUCC}, Q)$ for some Q. Thus as T_0, $c?v'.U_0$ satisfies the conditions.

($\Leftarrow$) Let $T' \equiv c?v'.U'$ where $v' \preceq d$ and $U' \lhd U$. Then $T_0 \equiv c?v_0.U_0$ and for $U' \lhd U_0$, $\top \in Obs(P, R, T_0)$. And for some P_1,P_2:

$$(P||T_0, \epsilon) \rightarrowtail^* (P_1||T_0, \epsilon) \rightarrowtail (P_2||U_0, < c?, t \frown v_0 >)$$

where $v_0 \preceq Val_R(t)$. Since $c? \in S(P)$, $\top \in Obs(P, R, U_0)$. Then by the induction hypothesis, $\top \in Res_I(P, R, U)$. From the definition of $\rightarrowtail_I$, $T \in Res_I(P, R, T)$ follows.

- $T \equiv c!d.U$
Similar to the case of $c?d.U$.
- $T \equiv 1.U$
U_0 suffices to be T_0.
- $T \equiv T_1 + T_2$

($\Rightarrow$) Since $Res_I(P, R, T) = Res_I(P, R, T_1) \cup Res_I(P, R, T_2)$, by the induction hypothesis:
For all $T_1' \lhd T_1$ there exists T_1'' such that
$T_1' \lhd T_1''$ and $\top \in Obs(P, R, T_1'')$ or
For all $T_2' \lhd T_2$ there exists T_2'' such that
$T_2' \lhd T_2''$ and $\top \in Obs(P, R, T_2'')$
Suppose the former case without loss of generality. T_0 should have the form $T_1'' + U$ where $U \lhd T_2$.

Assume $\top \notin Obs(P, R, T_0)$. Since $Obs(P, R, T_0) = Obs(Res(P, R, T_1'') \cup Res(P, R, U))$, there must exist $(\bot, Q', A') \in Res(P, R, U)$ such that $(\top, Q, A) \sqsubseteq_r (\bot, Q', A')$ and $(\bot, Q', A') \not\sqsubseteq_r (\top, Q, A)$ for all $(\top, Q, A) \in Res(P, R, T_1'')$. Thus, $A = A'$ and $Q \sqsubseteq_q Q'$ but $Q \neq Q'$. There are two possibilities to enter this state. First, deadlock at the last communication: in this case, the test goes on until deadlock, thus Q must be equal to Q' due to (I_I), (O_{I_1}) and (O_{I_2}). Second, value mismatch (O_{I_3}) at the end: But in this case, $Q' \sqsubseteq_q Q$. Then $\top \in Obs(P, R, T_0)$.

($\Leftarrow$) For $T' = T_1' + T_2'$, let $T_0 = T_{01} + T_{02}$. Since $\top \in Obs(P, R, T_0)$, $\top \in Obs(P, R, T_{01})$ or $\top \in Obs(P, R, T_{02})$. By the induction hypothesis, $\top \in Res_I(P, R, T_1)$ or $\top \in Res_I(P, R, T_2)$.

Now, we prove (ii).

- The base case is same as (i).
- $T \equiv c?d.U$

($\Rightarrow$) There are three cases to consider.
(1) If $c? \in S(P)$, then there is a sequence for some P_1 and P_2:
$(P||T) \rightarrowtail_I^* (P_1||T) \rightarrowtail_I (P_2||U)$ where $\bot \in Res_I(P_2, R, U)$.
(2) If $c? \notin S(P)$ and P is convergent, then P gets in deadlock after some internal actions. (3) P is divergent

Case(1) is proved by using the induction hypothesis as in (i). In Case(2) and (3), T' is suffice to be T_0.

($\Leftarrow$) Let $T' \equiv c?v'.U'$ where $v' \preceq d$ and $U' \lhd U$. Then $T_0 \equiv c?v_0.U_0$ and for $U' \lhd U_0$, $\bot \in Obs(P, R, T_0)$. There are also three cases. If $c? \in S(P)$ then there is a sequence for some P_1 and P_2:

$(P||T_0,\epsilon) \rightarrowtail^* (P_1||T_0,\epsilon) \rightarrowtail (P_2||U_0, < c?, t \frown v_0 >)$. By using the induction hypothesis, $\perp \in Res_I(P_2, R, U)$. Then from the definition of $\rightarrowtail_I$, $\perp \in Res_I(P, R, T)$. The case that $c? \notin S(P)$ and P is convergent and that P is divergent, any testing term beginning with $c?$ fails. Then $\perp \in Res_I(P, R, T)$

- $T \equiv c!d.U$

($\Rightarrow$) Similar to the case of $c?d.U$. But consider the case $d \preceq Val_R(t)$ and $d \neq Val_R(t)$. In this case, a value output test suspends. In this case, d must be in $T_{\Sigma_{vc}}$. Thus, T_0 exists.

($\Leftarrow$) This case is also similar to $c?d.U$ considering thee output suspension occurs only finite tests.

- $T \equiv 1.U$
  U_0 suffices to be T_0.
- $T \equiv T_1 + T_2$

($\Rightarrow$) As in (i), suppose for all $T_1' \lhd T_1$ there exists T_1'' such that $T_1' \lhd T_1''$ and $\perp \in Obs(P, R, T_1'')$. Let T_0 be $T_1'' + U$ where $U \lhd T_2$.

Assume $\perp \notin Obs(P, R, T_0)$. Since $Obs(P, R, T_0) = Obs(Res(P, R, T_1'') \cup Res(P, R, U))$, there must exist $(\top, Q', A') \in Res(P, R, U)$ such that $(\perp, Q, A) \sqsubseteq_r (\top, Q', A')$ and $(\top, Q', A') \not\sqsubseteq_r (\perp, Q, A)$ for all $(\perp, Q, A) \in Res(P, R, T_1'')$. Thus, $A = A'$ and $Q \sqsubseteq_q Q'$ but $Q \neq Q'$. The argument similar to that of (i) is applied.

($\Leftarrow$) Similar to that of (i).

Session 4

Constructive Semantics

Paul C. Brown
GE Corporate Research and Development
Schenectady, New York USA

Constructive Semantics is an approach to programming language semantics that treats a program as a constructive specification for an abstract state machine. This abstract machine is composed of a set of smaller "well-behaved" machines operating concurrently. The exact combination of machines is determined by the program, with each programming language construct appearing in the program defining a portion of the composition. The programming language itself specifies a number of primitive machines that form the basic building blocks of programs. These machines represent the basic operations and data types of the language. The resulting semantics is relatively easy to understand, and its relationship to the original program is clear.

Constructive semantics treats many higher level programming language abstractions also as specifications of state machines, where these machines serve as prototypes for entire sets of machines. For example, a basic data type in a programming language is modeled as a state machine, and each variable of the type is modeled as a copy of this machine. Behavioral equivalence of machines provides a basis for modeling abstract data types, in which behaviorally equivalent machines belong to the same abstract data type. Behavioral equivalence also provides a basis for modeling type hierarchies such as those found in object-oriented languages with multiple inheritance.

The formalism underlying constructive semantics is a process algebra known as the Hybrid Calculus of Communicating Systems (HCCS), since it contains elements of both Milner's Calculus of Communicating Systems (CCS) and Hoare's Communicating Sequential Processes (CSP). Behavioral equivalence in HCCS is based upon Hoare's failures equivalence.

Constructive semantics provides straightforward semantic models for other important aspects of programming languages, including concurrency (Ada tasks), elaboration and visibility computations.

1.0 Introduction

The purpose of this paper is to show how process algebras can be used to give a complete and understandable semantics for a programming language, including such higher level language constructs as data types, program blocks, subprograms, and Ada generics. The resulting semantics is complete in the sense that *all* of the features of the programming language can be modeled in a straightforward manner[1]. The simple relationship between the programming language constructs and the semantic model elements allows an understanding of the semantics to directly aid in understanding the programming language being modeled.

We shall not attempt within the scope of this paper to give a complete model for even a small programming language. Instead, we will lay the formal groundwork for the model, and give examples of simplified semantics for several programming language constructs to indicate the manner in which a complete semantics can be given. A more complete treatment of constructive semantics can be found in [5].

The following section provides an intuitive introduction to constructive semantics by giving a simple program fragment and a sketch of its corresponding semantic model. The following two sections then describe the formal underpinnings of constructive semantics, first by defining the process algebra HCCS (section 3), and then by describing the classes of machines that we will use directly in our semantics and the machine algebra that we will use for describing their composition to form larger machines (section 4). Finally, we give formal expressions for simplified semantics of some common programming language constructs.

2.0 An Overview of Constructive Semantics

The fundamental concept behind constructive semantics is that a program is simply the specification of an abstract state machine, and *all* constructs of the programming language in which the program is written relate, directly or indirectly, to the specification of this machine or its component machines. The specification for the entire program is, in turn, given as a composition of smaller state machine specifications. Each of these machines may, in itself, be a composition of still smaller machines. Since the programming language constructs themselves define the composition of machines from smaller machines, the structure of the semantic model very closely parallels the structure of the program.

Figure 1 shows some of the possible relationships between machines, using the notation of Rumbaugh et. al. [12]. Each machine is either a primitive machine or a composite machine. Composite machines are comprised of one or more machines (of either class), and, conversely, each machine is optionally a part of a composite machine. Our program itself is just a composite machine.

It is permissible to have many copies of a program executing simultaneously. Each of these copies is a state machine, separate and distinct from all of the others, yet sharing the same specification. When we ask a machine to perform an action, we must be specific as to which machine we are making the request of. To reflect this in our model, we give each action of a machine a "subscript" that uniquely identifies the machine that the action is associated with. The program is thus the specification of a *family* of *structurally isomorphic* machines (machines that are identical under a mapping replacing the action subscripts with subscripts indicating the other machine). We will call a family of structurally isomorphic machines a ***state type*** (for reasons that will become apparent later).

1. The only exception that we are aware of is that the concept of a time *interval*, such as the interval implicit in the Ada `delay` statement, is not directly definable, but must be defined with reference to some abstract clock, each of whose "ticks" is implies the passage of a certain time interval. This is a consequence of the underlying process algebra (as are all process algebras that we are aware of) being based upon a *point* ontology of time rather than an *interval* ontology. [13] contains an extensive discussion and comparison of these ontological structures.

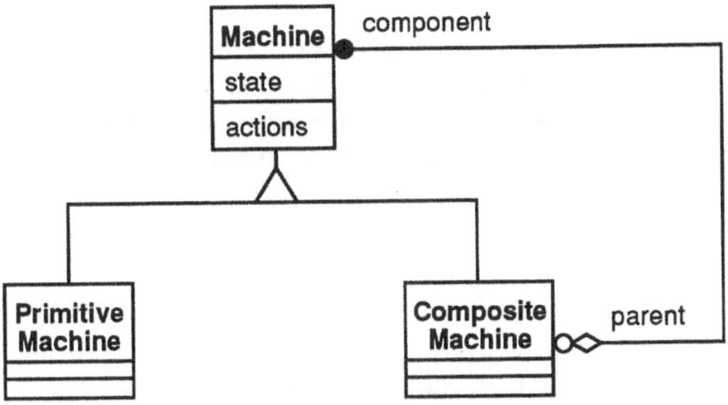

Figure 1 Composition of Machines

This concept of families of machines characterizes other programming language constructs as well. Subprograms and operators can be interpreted as families of state machines: each call of the subprogram or operation indicates an interaction with a different machine from the family. Data types are also families of machines, each of which duplicates the data storage and behavior associated with the type. Each variable belonging to a data type is associated with one member of the family of machines defined by the data type.

2.1 Behavioral Equivalence

As might be surmised, for any given behavior there are many models (machines or combinations of machines) that will exhibit that behavior. In constructive semantics, we use Hoare's failures equivalence [8][2] to determine when two machines are equivalent. Under failures equivalence, machines are considered equivalent if the sequences of actions that they will perform are the same and the actions that they can possibly refuse after a sequence of actions are also the same.

Since *implementations* of programs (as opposed to their abstract semantics) can also be modeled with process algebras, behavioral equivalence provides us with a formal means of comparing actual implementations of programs with the abstract semantics. Furthermore, since this behavioral equivalence is a congruence relation in the model, equivalence of two machines can be established by establishing that the component pieces of the two machines are pair-wise equivalent.

2.2 Types

There are at least three concepts frequently associated with the term type in programming languages: a set of values[2], a behavioral (interface) specification, or an implementation specification. In constructive semantics, these three concepts are not inconsistent, and are in fact closely related to one another. To see this, we must first realize that in a state machine model, there is no way to store a value directly. Instead, the value must be encoded as the state of some state machine, with each state of this

2. Or a representative of the set, as is the case in domain semantics

machine representing a different value. Reading and writing values then correspond to actions of the state machine in which the state of the machine is altered (writing) or its present state is revealed without altering the state (reading). Thus we see that values, per se, do not exist in our model. We must instead talk in terms of a machine capable of encoding these values. In particular, in order to model the storage or passing of values, we must explicitly model the machine that is used to hold the values. This explicit handling of value storage in the semantics makes clear the distinctions between the different parameter passing semantics used in subprograms, and provides a basis upon which to decide whether they are equivalent for a particular subprogram.

When a machine is interacting with a machine encoding values (we will call these machines ***value machines***), we are not *directly* observing the state of the machine (i.e. the encoding of the value being represented): we are only able to *infer* the intended value from the actions to which the value machine will respond. We conclude that the only characteristic we *can* observe about a machine is its *behavior*: its interactions with other machines. Returning to our discussion concerning a type as a set of values, it is the behavior of the machine that we use to store that values rather than the values themselves that we find in constructive semantics: values have thus been reduced to a special case of a type as a behavioral specification. We define an ***abstract type*** to be a set of actions and a family of machines that all exhibit the same behavior with respect to the actions belonging to the abstract type. Clearly more than one state type may be included in a given abstract type, and is relatively straightforward to show that every state type also defines an abstract type. Similarly, we define a ***state type*** to be a set of machines and an isomorphism between the actions of the machines. Thus all of the machines in a state type are identical up to the relabeling operation.

In constructive semantics, a data type declaration in the language is usually taken to be both the definition of an abstract type (a behavioral specification) and a state type belonging to that abstract type (an implementation) capable of encoding the values of the type. For built-in data types in the language, the state type and abstract type are taken to be part of the language specification itself. For user-defined types, the formal semantics of the language defines how these types are defined in terms of previously defined abstract and state types.

Abstract types provide a formal basis for defining type hierarchies based upon behavioral equivalence, in which the descendent types are required to exhibit the full abstract behavior of the parent[3]. This allows the implementation of the individual types to be different, and allows their functionality to be extended (new actions added) beyond that of the parent as long as the behavior with respect to the parent's actions is preserved. Abstract types even provide a model for multiple inheritance. In Figure 2 child type **C** performs all "`a`" actions and "`b`" actions as well as its own "`c`" actions. If the "`a`" actions and "`b`" actions are disjoint (a condition that we require of all unrelated types for both technical and philosophical reasons), the child type **C** will be of abstract type **A** (behaviorally equivalent to **A** when interaction is restricted to

3. These are frequently referred to as "is-a" type hierarchies in object oriented languages.

"a" actions) and will also be of abstract type *B* (behaviorally equivalent to *B* when interaction is restricted to "b" actions).

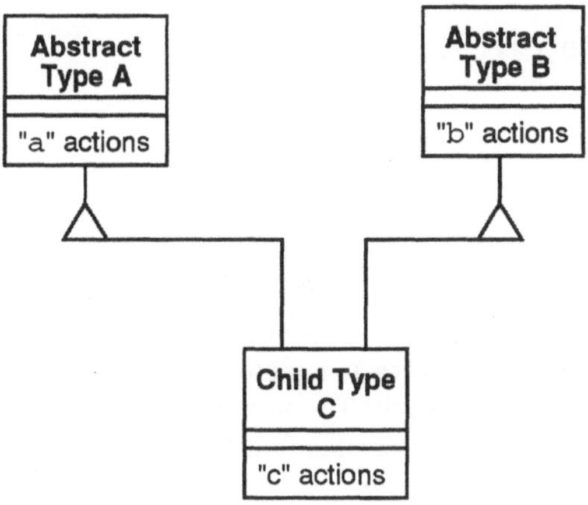

Figure 2 Multiple Inheritance

In constructive semantics, we will define families of machines by specifying a prototype machine for the family. The other machines in the family can then be generated by changing the subscripts of the actions as appropriate. This holds for abstract types as well as state types.

2.3 Classes of Machines

There are three classes of primitive machines that occur in constructive semantics. The first of these is the class of ***value machines***, which we have already encountered. Value machines are passive (by construction): they never ask other machines to perform actions. Thus value machines cannot interact.
Since value machines never request actions of other machines, and therefore cannot interact, we must have at least one other class of machines that can request actions of other machines. In fact, we have two classes of these machines. The first we call ***interaction machines***. Generally speaking, interaction machines do not retain encoded values in their internal state: instead, they act as intermediaries between other machines that can encode values. These are the machines that define the basic operations and relations: assignment, equality, arithmetic operations, etc.

The third class of machine correlates the activation and idling (starting and stopping) of other machines (we require that each machine has a unique activate and idle action). This class of machines we will call ***activation machines***.

2.4 An Example

Let us consider the following fragment as if it were a complete program, and examine its possible composition from smaller machines. The data type `integer` corresponds to a state type of value machines, and the machines representing the variables `a`, `b` and `c` are members of this state type. We will use V_a, V_b, and V_c to represent these three machines. In addition, we will use a fourth temporary variable $\mathbf{v}_t$ to hold the result of the addition prior to its assignment to the variable c.

```
declare
    a : integer := 2;
    b : integer := 3;
    c : integer;
begin
    c := a + b;
end;
```

Example 1 Simple Program Fragment

Constants are also represented by value machines (variants of the integer value machine), and we will use $\mathbf{c}_2$ and $\mathbf{c}_3$ to represent the machines corresponding to the integer constants 2 and 3 respectively. The operations of integer assignment and addition correspond to families of interaction machines. We will designate individual interaction machines with a subscript, and use a superscript notation to informally indicate the role that the machine plays in the program. For example, the first assignment machine (the one that assigns the constant 2 to a) would be designated by $=_1^{2\to a}$, and similarly the other assignment machines are designated by $=_2^{3\to b}$ and $=_3^{t\to c}$. The addition machine is designated by $+^{a,b\to t}$ (we omit the subscript here since there is only one addition machine involved).

Figure 3 shows the relationship between these machines using a reduced Petri net notation. The tokens in this Petri net are the individual machines that we have been discussing. Arcs originating at the heavy black bars indicate the activation of machines (the introduction of that machine or token into the network), and the labels indicate which machine is being activated. Arcs terminating at the heavy black bars indicate the idling of machines (the removal of machines or tokens from the network). The heavy black bars are, themselves, graphic representations of activation machines. In all of the other transitions, the tokens retain their individual identities as they pass through the transition.[4]

In this diagram, we see that the three value machines $\mathbf{v}_a$, $\mathbf{v}_b$, and $\mathbf{v}_c$, the two constant machines $\mathbf{c}_2$ and $\mathbf{c}_3$, and the first assignment machine $=_1^{2\to a}$ are all activated in parallel. When the first assignment machine is idled, the second assignment machine $=_2^{3\to b}$ is activated. When the second assignment machine is idled, both the temporary value machine $\mathbf{v}_t$ and the addition machine $+^{a,b\to t}$ are activated. The idling of the addition machine activates the final assignment machine $=_3^{t\to c}$, and finally the four value machines, two constant machines and the final assignment machine become idle simultaneously.

While this graphical notation is quite descriptive of the relationships between the machines, it is somewhat cumbersome. Consequently, we will use an algebraic notation that also describes the machine relationships. The expression for the above example is:

4. This extension to Petri nets in which token identity is preserved through the transition is due to Eaker [6][7]

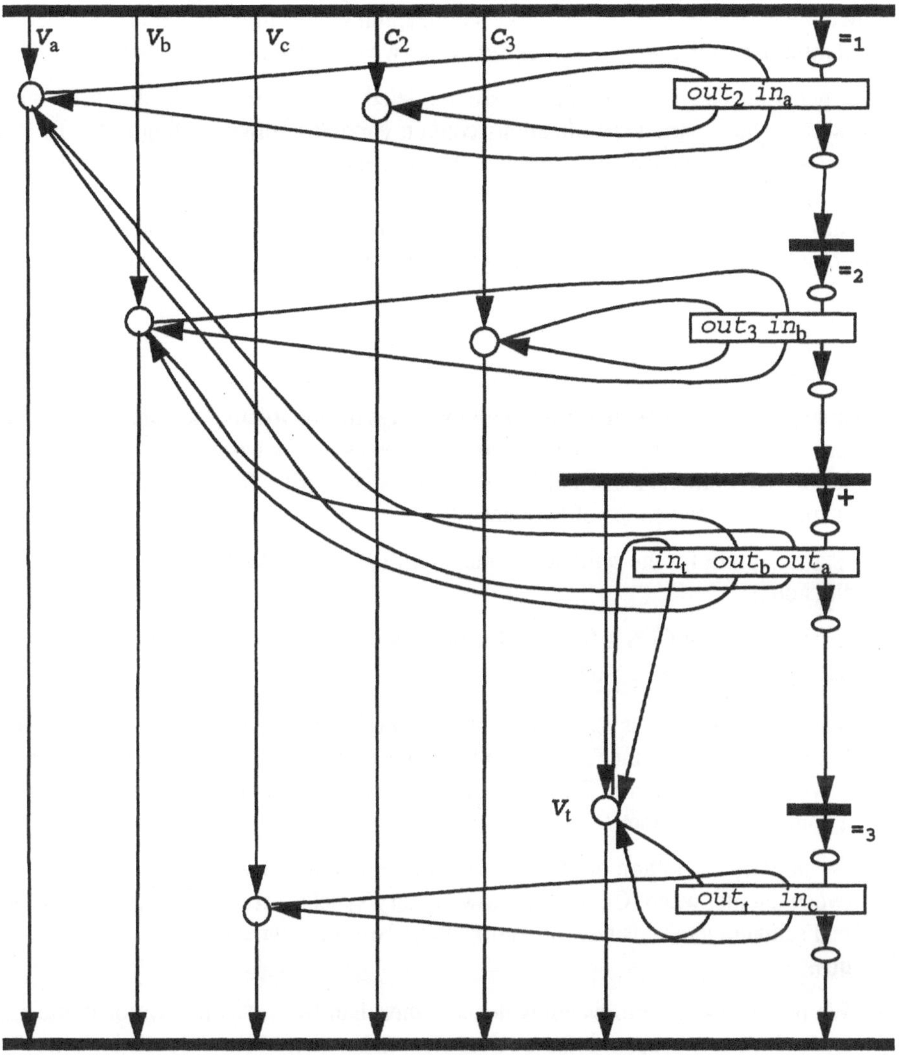

Figure 3 Reduced Petri Net of Example 1

$$\langle V_a \mid V_b \mid V_c \mid C_2 \mid C_3 \mid =_1^{2\to a} ;=_2^{3\to b} ;\langle V_t \mid +^{a,b\to t};=_3^{t\to c} \rangle\rangle \tag{1}$$

Here the vertical bars, angle brackets and semicolons represent specific kinds of activation machines. The vertical bars | indicates parallel composition, with the angle brackets ⟨•••⟩ indicate that the parallel machines within the brackets are both activated and idled simultaneously. The semicolon indicates that the idling of the machine before the semicolon is associated with the activation of the machine after the semicolon (we adopt the convention that the semicolon binds more tightly than the parallel composition bar, and thus avoid the need for parentheses to explicitly group terms). Note, however, that in this notation there is no explicit indication of which machines interact with which other machines. This information is contained in the definitions of the machines themselves, as we shall see in later sections.

Ada generics and C++ templates are modeled as partial specifications of state machines. These partial specifications contain variables corresponding to the formal parameters of the generic or template. The expected values for these variables are state machines. An instantiation of the generic or template is modeled as the state machine defined by replacing each variable with the state machine corresponding to the actual parameter (usually the prototype machine associated with a type or subprogram).

3.0 HCCS

We now proceed to give the formal underpinnings to constructive semantics. HCCS is the process algebra upon which constructive semantics is based, and is a hybrid of work done by Milner, Hoare and Brookes. The formal basis for this model is the synchronization tree semantics of Brookes [2].

The starting point for the semantic model is Milner's CCS [9][10], with the following modifications:

1 Instead of using CCS's flat set of actions, we use the abelian group of actions as used in Milner's SCCS and ASCCS.

2 Milner's + operator is replaced by the similar □ operation used by Hoare and defined in terms of Milner's synchronization trees by Brookes [2]. We will use ⊕ to represent this operation. The use of the Hoare operator makes failures equivalence a congruence relation.

3 We alter the definition of Milner's | operation to allow n-way synchronization between agents, where CCS only allows binary synchronization. We provide a formal semantics for this new operation using Brookes' synchronization tree semantics.

4 We use Brookes' failures equivalence rather than Milner's observational equivalence. Milner's observational equivalence makes non-observable distinctions between machines, while Brookes' failures equivalence only distinguishes between machines if the machines differ in observable behavior. The use of failures equivalence allows the proof of our parallelization theorem, which is not true using observational equivalence.

3.1 Actions

We first define `Act` the set of ***actions***. Actions can be thought of as the labels on the arcs of state machines. Actions are part of an abelian group ($Act, \times, 1, \overline{\ }$) freely generated by a set of *positive primitive actions* = {a,b,c,...}, a unique identity action 1, a binary operator × and a unary inverse operator $\overline{\ }$. The inverse of a positive primitive action a is the *negative primitive action* $\overline{a}$ (and vice-versa). We shall refer to the union of the positive and negative primitive actions and the identity action as the set of *primitives* = {...,$\overline{c}$,$\overline{b}$,$\overline{a}$,1,a,b,c,...}.

In our subsequent semantics, we shall see that the operator × is used as a parallel composition operator: when we write a×b, we mean that the actions a and b occur simultaneously. In writing actions, we shall frequently omit the parallel composition operator ×, writing st to represent s × t.

To facilitate the comparison of actions, we define *Primitives*(a) to be the set of primitive actions that an action a is comprised of. For example, if a and b are primitive actions, then *Primitives*(ab) = {a, b}. We note that this is a definition based on the *syntactic structure* of the action as a minimum length string of primitive actions. In particular, we do *not* want to infer that $\{a, \bar{a}\} \subseteq$ *Primitives*(x) just because $a\bar{a}=1$ and $1x = x$ for all actions $x \in Act$. However, we will always include the element 1 in the set of actions generated by *Primitives*(a). By extension, if *A* is a *set* of actions, we define *Primitives*(*A*) $= \cup_{a \in A}$ *Primitives*(a).

From time to time we will wish to define the set of actions a^+ that can be constructed from the primitive actions *Primitives*(a) of an action a. Formally, we define a^+ to be the submonoid of Act freely generated by *Primitives*(a) for any $a \in Act$.[5] As before, we extend this to sets of actions, defining A^+ to be the submonoid of Act freely generated by *Primitives*(*A*) for any $A \subseteq Act$. In a similar manner, we define the subgroup of actions a^* that can be constructed from the primitive actions *Primitives*(a) of an action a. Formally, we define a^* to be the subgroup of Act freely generated by *Primitives*(a) for any $a \in Act$.We extend this to sets of actions, defining A^* to be the subgroup of Act freely generated by *Primitives*(*A*) for any $A \subseteq Act$.

3.2 Agents and Agent Expressions: Syntactic State Machine Specifications

We now proceed to give a syntax for the specification of a state machine. We begin by defining ***E***, the set of ***agent expressions***. Agent expressions are syntactic representations of the states of state machines. Agent expressions may contain variables (whose values will be other agent expressions), in which case the agent expression is a *template* for a state that will be fully specified when the values of the variables appearing in the agent expression are given. An ***agent*** is an agent expression that contains no variables. An agent can be viewed as a fully specified state of a state machine. As we shall see, an agent, in conjunction with a collection of constant definitions and the HCCS inference rules relating one agent expression (state) to another, provides a complete specification of a state machine.

Returning to the syntactic specification of agent expressions, let ***K*** = {**0**,**1**,A,B,C,...} be a set of ***agent constants***, and ***X*** = {X,Y,Z,...} be a set of ***agent variables***. Then we define ***E***, the set of ***agent expressions***, to be the set generated by the following rules:

- $\boldsymbol{K} \cup \boldsymbol{X} \subset \boldsymbol{E}$

$\forall$ E,F $\in$ ***E***, $a \in Act$, $A \subseteq Act$:

Action:

5. Since a monoid does not include the inverse operator, the only primitives in A^+ are those that explicitly appear in *Primitives*(*A*).

$a.E \in \boldsymbol{E}$

where "." is a binary operator of type $Act \times \boldsymbol{E} \rightarrow \boldsymbol{E}$. Informally, this operator is used to provide the semantics of sequence: the agent a.E (under appropriate circumstances) performs the action a and then behaves like E.

Product:

$E \mid F \in \boldsymbol{E}$

where "|" is a binary operator of type $\boldsymbol{E} \times \boldsymbol{E} \rightarrow \boldsymbol{E}$. Informally, this operator is used to provide the semantics of parallel composition: E | F means that both E and F are operating in parallel.

Summation:

$\sum_{i \in I} E_i \in \boldsymbol{E}$

where " $\sum_{i \in I}$ " is an n-ary operator of type $\boldsymbol{E} \times \boldsymbol{E} \times ... \times \boldsymbol{E} \rightarrow \boldsymbol{E}$. Informally, this operator is used to provide the semantics of deterministic choice between the behaviors of the various E_i in the term. In the special case of a summation involving two machines, we shall write the summation as E⊕F.

It is important to note that the operation being defined here is *not* Milner's summation operator, but a generalization to an arbitrary number of terms of Hoare's ▯ operator. The behavior of the Milner and Hoare operators is the same for observable actions, but different for the non-observable action 1.

Restriction:

$E \restriction A \in \boldsymbol{E}$

where "$\restriction$" is a binary operator of type $\boldsymbol{E} \times A \rightarrow \boldsymbol{E}$, where A is a subset of Act. Informally, this is a restriction operator, with the semantics that $E \restriction A$ restricts the visible (available) actions of E to those present in A^* (i.e. hides all actions of E that are not in A^*).

From this operator we define a derived binary operator $\backslash : \boldsymbol{E} \times A \rightarrow \boldsymbol{E}$ as follows:

$$E \backslash A \equiv E \restriction \{a \in Act \mid \mathit{Primitive}(a) \cap \mathit{Primitive}(A^*) = 1\}$$

This operator hides any actions any of whose primitive actions are also primitive actions of A^*.

Morphism:

$E[\phi] \in \boldsymbol{E}$

where [] is a binary operator of type $\boldsymbol{E} \times (Act \rightarrow Act) \rightarrow \boldsymbol{E}$, and $\phi : Act \rightarrow Act$ is any mapping from Act to Act such that $\phi(\overline{a}) = \overline{\phi(a)}$ and $\phi(1) = 1$. Informally, [] is a relabeling operation such that E[ϕ] is the agent expression that results from replacing each action of E with the result of applying the mapping ϕ to that action.

Constant Definitions:

To allow recursive definitions, we use systems of equations involving constants, where a constant is simply a variable whose value has been fixed to be a particular agent expression. Recursive definitions can then be achieved by the appropriate use of constants on the right hand side. Each constant is defined to be an agent expression E:

$$A \equiv E$$

3.3 Semantics: Labeled Transition Systems

We now proceed to give semantics to agent expressions by defining *labeled transition systems*. Informally, a labeled transition system is a state machine in which the agent expressions are the "states" of the machines, and the transitions between agent expressions are labeled with actions, hence the term labeled transition system. Formally, a ***labeled transition system*** is a triple:

$$(\boldsymbol{E},\ Act,\ \{\xrightarrow{a}\ : a \in Act\})$$

where $\boldsymbol{E}$ is a set of agent expressions, Act is a set of actions, and each $\xrightarrow{a}$ is a relation between agent expressions.

In this system, we have the following inference rules[6]

Action:

$$\frac{}{a.E \xrightarrow{a} E} \qquad (2)$$

This rule states that the agent expression a.E becomes the agent E after it has performed the action a.

Summation:

$$\frac{E_j \xrightarrow{a} E'_j}{\sum_{i \in I} E_i \xrightarrow{a} E'_j} \quad (j \in I, a \neq 1) \qquad (3)$$

This rule states that if any element E_j of the summation can make a transition to E'_j by performing the observable action a, then the entire summation can also perform the a action and then behave like E'_j. In other words, by performing this observable action, all of the other alternative choices have been discarded.

$$\frac{E_j \xrightarrow{1} E'_j}{\sum_{i \in I} E_i \xrightarrow{1} \sum_{i \in I} E_i\,[E'_j/E_j]} \quad (j \in I) \qquad (4)$$

This second rule deals with the case in which the action is the non-observable action 1. In this case, the individual component of the summation may make this non-observable transition, *but the other choices in the summation are not discarded.* The notation $[E'_i/E_i]$ simply indicates that E_i has been replaced by E'_i.

6. While they are given here as axioms, these axioms are in fact derivable from synchronization tree semantics [5]

This pair of rules has the effect of preventing a non-observable transition of one element of a summation from eliminating the other possibilities in the summation. This is distinctly different from Milner's "+" semantics[7], which uses the first rule only and places no restriction on the observability of the action. Using Milner's semantics would prevent failures equivalence from being a congruence relation, which is a requirement of our model[8].

Product:

$$\frac{E \xrightarrow{a} E'}{E \mid F \xrightarrow{a} E' \mid F} \qquad (5)$$

$$\frac{F \xrightarrow{b} F'}{E \mid F \xrightarrow{b} E \mid F'} \qquad (6)$$

These two rules state that if one member of a parallel composition can make a transition on the action a, then that member can also make the transition as part of a parallel composition if the parallel composition performs the action a.

$$\frac{E \xrightarrow{a} E' \qquad F \xrightarrow{b} F'}{E \mid F \xrightarrow{ab} E' \mid F'} \qquad (7)$$

This rule says that both members of a parallel composition may make transitions simultaneously, provided that the appropriate actions occur simultaneously.

Restriction:

$$\frac{E \xrightarrow{a} E'}{E \upharpoonright S \xrightarrow{a} E' \upharpoonright S} \; (a \in S) \qquad (8)$$

Here we have the semantics of restricting the actions that a machine may perform to those present in a particular set S. If E makes a transition to E′ with the action a, and a is a member of S, then $E \upharpoonright S$ may also make a transition to $E' \upharpoonright S$ with the action a. Note that the absence of any other rule regarding restriction means that if a is not in S, no transition is possible. We also note that the non-observable action 1 is always a member of S.

Morphism:

7. [9] p. 271
8. Milner's observational equivalence is also not a congruence relation in CCS. It is conjectured that the use of Hoare's □ operator (which we are using here as our + operator) in lieu of Milner's + would make observational equivalence a congruence relation in the modified CCS system.

$$\frac{E \xrightarrow{a} E'}{E[\phi] \xrightarrow{f(a)} E'[\phi]} \tag{9}$$

This rule gives us the semantics of relabeling machines. Recall that ϕ is a homomorphism from actions to actions. The semantics are that if E can make a transition to E′ with the action a, then the relabeled version of E, E[ϕ], can make a transition to the relabeled version of E′, E[ϕ]′, with the mapped action ϕ(a).

Constants:

$$\frac{E \xrightarrow{a} E'}{C \xrightarrow{a} E'} \quad (C \equiv E) \tag{10}$$

The semantics of constant declarations is that if the constant C is defined to be the expression E, then any transition that E can make is also a transition of C.

4.0 Machines

In constructive semantics, a machine is a labeled transition system and an initial state of that system. The well-behaved machines used in constructive semantics are a restricted subset of the machines definable in HCCS. We now proceed to define the restrictions that we will place upon the machines, and then define the machine algebra in which machines may be composed in various serial/parallel combinations while preserving these well-behaved properties. Several theorems in machine algebra describe orthogonality (independence) conditions under which the serial/parallel relationships between machines in a composition may be altered, yielding behaviorally equivalent compositions.

4.1 Positive Primitive Actions Appear on Exactly One Machine

We now proceed to give an interpretation of actions. Positive actions represent actions that a machine can take. Since we wish such actions to be unique to a machine, we shall require that ***a positive primitive action may appear as a transition label on exactly one machine*** (although it may label any number of arcs on that machine). In contrast, the corresponding negative primitive action (a request made by another machine for this action) may appear on any number of machines. We will refer to the one machine on which a positive action appears as the ***defining machine*** of that action.

4.2 Well-Behaved Machines

We define a ***well-behaved machine*** to be a machine that has exactly one observable activate action and whose initial action is always that unique activate action, and has exactly one observable idle action and if this idle action occurs, the only observable action that may follow is the activate action[9].

9. Note that this does not imply that the machine must respond to the idle action in every state. This is similar to Milner's well-terminating property [10] p. 173

We note that ";" and "⟨⟩" both preserve the well-behaved property:

Lemma 1

Let $\mathcal{M}$ be the set of well-behaved machines. Then:

$$\mathbf{M}_1, \mathbf{M}_2 \in \mathcal{M} \Rightarrow \mathbf{M}_1 ; \mathbf{M}_2 \in \mathcal{M}$$

$$\mathbf{M}_1, \mathbf{M}_2, \ldots, \mathbf{M}_n \in \mathcal{M} \Rightarrow \langle \mathbf{M}_1 \mid \mathbf{M}_2 \mid \ldots \mid \mathbf{M}_n \rangle \in \mathcal{M}$$

We adopt the convention that the ";" operator binds more tightly than (takes precedence over) the | operator, and that ⟨⟩ binds more tightly than either.

4.3 Machine Algebra

Our ***machine algebra*** is then the system ($\mathcal{M}$, =, ;, |, ⟨⟩), where $\mathcal{M}$ is the set of well-behaved machines, = is behavioral equivalence, and ;, |, and ⟨⟩ are as defined above.

4.3.1 Parallelization Theorem

We wish to consider the circumstances under which the sequencing of machines in a composition may be altered. Consider the following abstracted model of a program or subprogram:

$$\langle \mathbf{V}_a \mid \mathbf{V}_b \mid \mathbf{M}_p ; \mathbf{M}_q ; \mathbf{M}_r ; \mathbf{M}_s \rangle \backslash S$$

where $S = Sort^{10}(\mathbf{M}_a) \cup \overline{Sort}(\mathbf{M}_a) \cup Sort(\mathbf{M}_b) \cup \overline{Sort}(\mathbf{M}_b) \cup Sort(\mathbf{M}_q) \cup \overline{Sort}(\mathbf{M}_q) \cup Sort(\mathbf{M}_r) \cup \overline{Sort}(\mathbf{M}_r)$, $\mathbf{V}_a$ and $\mathbf{V}_b$ are local variables of the program, $\mathbf{M}_p$ is a machine that copies actual parameter values into the local variables, $\mathbf{M}_q$ and $\mathbf{M}_r$ are machines that do the actual work of the subprogram by modifying the local variables, and $\mathbf{M}_s$ is a machine that copies the final values out into their target destinations. The restriction $\backslash S$ simply says that the internal workings of the program are not visible from outside the program (the variables are hidden and the machines that do the work cannot communicate with any machines outside of the program). The sequencing $\mathbf{M}_p ; \mathbf{M}_q ; \mathbf{M}_r ; \mathbf{M}_s$ says that the actual parameter values are copied in, then $\mathbf{M}_q$ does its work, then $\mathbf{M}_r$ does its work, and finally $\mathbf{M}_s$ copies the results back out of the program.

Now it seems intuitive that if 1) the variables cannot interact with each other, and 2) the machines that do the work ($\mathbf{M}_q$ and $\mathbf{M}_r$) cannot interact with each other, and 3) each working machine can only interact with *one* of the variables, then it should not matter which order $\mathbf{M}_q$ and $\mathbf{M}_r$ do their work. In fact, they could even operate in parallel! This is exactly what the parallelization theorem establishes:

Theorem 2 Parallelization Theorem

$$\mathbf{M}_a \perp \mathbf{M}_b, \mathbf{M}_a \perp \mathbf{M}_r, \mathbf{M}_b \perp \mathbf{M}_q, \mathbf{M}_q \perp \mathbf{M}_r \Rightarrow$$
$$\langle \mathbf{M}_a \mid \mathbf{M}_b \mid \mathbf{M}_p ; \mathbf{M}_q ; \mathbf{M}_r ; \mathbf{M}_s \rangle \backslash S =$$
$$\langle \mathbf{M}_a \mid \mathbf{M}_b \mid \mathbf{M}_p ; \langle \mathbf{M}_q \mid \mathbf{M}_r \rangle ; \mathbf{M}_s \rangle \backslash S =$$
$$\langle \mathbf{M}_a \mid \mathbf{M}_b \mid \mathbf{M}_p ; \mathbf{M}_r ; \mathbf{M}_q ; \mathbf{M}_s \rangle \backslash S$$

---

10. The sort of a machine is the set of primitive actions that appear on the machine. The inverse of the sort is the set of inverses of the actions appearing in the sort.

This is a valuable result, since it shows how to take a serial program and convert it into an equivalent parallel program with no analysis beyond simply determining the orthogonality (independence) of the component parts of the program. The proof of this theorem is in [5].

4.3.2 Change of Scope Theorem

Somewhat similar to the parallelization problem is the change of scope theorem. Consider the following configuration of machines:

$$\langle \mathbf{M}_a ;\langle \mathbf{M}_b \mid \mathbf{V}_x\rangle; \mathbf{M}_c\rangle\backslash S$$

where $S = Sort(\mathbf{V}_x)$. The situation we are modeling here is one in which $\mathbf{V}_x$ is a local variable used by $\mathbf{M}_b$ only. If $\mathbf{M}_a$ and $\mathbf{M}_c$ cannot interact with $\mathbf{V}_x$ and $\mathbf{V}_x$ is hidden from the outside, then it seems reasonable that the lifetime of $\mathbf{V}_x$ could be extended, yielding a behaviorally equivalent configuration:

$$\langle \mathbf{V}_x \mid \mathbf{M}_a ;\mathbf{M}_b; \mathbf{M}_c\rangle\backslash S$$

This leads to the following theorem:

Theorem 3 Change of Scope Theorem

$$\mathbf{M}_a \perp \mathbf{M}_x, \mathbf{M}_c \perp \mathbf{M}_x \Rightarrow$$
$$\langle \mathbf{M}_a ;\langle \mathbf{M}_b \mid \mathbf{M}_x\rangle; \mathbf{M}_c\rangle\backslash S = \langle \mathbf{M}_x \mid \mathbf{M}_a ;\mathbf{M}_b; \mathbf{M}_c\rangle\backslash S$$

where $S = Sort(\mathbf{M}_X)$. A proof of this theorem can be foundin [5].

5.0 Semantics

The intent of constructive semantics is to provide an interpretation of a program as the specification for an abstract state machine. We are now in a position to show how the constructive semantics of a programming language can be given using the results of the earlier sections. The style of this semantics will be denotational, showing how each construct in the language can be interpreted as a machine that is defined by the composition of the denotations of its component parts.

A programming language defines a number of basic data types and operations as given elements of the language. We assume that the machines denoted by these data types and operations are given as part of the formal semantics of the language.

We will take some example from the Ada programming language as the basis for showing how a constructive semantics for a language can be given. This subset includes many of the language features of Ada, including declarative blocks, composite types and exception handling. Packages and task's have been omitted because they differ little from declarative blocks in their semantics except for their extremely complex visibility computation rules, which are analyzed and modeled in [4].

In giving the semantics for each construct, we will not provide complete Ada semantics, but rather simplified semantics that highlight the strategy used in modeling each particular language construct. This is done because modeling the complex visibility rules of Ada would tend to divert attention away from the basic approach.

5.1 Declarations and References

In the previous sections we have laid the groundwork for modeling the semantics of a programming language in terms of machines. Each identifier in a program corresponds to a machine. We define an ***environment*** $\mathcal{E}: Ide \times \mathcal{M}$ to be a relation between identifiers and machines. We note that this environment is suitable for mapping identifiers into both machine instances and types if we use the "prototype machine" approach to defining machine types.

A ***declaration*** in a program usually results in both the definition of a machine and its entry in an environment relation in association with an identifier[11]. A ***reference*** is an occurrence of an identifier that must be associated with a machine through an environment relation[12]. *Each occurrence of an identifier in a program is either part of a declaration or it is part of a reference.* Consider the program in Example 2. In this example, we observe a single explicit declaration[13] of a variable named `a`, and a number of references: one to `integer` in the declaration itself; another to `a` in the assignment statement; and a third to the literal `1` in the assignment statement. The symbol `:=` is a reference to an assignment operator.

```
X: declare
      a : integer; -- a declaration of "a"
   begin
      a := 1; -- a reference to the "a"
   end X;
```

Example 2 Declarations and References

Taking this perspective of a program, one interesting problem is the determination of which machine a particular reference actually refers to. This problem can, in itself, be divided into two sub-problems: one is the determination of which machines are *visible* (available) at a given point in the program (we call this set of visible machines and their associated identifiers the ***direct environment***); the other is the determination of which of these is the one that is actually being referred to. In [4] we have investigated the computation of visibility, and at present have left the formalization of how a reference is selected for future work.

5.2 Elaboration: From Programs to Machines

If a program is the specification of a state machine, somewhere along the line the program must be converted into the machine itself. While one might be initially tempted to say that this is the role of the compiler, a compiler in simply generates the *initial state* of a machine, namely the computer in which the program will be executed. In constructive semantics, we wish to generate the specification of an abstract machine whose behavior is the "meaning" of the program. Of course, to be correct, the behavior

11. Note that this approach allows more than one identifier to be associated with the same machine. Declarations of aliases do not define new machines, but simply associate an existing machine with a new identifier.

12. There may be more than one entry in the relation with the same identifier

13. There is also an implicit declaration of the block X.

of the computer with the compiler generated initial state must be equivalent to the behavior of this abstract machine.

Converting a program into a state machine is not necessarily a one step process. The obvious counter-example is the interpreter, which does the conversion piecemeal. But even a compiled program may not be convertible into a complete state machine at load time. For example, some Ada data type declarations (definitions of machine types) are allowed to depend upon values computed previously in the program. Thus the machines defined by these type declarations are not even fully defined until part of the program has been executed.

In what follows, we shall have occasion to refer to the *process* of converting a program into a machine. For this purpose, we borrow a term from Ada and call this process ***elaboration***. Formally, elaboration is a mapping from a syntactic language term, a local declaration set, a direct environment and a type structure to a machine:

$$\mathbf{E}: \mathcal{L} \times \mathcal{E} \times \mathcal{E} \times \mathcal{T} \rightarrow \mathcal{M}$$

The first term is the language expression whose meaning is being determined. The second term is an environment that is to contain local declarations (typically, this environment is modified in the course of elaboration). The third term is an environment that contains machines that have been defined elsewhere that may be used in the course of performing the elaboration.The fourth term is a type structure, which may contain some type information initially and may also be added to during the course of elaboration.

5.3 Variables of a Simple Type

A variable belonging to a simple data type is modeled as a value machine belonging to the state type whose prototype machine is associated with the data type. The result of elaboration is a new machine "cloned" from the prototype machine of the data type.

$$\mathbf{E}(\llbracket \texttt{variablename: typename;} \rrbracket, \mathbf{LD}, \mathbf{DE}, \mathcal{T})$$
$$= \mathbf{M} \equiv \mathbf{New}(\mathbf{M}_T)$$

where

$$\mathbf{M}_T = \mathbf{Ref}(\mathit{ide}, \mathbf{DE})$$

Here **New** is an operation that takes a prototype machine and gets an unused member of the state type, and **Ref** is an operation that locates a machine in an environment by name.

The elaboration has side effects adding the new declaration to the set of local declarations and recording the type relationship in the *StateType* relation:

$$\mathbf{LD} = \mathbf{LD} \cup \{(\texttt{variablename}, \mathbf{M})\}$$

$$\mathit{StateType} = \mathit{StateType} \cup (\mathbf{M}, \mathbf{M}_T)$$

where *StateType* is a relation that records which state type a machine belongs to.

5.4 Subprogram Calls and Expression Evaluation

The semantics that we give for subprogram calls encompasses the invocation for both operators and subprograms. This treatment of operators as pre-defined subprograms allows the modeling of languages in which user-defined versions of these operators may be declared in a program. The newly defined operator affects the visibility of the original operator according to the visibility rules of the language.

We give several alternative representative formulations for calling two argument subprograms, each of which can be readily generalized to an arbitrary number of arguments. In defining these expressions, we will frequently need to know the syntactic name of the operation at the root of an expression. We define the syntactic function *Root*(expression) that returns this identifier.

The simplest form of subprogram call assumes that the arguments are simply references to existing machines (thus requiring no elaboration of the arguments) and further assumes that the arguments are of the correct type. We arrive at a relatively simple semantic for a subprogram call in which the prototype machine defining the subprogram is located and a new member of its associated state type is returned:

$$\mathbf{E}(\llbracket\, \texttt{subprogramName (<argument}_1\texttt{>, <argument}_2\texttt{>);}\, \rrbracket, \mathbf{LD}, \mathbf{DE}, \mathcal{T}) \\ = \mathbf{New}(\mathbf{Ref}(\texttt{subprogramName}, \mathbf{DE}))[\mathbf{M}_1/\mathbf{P}_1, \mathbf{M}_2/\mathbf{P}_2]\,\rangle$$

where $\mathbf{M}_1 = \mathbf{Ref}(\mathit{Root}(\texttt{<argument}_1\texttt{>}))$, $\mathbf{M}_2 = \mathbf{Ref}(\mathit{Root}(\texttt{<argument}_2\texttt{>}))$, and $[\mathbf{M}_1/\mathbf{P}_1, \mathbf{M}_2/\mathbf{P}_2]$ is a minor abuse of our morphism notation indicating that the actions of the formal parameter $\mathbf{P}_1$ are mapped to the actions of actual parameter $\mathbf{M}_1$, and similarly for $\mathbf{P}_2$ and $\mathbf{M}_2$. This assumes, of course, that $\mathbf{P}_1$ and $\mathbf{M}_1$ are of the same type, and similarly $\mathbf{P}_2$ and $\mathbf{M}_2$ are of the same type.

We now extend the semantics to include a modest amount of type checking. In this scheme, type information propagates only in one direction: upwards from actual arguments to functions.

$$\mathbf{E}(\llbracket\, \texttt{subprogramName (<argument}_1\texttt{>, <argument}_2\texttt{>);}\, \rrbracket, \mathbf{LD}, \mathbf{DE}, \mathcal{T}) \\ = \mathbf{New}(\mathbf{TypRef}(\texttt{subprogramName}, \mathbf{DE}, \\ \{\mathbf{RetSig}(\mathbf{M}_1) \times \mathbf{RetSig}(\mathbf{M}_2)\}))[\mathbf{M}_1/\mathbf{P}_1, \mathbf{M}_2/\mathbf{P}_2]$$

Here **RetSig** uses the type structure $\mathcal{T}$ to locate the return type signature of the indicated machine, which is, itself, a prototype machine representing some data type, and **TypRef** uses the same type structure to type qualify candidate machines found in the environment.

The semantics given for the previous two examples will work correctly if the argument is a variable, but if the argument is itself a subprogram reference (a function call), then the reference to the root of the argument will return the *prototype* machine of the subprogram, and we must then get a new member of its state type. We do this by elaborating the argument itself as part of the elaboration of the subprogram call. We thus get (ignoring type checking again):

$$\mathbf{E}([\![\texttt{subprogramName (<argument}_1\texttt{>, <argument}_2\texttt{>);}]\!], \mathbf{LD}, \mathbf{DE}, \mathcal{T})$$
$$= \langle\, \mathbf{New}(\mathbf{Ref}(\texttt{subprogramName}, \mathbf{DE}))[\mathtt{M}_1/\mathtt{P}_1, \mathtt{M}_2/\mathtt{P}_2] \mid$$
$$\text{if } \mathbf{Type}(\mathtt{M}_1) = \mathtt{M}_1 \text{ then } \mathbf{E}([\![\texttt{<argument}_1\texttt{>}]\!], \mathbf{LD}, \mathbf{DE}, \mathcal{T}) \text{ else } \mathbf{0} \mid$$
$$\text{if } \mathbf{Type}(\mathtt{M}_2) = \mathtt{M}_2 \text{ then } \mathbf{E}([\![\texttt{<argument}_2\texttt{>}]\!], \mathbf{LD}, \mathbf{DE}, \mathcal{T}) \text{ else } \mathbf{0} \,\rangle$$

Note that if $\mathtt{M}_1$ is a prototype machine, then $\mathbf{Type}(\mathtt{M}_1) = \mathtt{M}_1$. Here **0** is a degenerate machine with no actions.

This almost gives us the semantics that we want, except for a possible problem in the order of evaluation: we have not constrained the argument machines to ensure that they are evaluated (executed) before the subprogram itself is evaluated. To accomplish this, for each argument that requires elaboration, we introduce a temporary local variable to carry the result of the argument evaluation forward to the subprogram itself. Ignoring type checking again, and leaving out the conditionals (we assume that both arguments require elaboration) we now have:

$$\mathbf{E}([\![\textit{subprogramName}\ \texttt{(<argument}_1\texttt{>, <argument}_2\texttt{>);}]\!], \mathbf{LD}, \mathbf{DE}, \mathcal{T})$$
$$= \langle\, \mathbf{V_1} \mid \mathbf{V_2} \mid \mathbf{E}([\![\texttt{<argument}_1\texttt{>}]\!], \mathbf{LD}, \mathbf{DE}, \mathcal{T})[\mathbf{V_1}/\texttt{out}];$$
$$\mathbf{E}([\![\texttt{<argument}_2\texttt{>}]\!], \mathbf{LD}, \mathbf{DE}, \mathcal{T})[\mathbf{V_2}/\texttt{out}];$$
$$\mathbf{New}(\mathbf{Ref}(\textit{subprogramName}, \mathbf{DE}))[\mathbf{V_1}/\mathtt{P}_1, \mathbf{V_2}/\mathtt{P}_2] \,\rangle \backslash \mathit{Sort}(\mathbf{V_1}) \cup \mathit{Sort}(\mathbf{V_2})$$

where $[\mathbf{V_1}/\texttt{out}]$ relabels the output formal parameter with the actions of $\mathbf{V_1}$. Note that we have somewhat arbitrarily determined an order of evaluation for the actual parameter expressions.

In [3] we have explored event more complex type checking and overload resolution schemes. While these schemes add significantly to the type information that is passed around between references (the reference functions themselves become very complicated), the basic structure of the elaboration in terms of the structure of machines still remains the same.

A final note on subprogram calls – we have made no distinction between functions and procedures in this semantics, nor have we made any distinction between calls that occur as an actual statement and calls that occur as part of a subexpression. We note that if one of the actual arguments to the subprogram was accidentally a procedure, then the typed reference to the subprogram would fail to resolve to a machine. This further illustrates the generality of this approach to semantics.

5.5 Declarative Blocks

A declarative block is a collection of declarations followed by a sequence of statements and possibly an exception handler. There are a number of possible semantics for declarative blocks, each reflecting a different visibility semantics for the declarations that occur in the block. We show two possibilities here.

For let or let* visibility semantics[14], we have:

```
E([[ blockname:
     declare
        <declarative part>
     begin
        <sequence of statements>
     exception
        <exception handler list>
     end]], LD, DE, T)
```

$$
\begin{aligned}
= \quad \mathbf{M} \equiv \quad \langle\; & \mathbf{E}([\![\texttt{<declarative part>}]\!], \mathbf{M{:}LD}, \mathbf{DE}, \mathcal{T}) \mid \\
& \mathbf{E}([\![\texttt{<sequence of statements>}]\!], \mathbf{M{:}LD}, \mathbf{M{:}DE}, \mathcal{T}) \mid \\
& \mathbf{E}([\![\texttt{<exception handler list>}]\!], \mathbf{M{:}LD}, \mathbf{M{:}DE}, \mathcal{T}) \rangle
\end{aligned}
$$

lere **M:LD** is a new local declaration set associated with the newly created machine [, and **M:DE** is a new direct environment associated with the same machine.

Ve have the following side effects:

$$\mathbf{LD} = \mathbf{LD} \cup \{(\mathit{blockname}, \mathbf{M})\}$$

$$\mathbf{M{:}DE} = \mathbf{M{:}LD} \underset{m}{\cup} \mathbf{DE}$$

lere $\underset{m}{\cup}$ denotes a modified set union [4][5] that formalizes the hiding of some mem- er of the second set (**DE**) by members of the first set (**M:LD**).

'or letrec[15] visibility semantics the direct environment passed to the declarative part /ould be different, giving:

$$
\begin{aligned}
\mathbf{M} \quad \equiv \quad \langle\; & \mathbf{E}([\![\texttt{<declarative part>}]\!], \mathbf{M{:}LD}, \mathbf{M{:}DE}, \mathcal{T}) \mid \\
& \mathbf{E}([\![\texttt{<sequence of statements>}]\!], \mathbf{M{:}LD}, \mathbf{M{:}DE}, \mathcal{T}) \mid \\
& \mathbf{E}([\![\texttt{<exception handler list>}]\!], \mathbf{M{:}LD}, \mathbf{M{:}DE}, \mathcal{T}) \rangle
\end{aligned}
$$

.5.1 *Declarative Part*

'he elaboration of the declarative part simply elaborates each of the declarations, com- osing any machines that result in parallel. It is important to note that the elaboration f a variable will return a machine. The elaboration of a function declaration or type eclaration *will not return a machine* - the created machine will be associated with the eclared name in the local declaration set **LD**, but no machine is actually instantiated s part of the elaboration.

or let or letrec visibility semantics, we would have

14. Let and let* [11] are constructs arising in the language scheme in which declarations in a block are not visible at all to each other (let semantics) or earlier declarations are visible to later declarations (let* semantics). [4][5] cover the variations in visibility semantics in more detail.
15. In letrec semantics [11] declarations in a block are all mutually visible, thus allowing recursive declarations.

$$\mathbf{E}([\![\texttt{<declarative part>}]\!], \mathbf{LD}, \mathbf{DE}, \mathcal{T})$$
$$= \quad \langle \; \mathbf{E}([\![\texttt{<declaration}_1\texttt{>}]\!], \mathbf{LD}, \mathbf{DE}, \mathcal{T}) \mid$$
$$\mathbf{E}([\![\texttt{<declaration}_2\texttt{>}]\!], \mathbf{LD}, \mathbf{DE}, \mathcal{T}) \mid \ldots \mid$$
$$\mathbf{E}([\![\texttt{<declaration}_n\texttt{>}]\!], \mathbf{LD}, \mathbf{DE}, \mathcal{T}) \; \rangle$$

It is important to note that for letrec visibility semantics, the enclosing declarative block has included **LD** in the computation of **DE**. Thus the elaboration of one declaration could well affect the meaning of a reference in another. This points out the importance of the order of elaboration in determining the meaning of a program. Some languages put such severe constraints upon the relative positions of declarations with respect to references that the order of elaboration is not an issue. Other languages, like Ada, provide mechanisms to specify the order of elaboration in cases where the order may not be sufficiently constrained[16]. In [4] we show that an appropriate ordering, if one exists, may be determined through the construction of a dependency graph relating declarations and references. Cycles in this graph indicate that no proper elaboration ordering exists.

For let* visibility semantics, we would have

$$\mathbf{E}([\![\texttt{<declarative part>}]\!], \mathbf{LD}, \mathbf{DE}_0, \mathcal{T})$$
$$= \quad \langle \; \mathbf{E}([\![\texttt{<declaration}_1\texttt{>}]\!], \mathbf{LD}_1, \mathbf{DE}_0, \mathcal{T}) \mid$$
$$\mathbf{E}([\![\texttt{<declaration}_2\texttt{>}]\!], \mathbf{LD}_2, \mathbf{DE}_1, \mathcal{T}) \mid \ldots \mid$$
$$\mathbf{E}([\![\texttt{<declaration}_n\texttt{>}]\!], \mathbf{LD}_n, \mathbf{DE}_{n-1}, \mathcal{T}) \; \rangle$$

Here the order of elaboration is defined to be the order of declaration. For these elaborations, we have:

$$\mathbf{LD}_0 = \varnothing$$

Prior to the ith elaboration, we have:

$$\mathbf{LD}_i = \mathbf{LD}_{i-1}$$

$$\mathbf{DE}_i = \mathbf{LD}_{i-1} \cup \mathbf{DE}_0$$

After the ith elaboration, $\mathbf{LD}_i$ also contains the declaration resulting from the elaboration. After the last elaboration, we compute the returned set of local declarations:

$$\mathbf{LD} = \mathbf{LD}_n$$

5.5.2 Sequence of Statements

Because statements may have labels on them and references to them, elaboration order is important here as well. We show the semantics for letrec style visibility[17] (for se-

16. [1] p. 10-11
17. For sequential elaboration order and let* visibility, the computation of the local and direct environments is exactly the same as for the let* visibility of declarations.

quential or let* visibility, the local and direct environments are computed exactly as for the let* declarative items).

$$\mathbf{E}(\llbracket \texttt{<sequence of statements>} \rrbracket, \mathbf{LD}, \mathbf{DE}, \mathcal{T})$$
$$= \quad \langle \ \mathbf{E}(\llbracket \texttt{<statement}_1\texttt{>} \rrbracket, \mathbf{LD}, \mathbf{DE}, \mathcal{T}) \ ;$$
$$\mathbf{E}(\llbracket \texttt{<statement}_2\texttt{>} \rrbracket, \mathbf{LD}, \mathbf{DE}, \mathcal{T}) \ ; \ldots;$$
$$\mathbf{E}(\llbracket \texttt{<statement}_n\texttt{>} \rrbracket, \mathbf{LD}, \mathbf{DE}, \mathcal{T}) \ \rangle$$

6.0 Summary

We have given an outline of how the semantics of a programming language can be constructively given in terms of primitive state machines and compositions of state machines. Thus the semantics of a program is given as an abstract state machine whose structure is constructively specified by the program itself. We have shown that the dominant concepts of a programming language are readily understood in terms of three basic semantic concepts: state machines, state types (sets of isomorphic state machines), and generic machines (parameterized specifications of state types.) We have shown that programs, subprograms and data types all have a uniform interpretation as state types. We have described the relationship between the identifiers in the language and the semantic model elements that they correspond to, and in [4] and [5] we have provided a set-theoretic description of the computation of visibility in programming languages.

Constructive semantics is fully abstract in the sense that behavioral equivalence defines equivalence classes of semantic expressions, and these equivalence classes can be taken to be the fully abstract semantics of the expression. We have left two interesting questions open in this area. Is there a normal form for machine algebra expressions that would ease their syntactic comparison? Is behavioral equivalence decidable in the restricted classes of machines used in our semantics? Brookes' work on normal forms of synchronization trees[18] (which underlie HCCS) leads us to believe that for *finite* machines a unique (up to the ordering of terms) normal form exists in HCCS for each equivalence class, but we suspect that the existence of a normal form of machine algebra expressions is precluded by the constraints that our machine algebra places on the form of HCCS expressions.

References

1. Ada Programming Language, ANSI/MIL-STD-1815A (1983)

2. Brookes, Stephen D, A Model for Communicating Sequential Processes, Ph.D. thesis, University College, Oxford University (1983)

18. [2] pp. 99-100.

3. Brown, Paul C., Oconnor, D.M., and Kelliher, Tim, "An Extended Overload Resolution Algorithm that allows Types and Subprograms as First Class Objects," internal document, GE Corporate Research and Development Center, Schenectady, New York (1989)

4. Brown, Paul C., Computing Visibility in Programming Languages, Technical Report 90CRD098, GE Research and Development Center, Schenectady, New York 12301 (1990)

5. Brown, Paul C., Constructive Semantics, Ph.D. thesis, Rensselaer Polytechnic Institute, Troy, New York (1992)

6. Eaker, Charles E., "Creating Software Should Be Easy," (unpublished) GE Corporate Research and Development Center, Schenectady, New York (1991)

7. Eaker, Charles E., "How to Create Software: A Guide to the Perplexed," (unpublished) GE Corporate Research and Development Center, Schenectady, New York (1991)

8. Hoare, C.A.R., Communicating Sequential Processes, Prentice Hall International (1985)

9. Milner, Robin, "Calculi for Synchrony and Asynchrony," Journal of Theoretical Computer Science, Vol. 25, pp 267-310 (1983)

10. Milner, Robin, Communication and Concurrency, Prentice Hall International (1989)

11. Rees, Jonathan and Clinger, William (eds), "Revised3 Report on the Algorithmic Language Scheme," SIGPLAN Notices, Vol. 21, No. 12 (1986).

12. Rumbaugh, James et. al., Object Oriented Modeling and Design, Prentice Hall (1991)

13. Van Benthem, Johan, The Logic of Time: A Model-Theoretic Investigation into the Varieties of Temporal Ontology and Temporal Discourse, D. Reidel Publishing Company (1982),

A Causality-based Semantics for CCS

R. J. Coelho da Costa* J.-P. Courtiat

LAAS/CNRS – 7, Av. du Colonel Roche
31077 Toulouse Cedex – France
E-mail: {rosvelte,courtiat}@laas.fr

Abstract

In this paper we present an original approach providing a causality-based semantics for CCS based on preliminary results appeared in [6, 7, 8, 11]. This approach relies on a simple intuition related to the expression of causal relationships among occurrences of actions, in a way closely related to the notion of variable abstraction of Lambda-Calculus. Main results deal with the definition of an operational semantics, as well as with the characterization of strong and weak bisimulation equivalences within the proposed causality framework. Discussion is provided aiming at relating the proposed approach with other approaches, like Causal Trees and location-based semantics, presenting some degree of similarity with ours, at least at an intuitive level.

Introduction

Milner's CCS (Calculus of Communicating Systems) remains one of the most important formalisms for studying concurrent systems. Originally, only an interleaving-based semantics has been proposed for CCS. Nevertheless, although interleaving-based semantics permit a relatively simple mathematical treatment when analyzing many interesting properties of concurrent systems, it is well known that this approach is not always appropriate; for instance, when the actions are assumed to be (temporally or spatially) nonatomic. On the other hand, causality-based semantics (also called "true concurrency"

* On leave from the Dept. of Computer Science of the Universidade Federal de Santa Catarina.
Current address: Universidade Federal de Santa Catarina
Dept. of Computer Science (CEC/CTC)
Trindade — Caixa Postal 476
88049 Florianópolis-SC, Brazil
E-mail: cec1cdc@brufsc.bitnet

semantics) seem to be immune to many drawbacks of interleaving-based semantics. Causality-based semantics postulate that two events are concurrent iff they are not causally related. In the past few years, several works have been devoted to the definition of causality-based semantics for CCS and related languages (see [15, 24, 22, 21, 25, 1, 4, 2, 23, 12, 16, 14, 17] for a representative sample and further references). Unfortunately, studying concurrent systems using causality-based semantics seems to be much more difficult than using interleaving-based approaches. Therefore, although several interesting results have already been achieved, many useful notions need to be carefully investigated.

The approach, presented in this paper, provides a causality-based operational semantics for CCS, as well as the characterization of strong and weak bisimulation equivalences within this causality framework. The main advantage of the proposal is that it actually provides a simple and straightforward generalization of the classical interleaving-based semantics making it possible to adapt and reuse many results achieved for the interleaving case.

In this paper, more emphasis has been given to the underlying intuition of our proposal than to the technical details. For this reason, most proofs have been omitted, but are available in [5]. The paper is organized into five sections. The first section presents the basic insights behind the proposal by means of simple examples. The second section deals with the presentation of the causality-based operational semantics. The third and fourth sections introduce respectively strong and weak bisimulation equivalences. Finally, the last section presents an extensive discussion on the relationships of the proposed approach with other related approaches, in particular causal trees [12] and location-based semantics [3].

1 Basic Insights

The approach presented in this paper is fundamentally based on the notion of *configuration*. A configuration of a concurrent system is just defined as a collection of *local states*. A local state is a pair $\langle E, M \rangle$, where M, called the *determinant* of the local state, is a finite non-empty set of event names which identifies the causative events enabling local behavior E (here, a CCS expression). Note that local behavior and local state are closely related notions. In fact, most concurrent models do not make any difference between them. In our approach, however, the notion of local state is contextual, whereas the notion of local behavior is only structural.

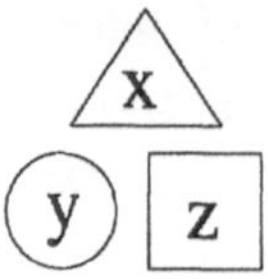

Figure 1

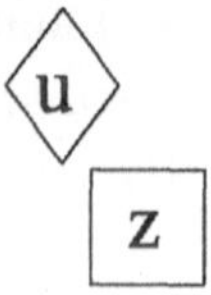

Figure 2

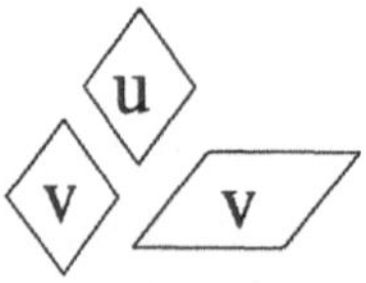

Figure 3

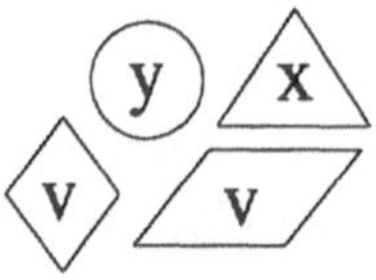

Figure 4

The notion of configuration leads to another important notion of our approach: the notion of *atom.* An atom intends to capture the notion of local occurrence of an action. For instance, let us assume that at configuration of Figure 1, a concurrent system can locally perform action a by enforcing a cooperation between local behaviors $\triangle$ and $\bigcirc$. Let us name u ($u \not\equiv z$) this occurrence of a and assume further that it enables local behavior $\Diamond$ (Figure 2). u is the determinant of the new local state. In our approach, this local occurrence of a is represented by atom ${}_{\{x,y\}}a_u$, where set $\{x, y\}$ is the *motive* of a, i.e. the causative events enabling this local occurrence of a. Now, let us assume that local behavior $\Box$ can perform action b, and consequently enables two local behaviors: $\Diamond$ and ▱. This local occurrence of b can be represented by atom ${}_{\{z\}}b_v$, where v ($v \not\equiv u$) is the name given to this local occurrence of b. The enabled configuration is pictured in Figure 3. Of course, as this local occurrence of b does not depend on the previous occurrence of a (i.e. they occur concurrently), this occurrence of b must also be possible at configuration of Figure 1. For instance, an occurrence of b, represented by atom ${}_{\{z\}}b_v$ ($v \not\equiv x, y$), from configuration of Figure 1 enables configuration of Figure 4.

With each local occurrence of action, a transition $\mathcal{E} \xrightarrow{\alpha} \mathcal{F}$ is associated, where $\mathcal{E}$ and $\mathcal{F}$ are configurations and α is an atom. In this way, the behavior of a concurrent system can be represented by a *configuration graph* as shown in Figure 5.

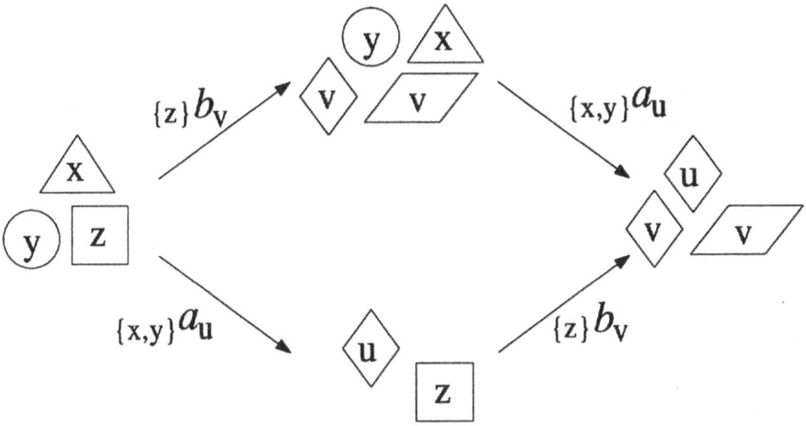

Figure 5

Our notion of configuration is closely related to the notion of marking in Petri nets [10], where each local state $\langle E, M \rangle$ is interpreted as a token at the place corresponding to local behavior E.

As usual, we are mainly interested in the event-based behavior of concurrent systems. For this reason, we represent a concurrent system using a kind of *concurrent labeled transition systems*, where configurations are interpreted as states (the vertices) and transitions (the arcs) are labeled with atoms. For instance, the concurrent labeled transition system of Figure 6 intends to represent the behavior of configuration $\{\langle a; b; \text{Nil}|c; \text{Nil}, \{x\}\rangle\}$, where, similarly to the usual graphic notation of labeled transition system, the little arrow points to the initial configuration.

Infinite behaviors can be nicely represented in terms of concurrent labeled transition systems. For instance, configuration $\{\langle \text{rec}\, X.a; X + b; \text{Nil}, \{x\}\rangle\}$ corresponds to the concurrent labeled transition system of Figure 7.

A detailed presentation of concurrent labeled transition systems is given in [5]. In this paper, concurrent labeled transition system are only implicitly considered.

2 Causality-Based Operational Semantics for CCS

First of all, let us briefly remind the syntax of CCS behavior expressions

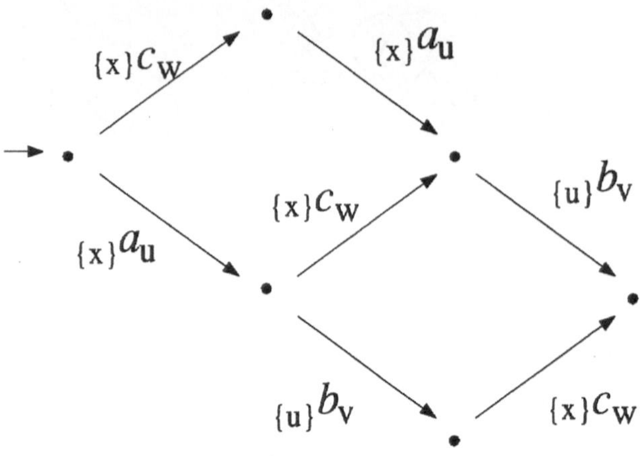

Figure 6

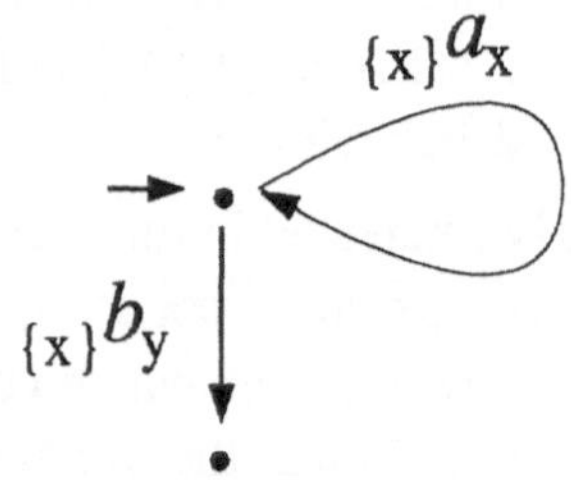

Figure 7

(for the sake of simplicity, relabeling is not considered). For this purpose, let us assume a set $\Delta = \{l, \ldots\}$ of *labels* and a set $\overline{\Delta} = \{\bar{l}, \ldots\}$ of *complementary labels* such that $\bar{\bar{l}} = l$ for all $l \in \Delta$. Let $\Lambda = \Delta \cup \overline{\Delta}$, ranged over by o, denote the set of all *visible labels*. Thus, $\mathcal{A}ct = \Lambda \cup \{\tau\}$, where $\tau \notin \Lambda$ and $\bar{\tau} \stackrel{def}{=} \tau$, denotes the usual set of *actions*, ranged over by $a, b, \ldots$. Finally, let Var ranged over by X, Y, Z, be a denumerable set of variables. Then, the syntax of *behavior expressions* is defined by the following BNF-like grammar:

$$P ::= \mathrm{Nil} \mid X \mid a; E \mid E + E' \mid E|E' \mid E \backslash l \mid \mathrm{rec}\, X.E$$

Let $\mathfrak{B}$, ranged over by $E, F, \ldots$, denote the set of all behavior expressions. Let also $\mathfrak{B}_0$, ranged over by $P, Q, \ldots$, denote the set of all *closed behavior expressions*, in short *behaviors*. In the sequel, the usual priority order for the above operators is assumed and 'Nil' is omitted whenever there is no danger of confusion.

Let us now introduce formally our concept of configuration. First, we assume a denumerable set $\mathfrak{X}$ of event names, ranged over by $\ldots, x, y, z$. The syntax of *configuration expressions* is then defined by the following grammar:

$$\mathcal{E} ::= {}_M[E] \mid \mathcal{E}|^{\bullet}\mathcal{E}' \mid \mathcal{E}\backslash^{\bullet}l$$

where, for any behavior expression E, ${}_M[E]$ denotes a local state with behavior E and determinant M. Operations over configurations $\mathcal{E}|^{\bullet}\mathcal{E}'$ and $\mathcal{E}\backslash^{\bullet}l$ are intuitively similar to the corresponding operations over behavior expressions. We use $\mathfrak{P}$, ranged over by $\mathcal{E}, \mathcal{F}, \ldots$, to denote the set of all configuration expressions. Furthermore, $\mathcal{E}$ is said to be a *closed configuration expression*, in short *configuration*, iff, for all terms ${}_M[E]$ in $\mathcal{E}$, E is a closed behavior expression. We use $\mathfrak{P}_0$, ranged over by $\mathcal{P}, \mathcal{Q}, \ldots$, to denote the set of all configurations.

Before providing a semantics for configurations, some preliminary definitions are needed.

2.1 Definition The set of all event names occurring in $\mathcal{E}$, in symbols $\phi(\mathcal{E})$, is defined inductively as follows:

(i) $\phi({}_M[E]) = M$ (ii) $\phi(\mathcal{E}|^{\bullet}\mathcal{E}') = \phi(\mathcal{E}) \cup \phi(\mathcal{E}')$

(iii) $\phi(\mathcal{E}\backslash^{\bullet}l) = \phi(\mathcal{E})$ □

2.2 Definition Substitution $M/x\,\mathcal{E}$ is defined inductively as follows:

(i) $M/x\ {}_N[E] \equiv {}_{N'}[E]$

where either $N' = (N - \{x\}) \cup M$ if $x \in N$, or else $N' = N$

(ii) $M/x\,(\mathcal{E}|^{\bullet}\mathcal{E}') \equiv M/x\,\mathcal{E}|^{\bullet}M/x\,\mathcal{E}'$ (iii) $M/x\,(\mathcal{E}\backslash^{\bullet}l) \equiv M/x\,\mathcal{E}\backslash^{\bullet}l$ □

2.3 Proposition For all $\mathcal{E} \in \mathfrak{P}$, $\phi(M/x\,\mathcal{E}) = M/x\,\phi(\mathcal{E})$. □

2.4 Definition An atom is a triple ${}_M a_x \in 2^{\mathfrak{X}}_{\mathrm{fn}} \times A \times \mathfrak{X}$, where: M is a finite non-empty set of event names; a is an action of some set of actions A; and x is an event name. We note by Atm_A the set of all atoms with respect to A; in particular, we note by Atm, ranged over by $\alpha, \beta, \ldots$, the set of all atoms with respect to Act. □

We are now ready to provide an operational semantics for configurations, given in Plotkin's SOS style.

2.5 Definition Set $\rightarrow \subseteq \mathfrak{P} \times \mathrm{Atm} \times \mathfrak{P}$ of *transitions* is defined to be the least set satisfying the following laws:

L1. ${}_M[a; E] \overset{{}_M a_z}{\rightarrow} {}_z[E]$ for any event name z.

L2. If ${}_M[E] \overset{\alpha}{\rightarrow} \mathcal{F}$ then ${}_M[E + F] \overset{\alpha}{\rightarrow} \mathcal{F}$ and ${}_M[F + E] \overset{\alpha}{\rightarrow} \mathcal{F}$.

L3. If ${}_M[E]|^{\bullet}\, {}_M[F] \overset{\alpha}{\rightarrow} \mathcal{G}$ then ${}_M[E|F] \overset{\alpha}{\rightarrow} \mathcal{G}$.

L4. If $\mathcal{E} \overset{{}_M a_x}{\rightarrow} \mathcal{E}'$ then $\mathcal{E}|^{\bullet}\mathcal{F} \overset{{}_M a_z}{\rightarrow} z/x\, \mathcal{E}'|^{\bullet}\mathcal{F}$ and $\mathcal{F}|^{\bullet}\mathcal{E} \overset{{}_M a_z}{\rightarrow} \mathcal{F}|^{\bullet} z/x\, \mathcal{E}'$
for any $z \notin (\phi(\mathcal{E}) - \{x\}) \cup \phi(\mathcal{F})$.

L5. If $\mathcal{E} \overset{{}_M o_x}{\rightarrow} \mathcal{E}'$ and $\mathcal{F} \overset{{}_N \bar{o}_y}{\rightarrow} \mathcal{F}'$ then $\mathcal{E}|^{\bullet}\mathcal{F} \overset{{}_{M \cup N} \tau_z}{\rightarrow} z/x\, \mathcal{E}'|^{\bullet} z/y\, \mathcal{F}'$
for any $z \notin (\phi(\mathcal{E}') - \{x\}) \cup (\phi(\mathcal{F}') - \{y\})$.

L6. If ${}_M[E]\backslash^{\bullet} l \overset{\alpha}{\rightarrow} \mathcal{F}$ then ${}_M[E\backslash l] \overset{\alpha}{\rightarrow} \mathcal{F}$.

L7. If $\mathcal{E} \overset{{}_M a_x}{\rightarrow} \mathcal{F}$ then $\mathcal{E}\backslash^{\bullet} l \overset{{}_M a_x}{\rightarrow} \mathcal{E}'\backslash^{\bullet} l$ if $a \not\equiv l, \bar{l}$.

L8. If ${}_M[[\mathrm{rec}\, X.E/X]E] \overset{\alpha}{\rightarrow} \mathcal{G}$ then ${}_M[\mathrm{rec}\, X.E] \overset{\alpha}{\rightarrow} \mathcal{G}$.

where $\mathcal{E} \overset{{}_M a_x}{\rightarrow} \mathcal{F}$ is a shorthand for $\langle \mathcal{E}, {}_M a_x, \mathcal{F} \rangle \in \rightarrow$. □

The remainder of this section is devoted to the presentation of some basic results of this causality-based operational semantics of CCS.

2.6 Proposition For all $\mathcal{E} \in \mathfrak{P}$,

$$\bigcup_{\mathcal{E} \overset{{}_M a_x}{\rightarrow} \mathcal{F}} (M \cup (\phi(\mathcal{F}) - \{x\})) \subseteq \phi(\mathcal{E})$$

□

2.7 Proposition For all $\mathcal{E} \in \mathfrak{P}$ and $\langle N, u \rangle \in 2^{\mathfrak{X}}_{\mathrm{fn}} \times \mathfrak{X}$,

if $\mathcal{E} \overset{{}_M a_x}{\rightarrow} \mathcal{F}$ then

$$N/u\, \mathcal{E} \begin{cases} \overset{{}_{M'} a_x}{\rightarrow} \mathcal{F} & \text{if } x \equiv u \\ \overset{{}_{M'} a_x}{\rightarrow} N/u\, \mathcal{F} & \text{if } x \not\equiv u \wedge (x \notin N \vee u \notin \phi(\mathcal{F})) \\ \overset{{}_{M'} a_z}{\rightarrow} N/u\, z/x\, \mathcal{F} & \text{if } x \not\equiv u \wedge x \in N \wedge u \in \phi(\mathcal{F}) \\ & \text{for some } z \notin N, \phi(\mathcal{F}) \end{cases}$$

where $M' = N/u\, M$. □

2.8 Proposition For all $\mathcal{E} \in \mathfrak{P}$, $\mathcal{E} \overset{M^{a_x}}{\rightarrow} \mathcal{F}$ iff $\mathcal{E} \overset{M^{a_z}}{\rightarrow} z/x\,\mathcal{F}$, for any $z \notin \phi(\mathcal{F}) - \{x\}$. □

The so-called *Diamond Property* which expresses the fact that the behavior of a concurrent system is confluent under concurrent executions of actions, may be stated in our approach as follows:

2.9 Proposition $\forall \mathcal{E}, \mathcal{F}, \mathcal{H} \in \mathfrak{P}$ such that $\mathcal{E} \overset{M^{a_x}}{\rightarrow} \mathcal{F} \overset{N^{b_y}}{\rightarrow} \mathcal{H}$, if $x \notin N$ (i.e. the occurrences of a and b are concurrent) then $\exists \mathcal{G} \in \mathfrak{P}$ such that $\mathcal{E} \overset{N^{b_z}}{\rightarrow} \mathcal{G} \overset{M^{a_x}}{\rightarrow} z/y\,\mathcal{H}$ for some $z \notin (\phi(\mathcal{H}) - \{x, y\}) \cup M \wedge z \equiv x \Rightarrow y \notin \phi(\mathcal{H})$. □

This proposition is useful because it permits to obtain a reduced representation of the transition relation (saving memory), and it can therefore be used as a palliative solution to the state space explosion problem.

3 Causality-based Strong Bisimulation

Now, Park's and Milner's notion of bisimulation is introduced in our framework.

3.1 Definition Let $\mathbf{R} \subseteq \mathfrak{P}_\circ \times \mathfrak{P}_\circ$ be a symmetric binary relation between configurations. Then, let $\mathbb{F} : \mathrm{Rel}(\mathfrak{P}_\circ) \rightarrow \mathrm{Rel}(\mathfrak{P}_\circ)$ be a function defined as follows: $\langle \mathcal{P}, \mathcal{Q} \rangle \in \mathbb{F}(\mathbf{R})$ iff

whenever $\mathcal{P} \overset{M^{a_x}}{\rightarrow} \mathcal{P}'$, there exists $\mathcal{Q} \overset{M^{a_y}}{\rightarrow} \mathcal{Q}'$ such that $\langle z/x\mathcal{P}', z/y\mathcal{Q}' \rangle \in \mathbf{R}$ for some $z \notin (\phi(\mathcal{P}') - \{x\}) \cup (\phi(\mathcal{Q}') - \{y\})$.

$\mathbf{R}$ is called a *causality-based strong bisimulation* iff $\mathbf{R} \subseteq \mathbb{F}(\mathbf{R})$. If $\langle \mathcal{P}, \mathcal{Q} \rangle \in \mathbf{R}$ for some causality-based strong bisimulation $\mathbf{R}$, then $\mathcal{P}$ and $\mathcal{Q}$ are said to be *causality-based strongly bisimilar*, in symbols $\mathcal{P} \sim \mathcal{Q}$. □

Our causality-based strong bisimilarity is rather similar to the classical interleaving-based strong bisimilarity. Nevertheless, and differently to the classical interleaving-based approach, we are here concerned with *local occurrences* of actions, and we can then distinguish, for instance, $a|b$ from $a; b+b; a$. Note however that our causality-based strong bisimulation does not consider action synchronization as an unique atomic event (as it is in general the case), but as two interdependent distinct activities. For instance, $P = a; (b|\bar{b}; \mathrm{c})\backslash b$ and $Q = (a; b|\bar{b}; \mathrm{c})\backslash b$ are not causality-based strongly bisimilar, as action $\bar{b}$ in

Q can be performed before the complete occurrence of action a. Nevertheless, it is also possible to slightly modify the previous definition of causality-based strong bisimulation, by introducing some notion of transitivity among transitions (omitted here) in order to permit identifying behaviors such that P and Q.

In [5], several properties of the causality-based strong bisimilarity are discussed. In particular, we prove that $\sim$ is an equivalence relation. In this paper, however, we present only a basic property of the causality-based strong bisimilarity which establishes that $\sim$ is preserved under name substitution.

3.2 Proposition For any $\langle M, x\rangle \in 2^{\mathfrak{X}}_{\mathrm{fn}} \times \mathfrak{X}$,

if $\mathcal{P} \sim \mathcal{Q}$ then $M/x\,\mathcal{P} \sim M/x\,\mathcal{Q}$. □

This proposition is important because it makes it possible to consider finite bisimulations whenever possible.

4 Observation

In this section a causality-based weak bisimilarity is defined following Milner's notion of observational bisimilarity. We start by introducing the notion of weak transition. For this purpose, let $\mathrm{Act}_{\triangleleft} = \mathrm{Act} - \{\tau\}$ be the set of *visible actions* and let $\mathrm{Atm}_{\triangleleft}$, ranged over by ν, denote the set of all atoms with respect to $\mathrm{Act}_{\triangleleft}$. Finally, we define $\mathrm{Atm}_{\epsilon} = \mathrm{Atm}_{\triangleleft} \cup \{\epsilon\}$, where $\epsilon \notin \mathrm{Atm}_{\triangleleft}$ stands for an *invisible evolution*, and we use θ to range over Atm_{ϵ}.

4.1 Definition Set $\twoheadrightarrow \subseteq \mathfrak{P} \times \mathrm{Atm}_{\epsilon} \times \mathfrak{P}$ of *weak transitions* is defined to be the least set satisfying the following laws:

(i) $\forall \mathcal{E}, \mathcal{F} \in \mathfrak{B}.\ \mathcal{E} \overset{M^{\tau_x}}{\rightarrow} \mathcal{F} \Rightarrow \mathcal{E} \overset{\epsilon}{\twoheadrightarrow} M/x\ \mathcal{F}.$

(ii) $\forall \mathcal{E}, \mathcal{F} \in \mathfrak{B}.\ \mathcal{E} \overset{\nu}{\rightarrow} \mathcal{F} \Rightarrow \mathcal{E} \overset{\nu}{\twoheadrightarrow} \mathcal{F}.$

(iii) $\forall \mathcal{E} \in \mathfrak{B}.\ \mathcal{E} \overset{\epsilon}{\twoheadrightarrow} \mathcal{E}.$

(iv) $\forall \mathcal{E}, \mathcal{F}, \mathcal{G} \in \mathfrak{B}.\ \mathcal{E} \overset{\epsilon}{\twoheadrightarrow} \mathcal{F} \overset{\theta}{\twoheadrightarrow} \mathcal{G} \vee \mathcal{E} \overset{\theta}{\twoheadrightarrow} \mathcal{F} \overset{\epsilon}{\twoheadrightarrow} \mathcal{G} \Rightarrow \mathcal{E} \overset{\theta}{\twoheadrightarrow} \mathcal{G}.$

where $\mathcal{E} \overset{\theta}{\twoheadrightarrow} \mathcal{F}$ is a shorthand for $\langle \mathcal{E}, \theta, \mathcal{F}\rangle \in \twoheadrightarrow$. In the sequel, $\mathcal{E} \overset{\epsilon}{\twoheadrightarrow} \mathcal{F}$ is shortly noted as $\mathcal{E} \twoheadrightarrow \mathcal{F}$. □

For instance, from configuration $\mathcal{P} = {}_x[(a;(b+m)|c;\overline{m};d)\backslash m]$ the following weak transition derivation is possible:

$$\mathcal{P} \overset{x^{a_u}}{\twoheadrightarrow} \overset{x^{c_v}}{\twoheadrightarrow} \mathcal{Q} \overset{u,v^{d_y}}{\twoheadrightarrow} \mathcal{R}$$

i.e. after two causally independent occurrences of actions a and c, it is possible to reach a configuration $\mathcal{Q}$ where only an occurrence of action d, causally dependent to both occurrences of actions a and c, is possible.

We are now ready to introduce the notion of causality-based weak bisimulation.

4.2 Definition Let $\mathbf{R} \subseteq \mathfrak{P}_o \times \mathfrak{P}_o$ be a symmetric binary relation between configurations. Then, let $\mathbb{F}^w : \mathrm{Rel}(\mathfrak{P}_o) \to \mathrm{Rel}(\mathfrak{P}_o)$ be a function defined as follows: $\langle \mathcal{P}, \mathcal{Q} \rangle \in \mathbb{F}(\mathbf{R})$ iff

(i) whenever $\mathcal{P} \twoheadrightarrow \mathcal{P}'$, there exists $\mathcal{Q} \twoheadrightarrow \mathcal{Q}'$ such that $\langle \mathcal{P}', \mathcal{Q}' \rangle \in \mathbf{R}$.

(ii) whenever $\mathcal{P} \overset{M^{o_x}}{\twoheadrightarrow} \mathcal{P}'$, there exists $\mathcal{Q} \overset{M^{o_y}}{\twoheadrightarrow} \mathcal{Q}'$ such that $\langle z/x\mathcal{P}', z/y\mathcal{Q}' \rangle \in \mathbf{R}$ for some $z \notin (\phi(\mathcal{P}') - \{x\}) \cup (\phi(\mathcal{Q}') - \{y\})$.

$\mathbf{R}$ is called a *causality-based weak bisimulation* iff $\mathbf{R} \subseteq \mathbb{F}^w(\mathbf{R})$. If $\langle \mathcal{P}, \mathcal{Q} \rangle \in \mathbf{R}$ for some causality-based weak bisimulation $\mathbf{R}$, then $\mathcal{P}$ and $\mathcal{Q}$ are said to be *causality-based weakly bisimilar*, in symbols $\mathcal{P} \approx \mathcal{Q}$. □

Finally, let us remark that the concept of unique fixed point up to strong/weak bisimulation equivalence for strongly/weakly guarded behavior expressions can be naturally extended to configuration expressions.

5 Conclusion and Related Works

In this paper, we have been concerned with the development of an original approach aiming at defining a causality-based semantics for CCS. The approach relies on a simple intuition related to the expression of causal relationships among occurrences of actions. This intuition has been reflected by two basic notions, namely atoms and configuration expressions. It has been shown how to define an operational semantics, as well as strong and weak bisimulation equivalences for CCS within the proposed causality framework.

There have been earlier attempts for providing causality-based semantics for CSP/CCS-based calculi, see for instance [15, 24, 22, 21, 25, 1, 4, 2, 23, 12, 16, 14, 17]. Many of these attempts have been developed using nonstandard (sometimes complicated) mathematical formalization, and consequently they are far from exhibiting the elegance and simplicity of interleaving-based approaches. Two existing theories however, namely the causal trees of Darondeau & Degano and the location-based semantics of Boudol & Castellani & Kiehn & Hennessy present some similarities with the ideas presented in this paper, and are consequently discussed in more details in the following paragraphs.

5.1 Causal Trees

Causal trees [12, 13] are labeled trees generalizing synchronization trees [20]. In causal trees, a pair $\langle a, K \rangle$ is associated with each arc, let say arc n, where a is an action and K is a set of natural numbers. Each $i \in K$ acts as a backward pointer to the ith predecessor arc of n. Intuitively, arc n is interpreted as an event (an occurrence of action a) and K represents the set of its causative events. In this way, each path $\langle a_0, K_0 \rangle \langle a_1, K_1 \rangle \ldots \langle a_i, K_i \rangle \ldots$ represents a pomset (partial order multi-set) of actions $\{a_0, \ldots, a_i, \ldots\}$ corresponding to a possible execution of the concurrent system. The transitivity of the partial order is obtained by the transitive closure of sets K_i's. See for instance the causal tree of Figure 8 corresponding to CCS term $ab|c$.

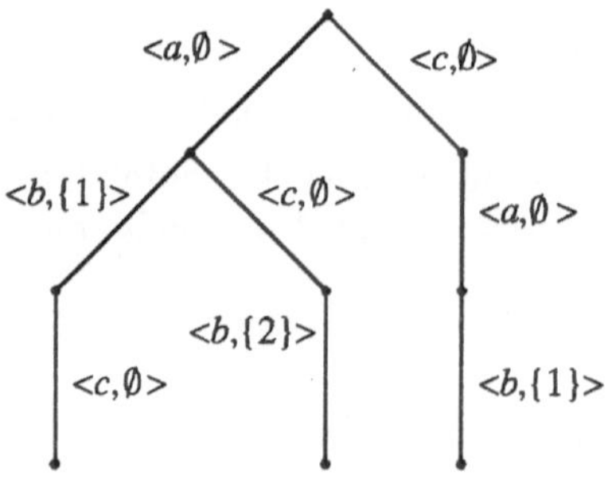

Figure 8

The approach followed in the paper is technically different from the causal tree approach, for two main reasons: the underlying algebraic structure is different, as we consider labeled transition systems instead of synchronization trees, and the way, the causal relationships among events are expressed, is also different, as we use event name abstractions (closely related to variable abstraction of Lambda-Calculus) instead of backward pointers. In spite of these differences, it is easy to be convinced, that both approaches are, as expected,

meaning that action a has been performed at a location named u. In this way, all possible occurrences of actions in sub-process $u :: P$ will be performed at some sub-location of u. Now, let $P = b|c$. Then, we have

$$u :: P \overset{b}{\underset{uv}{\rightarrow}} u :: (v :: \mathrm{Nil}|c) \overset{c}{\underset{uw}{\rightarrow}} u :: (v :: \mathrm{Nil}|w :: \mathrm{Nil})$$
$$u :: P \overset{c}{\underset{uw}{\rightarrow}} u :: (b|w :: \mathrm{Nil}) \overset{b}{\underset{uv}{\rightarrow}} u :: (v :: \mathrm{Nil}|w :: \mathrm{Nil})$$

as possible transition sequences. Note that b and c are performed at different spatial locations.

In spite of the fact that the notions of location and sub-process are technically rather different from our notion of determinant and configuration, location transitions are intuitively quite close with our notion of atom transitions. In the approach presented in this paper, however, no "spatial locality" for actions is assumed, i.e. all actions dwell in the same space location. Unfortunately, location semantics, as they have been currently formulated, present some difficulties when dealing with infinite behaviors.

Nevertheless, initial results have shown that it could be possible to extend our causality-based bisimulations in order to preserve the spatial distribution of processes, thus distinguishing behaviors such that R and S above.

Recently, A. Kiehn [18] has proposed a noninterleaving semantics for finite CCS using a method closely related to the one used in location-based semantics. In his approach, CCS terms have been extended with two additional operators: *local cause prefixing* $l :: P$ and *global cause prefixing* $A :: P$, where l is an event name, called *local cause*, and A is a set of event names, called *global causes*. Term $l :: P$ has the same meaning as in the location-based semantics, and term $A :: P$ is intuitively similar to our notion of local state. For this extended CCS calculus, an operational semantics based on the notion of local/global cause transitions has been proposed. A local/global cause transition is defined as $P \overset{a}{\underset{L,G,l}{\rightarrow}} P'$ and represents a possible occurrence of action a, with local causes L and global causes G. l is the event name associated with this occurrence of action a. Intuitively, the corresponding transition in our approach would be $P \overset{G \, a_l}{\rightarrow} P'$ (as $L \subseteq G$), which illustrates the technical similarity of the approaches. In fact, the mechanism by which the causal relationships among events are expressed in local/global cause-based semantics is closely related to the notion of atoms used in this paper and introduced in previous works of the authors [6, 7, 8, 11]. The consideration of local causes in the transition relation has permitted however to define a new equivalence notion, *local/global cause bisimilarity*, which is stronger than both causality-based and location-based bisimilarities. Nevertheless, presently, local/global cause-based semantics do not consider infinite behaviors; at least no treatment for dealing with recursive behaviors has been mentioned or presented in Kiehn's paper. For this, the approach can not yet be evaluated with respect to possible drawbacks in handling infinite behaviors, as it is the case for causal trees and location-based semantics. Normally, the same treatment for infinite behaviors used in this paper could also be applied to local/global cause-based approach without too many difficulties.

With respect to both approaches, location and local/global cause-based semantics, our claim is that the approach presented in this paper is simpler, and provides a purer (less syntax-oriented) mathematical treatment for dealing with the notion of concurrent labeled transition systems, and it has been successfully applied to all CCS/CSP constructs. Furthermore, as it has been proposed in previous authors' works [7, 11], CCS/CSP-based calculi can be extended with a new operator, closely related to the causality-based semantics presented in this paper, called *partial prefixing*, which permits to reduce the expressiveness gap between the syntactic and semantic models.

Although the results presented in this paper are far from being complete, it should already appear that the proposed approach is rather attractive as it provides a direct generalization of results obtained within an arbitrary interleaving framework. Future works will deal with the exploitation of the results introduced in this paper, the objective being to assess the advantages in considering true concurrency semantics for CCS/CSP-based Formal Description Techniques, like LOTOS [9].

References

[1] Boudol, G., and Castellani, I. Concurrency and atomicity. *TCS 59* (1988), 1–60.

[2] Boudol, G., and Castellani, I. Permutation of transitions: An event structure semantics for CCS and SCCS. In *Linear Time, Branching Time and Partial Order in Logics and Models for Concurrency*, J. de Bakker, W.-P. de Roever, and G. Rozenberg, Eds., vol. 354 of *LNCS*. Springer-Verlag, 1989, pp. 411–427.

[3] Boudol, G., Castellani, I., Kiehn, A., and Hennessy, M. Observing localities. In *Proc. of Mathematical Foundations of Computer Science*

1991, A. Tarlecki, Ed., vol. 520 of *LNCS*. Springer-Verlag, 1991, pp. 93–102.

[4] Castellani, I., and Hennessy, M. Distributed bisimulation. *J. ACM 36* (1989), 887–911.

[5] Coelho da Costa, R. J. PhD thesis, Université Paul Sabatier, Toulouse, 1992. To appear.

[6] Coelho da Costa, R. J., and Courtiat, J.-P. Definition of a new notation for representing pomsets (preliminary report). Rapport de Recherche 91121, LAAS, Mar. 1991.

[7] Coelho da Costa, R. J., and Courtiat, J.-P. POC: A partial order calculus for modeling concurrent systems. Rapport de recherche 91179, LAAS, May 1991.

[8] Coelho da Costa, R. J., and Courtiat, J.-P. SPOC – A simple partial order calculus for representing pomsets. In *Proceedings of the XVIII Integrated Seminar on Software and Hardware* (Santos/São Paulo - Brazil, Aug. 1991). Also available as Rapport de Recherche LAAS No. 91156 (May/1991).

[9] Coelho da Costa, R. J., and Courtiat, J.-P. A true concurrency semantics for LOTOS. In *Proc. of the IFIP TC6/WG6.1 5th Int. Conf. on Formal Description Techniques for Distributed Systems and Communication Protocols, FORTE'92* (Perros-Guirec, France, Oct. 1992), M. Diaz and G. R, Eds., North-Holland. Also available as Rapport de Recherche LAAS No. 92017, January/1992.

[10] Coelho da Costa, R. J., and Courtiat, J.-P. Using Petri nets as a model for Petri nets. In *Proceedings of Third IEEE Workshop on Future Trends of Distributed Computing Systems in the 1990's* (Taipei, Apr. 1992), IEEE Computer Society Press.

[11] Courtiat, J.-P., and Coelho da Costa, R. J. A LOTOS based calculus with true concurrency semantics. In *Proc. of the IFIP TC6/WG6.1 Fourth Int. Conf. on Formal Description Techniques for Distributed Systems and Communication Protocols, FORTE'91* (Sydney, Australia, Nov. 1991), G. A. Rose and K. R. Parker, Eds., North-Holland.

[12] Darondeau, P., and Degano, P. Causal trees. In *Proc. of 16th ICALP*, G. Ausiello, M. Dezani-Ciancaglini, and S. R. D. Rocca, Eds., vol. 372 of *Springer LNCS*. 1989, pp. 234–248.

[13] Darondeau, P., and Degano, P. Causal trees: interleaving + causality. In *Semantics of Systems of Concurrent Processes, Proc. of LITP Spring*

School on Theor. Comp. Sci., I. Guessarian, Ed., vol. 469 of *LNCS*. Springer-Verlag, 1990, pp. 239–255.

[14] Degano, P., De Nicola, R., and Montanari, U. A partial ordering semantics for CCS. *Theor. Comp. Sci. 75* (1990), 223–262.

[15] Degano, P., and Montanari, U. Concurrent histories: a basis for observing distributed systems. *J. of Comput. System Sci. 34* (1987), 422–461.

[16] Ferrari, G. L., and Montanari, U. Towards the unification of models for concurrency. In *Coll. on Algebra and Trees in Prog. (CAAP'90)* (Copenhagen, 1990). (LNCS 431, 162–176).

[17] Ferrari, G. L., and Montanari, U. The observation algebra of spatial pomsets. In *CONCUR'91 — 2nd Int. Conference on Concurrency Theory*, J. C. M. Baeten and J. F. Groote, Eds., vol. 527 of *LNCS*. Springer-Verlag, 1991, pp. 188–202.

[18] Kiehn, A. Local and global causes. Research Report 342/23/91, Technishe Universität Müchen, Institut für Informatik, Aug. 1991.

[19] Kiehn, A. Distributed bisimulation for finite CCS. CS Report 7/89, University of Sussex, Apr. (December) 1989.

[20] Milner, R. *A Calculus of Communicating Systems*, vol. 92 of *LNCS*. Springer-Verlag, 1980.

[21] Nielsen, M. CCS- and its relationship to net theory. In *Advanced Course in Petri Nets*, W. Brauer, W. Reisig, and G. Rozenberg, Eds., vol. 255 of *LNCS*. Springer-Verlag, Bad Honeff 1986, 1987, pp. 393–415.

[22] Olderog, E.-R. Operational Petri net semantics for CSP. In *Advances in Petri Nets 1987*, G. Rozenberg, Ed., vol. 266 of *LNCS*. Springer-Verlag, 1987, pp. 196–223.

[23] Taubner, D. The representation of CCS programs by finite predicate/transition nets. In X^{th} *International Conference on Application and Theory of Petri Nets* (Bonn-Germany, 1989), pp. 348–370.

[24] van Glabbeek, R., and Vaandrager, F. Petri net models for algebraic theories of concurrency. In *Proc. PARLE Conf.* (1987), A. N. J. W. de Bakker and P. Treleaven, Eds., vol. 259 of *LNCS*, Springer-Verlag.

[25] Winskel, G. Event structures. In *Advanced Course on Petri Nets*, vol. 255 of *LNCS*. Springer-Verlag, 1987, pp. 325–392.

A π–calculus Semantics of Logical Variables and Unification

Brian J. Ross
Department of Computer Science
Brock University
St. Catharines, Ontario, Canada L2S 3A1

Abstract

A π–calculus semantics of terms and logical variables, environment creation *visavis* term copying and variable refreshing, and sequential unification is presented. The π–calculus's object–oriented approach to modelling evolving communication structures is used to model the evolving communication environment found in concurrent logic program computations. The novelty of this semantics is that it explicitly models logic variables as active channels. These channels are referenced by π–calculus channel labels, and when used in concert with the ν restriction operator, model variable scopes and environments. Sequential unification without occurs check is modelled by traversing term expressions, and binding variables to terms as appropriate. The π–calculus is well-suited for this, as its object–oriented view of concurrency permits the modelling of the object passing and variable redirection that occurs during unification. This semantics is a central component of a more comprehensive operational semantics of concurrent logic programming languages currently being developed.

1 Introduction

The π–calculus is a process algebra suitable for modelling concurrent networks with evolving communicative structures [MPW89a, MPW89b, Mil91]. The π–calculus is similar to Milner's earlier CCS [Mil89], except that it is embellished with channel-label passing, which gives an object–oriented view of concurrency. The ability to treat channels as objects greatly enriches the descriptive power of the formalism, which can be seen by its modelling of the λ–calculus in [MPW89a].

This paper introduces a π–calculus semantics of logical variables and unification. The π–calculus is ideal for modelling concurrent logic programming languages. Concurrent logic programs are characterised as networks of concurrent processes which communicate with one another using shared logical variables. As computations proceed, the unification procedure binds logical variables to various term structures. These bindings alter the communication environment on which the interacting processes depend. The π–calculus concept of dynamic communication topologies is used to model the evolving communication environment found in concurrent logic program computations. This model serves as a basis on which more complex concurrent logic program control semantics can be built upon.

Section 2 reviews the logic programming domain being modelled in the paper, as well as some basic ideas behind the π–calculus. A π–calculus semantics of terms and logical variables, environments, and sequential unification is given in section 3. Some examples are in section 4. A discussion concludes the paper in section 5.

2 Review

2.1 Some logic programming concepts

The semantics in this paper models the data domain used in logic programming and term rewriting. This section will briefly review the essentials needed for developing an intuition of the domain being modelled. The following is found in greater depth in [Llo84].

Terms are defined recursively as follows: (i) a variable is a term; (ii) a constant is a term; (iii) if f is an n-ary function, and $t_1, ..., t_n$ are terms, then $f(t_1, ..., t_n)$ is a term. Constants are often considered to be functions of arity 0. Variables are written in upper-case, and constants and functions in lower-case. A shorthand notation for a term of arity > 0 is $f(\tilde{t})$. The variables in a term $\tilde{t}$ are denoted $Vars(\tilde{t})$.

A substitution θ is a finite set of the form $\{v_1 \leftarrow t_1, ..., v_n \leftarrow t_n\}$, where the v_i's are distinct variables and the t_i's are terms. Each $v_i \leftarrow t_i$ is called a binding for v_i. If E is a term, then $E\theta$ is the term obtained from E by simultaneously replacing each occurence of the variable v_i in E by the term t_i as found in θ. A variable-pure substitution is one where all the t_i's are variables.

Let $\theta = \{u_1 \leftarrow s_i, ..., u_m \leftarrow s_m\}$ and $\sigma = \{v_1 \leftarrow t_i, ..., v_m \leftarrow t_m\}$. The composition $\theta\sigma$ is the substitution obtained from the set

$$\{u_1 \leftarrow s_1\sigma, ..., u_m \leftarrow s_m\sigma, v_1 \leftarrow t_1, ..., v_n \leftarrow t_n\}$$

by deleting any binding $u_i \leftarrow s_i\sigma$ for which $u_i = s_i\sigma$ and deleting any binding $v_j \leftarrow t_j$ for which $v_j \in \{u_1, ..., u_m\}$.

Let E be a term and V be the set of variables occurring in E. A renaming substitution for E is a variable-pure substitution $\{x_1 \leftarrow y_1, ..., x_n \leftarrow y_n\}$ such that $\{x_1, ..., x_n\} \subseteq V$, the variables y_i are distinct, and $(V - \{x_1, ..., x_n\}) \cap \{y_1, ..., y_n\} = \emptyset$.

Let S be a finite set of terms. A substitution θ is called a unifier for S if $S\theta$ is a singleton. A unifier θ for S is called a most general unifier (mgu) for S if, for each unifier σ of S, there exists a substitution γ such that $\sigma = \theta\gamma$. For example, letting X, Y be variables, $\{p(f(X), Z),\ p(Y, a)\}$ has the mgu $\theta = \{Y \leftarrow f(X), Z \leftarrow a\}$, and their common instance is $p(f(X), a)$. On the other hand, $\{p(f(X), a),\ p(Y, f(c))\}$ is not unifiable, as the second arguments cannot unify.

When computing a unifier for a set S, the disagreement set of S is computed in the following way. Locate the leftmost symbol position at which not all expressions in S have the same symbol, and extract from each expression in S the subexpression beginning at that symbol. The set of all such subexpressions is the disagreement set.

An algorithm for determining the most general unifier is the following. S denotes a finite set of terms.

1. Put $k = 0$ and $\sigma_0 = \epsilon$ (the empty subsitution).

2. If $S\sigma_k$ is a singleton, then stop with σ_k as the mgu of S. Otherwise find the disagreement set D_k of $S\sigma_k$.

3. If there exist v and t in D_k such that v is a variable that does not occur in t, then put $\sigma_{k+1} = \sigma_k\{v \leftarrow t\}$, increment k, and go to step 2. Otherwise S is not unifiable, so stop.

The test in step 3 for occurence of a variable v in the term t is called an *occur check*. This step is necessary for preserving the soundness of unification. However, it is computationally expensive, and is normally ignored in Prolog [CM81].

2.2 The π–calculus

The monadic π–calculus of [MPW89a, MPW89b] will be used. We will embellish it with some devices from a later incarnation in [Mil91], and with some notational conveniences from CCS [Mil89]. The reader is referred to these sources for detailed treatments.

Like CCS, the π–calculus is a process algebra which models concurrency via interleaving. The π–calculus differs from CCS in its ability to pass labels as objects along specified channels. These labels can be either constants or other channel labels; the π–calculus does not distinguish label usage. For example, the expression

$$(\nu s)\ s(x).P \mid \bar{s}y.Q$$

passes label y (on the right) through channel s to the prefix of P. After applying a reduction rule (described below), the expression reduces to:

$$(\nu s)\ P\{y/x\} \mid Q$$

The label y replaces x in the expression P. The ν term hides s in the expression. The power of the π–calculus comes from the ability to pass channels themselves. For example, the expression

$$\begin{array}{lll} & (\nu s\ y)\ s(x).\bar{x}z.P \mid \bar{s}y.Q \mid y(w).R & (1) \\ \text{reduces to} & (\nu s\ y)\ \bar{y}z.P' \mid Q \mid y(w).R & (2) \\ \text{and then to} & (\nu s\ y)\ P' \mid Q \mid R' & \end{array}$$

where $R' \equiv R\{z/w\}$. This represents the passing of channel label y as an object in (1), and then using it to transfer label z in (2).

The basic syntax and semantics of the π–calculus is as follows. Let the names $x, y ... \in \mathcal{X}$, the agents $P, Q, ... \in \mathcal{P}$, and K range over agent identifiers. Then an agent P is inductively defined by

$$P ::= \sum_{i \in I} \pi_i.P_i \mid P \mid Q \mid\ !P \mid (\nu x)P \mid [x = y]P \mid K(\tilde{y})$$

[Mil91] defines a transition relation $\rightarrow$ using one reduction rule and three inference rules (figure 1). This lean semantic definition is made possible by using

$$\textbf{Comm}: \quad (\cdots + x(y).P) \mid (\cdots + \overline{x}z.Q) \rightarrow P\{z/y\} \mid Q$$

$$\textbf{Par}: \quad \frac{P \rightarrow P'}{P|Q \rightarrow P'|Q} \qquad \textbf{Res}: \quad \frac{P \rightarrow P'}{(\nu x)P \rightarrow (\nu x)P'}$$

$$\textbf{Struct}: \quad \frac{Q \equiv P \quad P \rightarrow P' \quad P' \equiv Q'}{Q \rightarrow Q'}$$

Figure 1: Reduction and inference rules for π–calculus

a structural congruence relation $\equiv$ over expressions. Some examples of $\equiv$ are

$$\begin{array}{rcll} !P & \equiv & P \mid !P & \\ (\nu x)\,\mathbf{0} & \equiv & \mathbf{0} & \\ (\nu x)\,(\nu y)\,P & \equiv & (\nu y)\,(\nu x)\,P & \\ (\nu x)\,(P \mid Q) & \equiv & P \mid (\nu x)\,Q & : if \ x \ not \ free \ in \ P \end{array}$$

A detailed semantic account of these π–calculus operators is beyond the scope of this review. An informal description follows.

A summation expression $\sum_{i \in I} \pi_i.P_i$ denotes a choice of possible behaviours in an expression. For example, in $P + Q$, both P and Q are alternate behaviors. I is a finite indexing set. If $I = \emptyset$, then the sum is $\mathbf{0}$, which is the empty process denoting inactivity. The term π is an atomic action, and has one of two forms:

$x(y)$	binds the input from link x to y
$\overline{x}y$	output y on link x

Channel communications which contain empty data result in CCS–style actions. For example, in

$$x.P \mid \overline{x}.Q$$

the action x is treated as a channel communication having no object. The $\mid$ operator denotes concurrent composition, and it models interleaved streams with message passing as generated by its constituent agent arguments. As with CCS, a hidden τ action denotes a handshake between two processes whose matching π prefixes are of opposite polarities. The semantics of $\mid$ accounts for the passing of labels on channels within π atoms. Infinite replication is denoted by $!P$. This expression denotes the infinite generation of P:

$$!P \; = \; P \mid P \mid \cdots \mid P \mid !P$$

Label restriction is denoted by (νx), where x is one or more label names. Matching is denoted by $[x = y]P$. The agent P proceeds only as long as the labels contained in x and y are identical. A useful abbreviation for a set of matches is:

$$x : [y_1 \Rightarrow A, y_2 \Rightarrow B, ..., else \Rightarrow Z] \quad \equiv \quad [x = y_1]A \mid [x = y_2]B \mid \cdots \mid \forall_i[x \neq y_i]Z$$

where all y_i are unique. An agent definition is invoked via the expression $K(\tilde{y})$, where K ranges over agent labels. Associated with each agent identifier K is an expression

$$K(x_1, ..., x_k) \stackrel{\text{def}}{=} E$$

where the x_i labels are distinct.

A notational convenience is the use of a CCS–style action renaming function. This can be implemented in the π–calculus as:

$$A[x/y] \equiv (\nu y)\,(A \mid !(y(w).\overline{x}w))$$

In addition, some useful abbreviations are:

$$\begin{array}{lcl} \pi & \equiv & \pi.0 \\ x(y_1 \cdots y_n) & \equiv & x(w).w(y_1).\cdots.w(y_n) \\ \overline{x}y_1 \cdots y_n & \equiv & (\nu w)\,\overline{x}w.\overline{w}y_1.\cdots.\overline{w}y_n \end{array}$$

Finally, two sequencing operators will be used in the paper:

$$\begin{array}{rcl} P\ Before\ Q & \stackrel{\text{def}}{=} & (\nu\, d)\,(P[d/done] \mid d.Q) \\ P \bullet Q & \stackrel{\text{def}}{=} & (\nu\, t')\,(P[t'/t] \mid t'.Q) \end{array}$$

In $Before$ (used in the appendix), P must produce the action $\overline{done}$ as the last action before it terminates, which triggers Q's execution. The $\bullet$ operator is like $Before$, but sequences on t.

3 Semantics

3.1 Terms

Conceptually, the labels used by the semantics fall into one of three categories:

(i) User–defined constants $\mathcal{C}$
(ii) Reserved constants $\mathcal{R} = \{\phi, \xi, t, f, get, free\}$
(iii) Channel labels $\mathcal{X}$

The set of user-defined constants $\mathcal{C}$ is a finite predefined set built from the constants in a logic program. The reserved labels $\mathcal{R}$ are distinguished labels, and can be considered to be uniquely coloured to separate them from the elements of $\mathcal{C}$. The ϕ label is an argument tuple delimiter, and ξ is used in the definition of logical variables. The labels t and f denote logical true and false respectively, and in the context of this paper can be taken to mean success and failure. get and $free$ are semaphore signals. Channel names are generic labels. Labels are often indexed (eg. u_2, t').

The denotation of terms and data structures is similar to that used in [MPW89a], except that, instead of using a Lisp–style list structure, a flat static–length graph data structure is adopted. This static structure suffices because pure logic program terms have static arities. Letting $\mathrm{k} \in \mathcal{C}$, t_i be terms, x be a logical variable term, and $[\![\ ]\!]_t$ be a term translation function, then terms are recursively modelled in figure 2. The u argument is used as a channel on which to transmit the term structure. The communication of ϕ delimits the end of the term. The replication "!" is used so that terms can be repeatedly read by other expressions; otherwise, it is ephemerally read once. The figure shows a partial translation of logical variables that is completed in section 3.2.

Let E be an agent identifier, and let the term t be an intended argument. The following abbreviation is convenient.

$$E(\ldots,\underline{t},\ldots) \equiv (\nu u)\, E(\ldots,u,\ldots) \mid [\![t]\!]_t(u) \qquad \textit{for some new } u$$

3.2 Logical Variables and Environments

An environment is a collection of computational mechanisms which affect the course of a computation. In this paper, an environment incorporates two aspects: (i) general computational mechanisms (processes, programs), and (ii) the context within which logical variables are defined. Environments therefore include a representation of process memory, and determine the scope or communication limits of logical variables. Semantically, an environment is an expression which contains logical variable definitions and their scopes, as well as any other mechanisms which can access these variables.

The view taken here is that logical variables are channels on which reading and writing can take place. These communications are restrained according to the state of the variable, and variables are considered to be "write–once". If one reads from an unbound logical variable, then some sort of communication will be given by the variable to indicate its unbound state. On the other hand, once a variable is bound to a term, then reading from it results in a communication of that term. Similarly, writing a term to an unbound variable results in its being bound to it. Such a term can be a data structure or another variable, and if the latter, a notion of memory or structure sharing is required.

A logical variable has two states, unbound and bound, which correspond respectively to the two agents Var and Set in figure 3. Variable channels define an active environment, in the sense that they actively communicate with other mechanisms. Variables are initialized in the Var state, which occurs with the use of $NewVar$ during the environment definition. They stay in Var until they are bound to a non–variable term, at which time they are managed by Set. In both states, the label x is a bidirectional channel. Reading from the channel will result in either the "unbound" signal ξ being emitted in Var, or

$$[\![\mathrm{k}(t_1,\ldots,t_n)]\!]_t(u) = (\nu v x_1 \ldots x_n)\, !\,\overline{u}v \mid !\overline{v}\,\mathrm{k} x_1 \ldots x_n \phi \mid [\![t_1]\!]_t(x_1) \mid \cdots \mid [\![t_n]\!]_t(x_n) \quad (n > 0)$$

$$\textit{where } x_i \equiv \begin{cases} \textit{new label } v_i & : t_i \textit{ is non-variable term} \\ t_i & : t_i \textit{ is a variable} \end{cases}$$

$$[\![\mathrm{k}]\!]_t(u) = (\nu v)\, !\,\overline{u}v \mid !\overline{v}\,\mathrm{k}\phi$$

$$[\![x]\!]_t(u) = \begin{cases} NewVar(x) & : \textit{variable } x \textit{ not translated yet} \\ (\textit{nothing}) & : \textit{otherwise} \end{cases}$$

Figure 2: Term translation

$$
\begin{array}{lcl}
NewVar(x) & \stackrel{\mathrm{def}}{=} & (\nu w)\, Var(x,w) \mid !\overline{w}\xi \\
Var(x,y) & \stackrel{\mathrm{def}}{=} & \overline{x}y.Var(x,y) + x(w).w(z).z : [\xi \Rightarrow Var(x,w), \\
 & & \qquad\qquad\qquad\qquad\qquad\quad else \Rightarrow Set(x,w)] \\
Set(x,y) & \stackrel{\mathrm{def}}{=} & \overline{x}y.Set(x,y)
\end{array}
$$

Figure 3: Logic variable channels

the term structure being communicated in *Set*. When in the *Var* state, a variable can be reset to another unbound variable an indefinite number of times. It can only be set to a non-variable term structure once. This can be seen in the definition of *Set*, which disallows any further writing. An important feature of the denotation of logical variables is that the variable channels indirectly point to their bound structures. This will be the basis for unification in section 3.3.

When translating a term, each logical variable will have a unique bidirectional channel defined for it, which is denoted a private channel label. Each of these channels has an associated agent expression which suitably defines the communication characteristics of logical variables in the language being modelled. Consider a term $\mathtt{k}(\tilde{t})$ that contains within it a set of logical variables $\{v_1, ..., v_n\}$. The semantic translation of $\mathtt{k}(\tilde{t})$ is:

$$[\![\, \mathtt{k}(\tilde{t})\,]\!](u) \;=\; (\nu v_1 ... v_n)\, [\![\, \mathtt{k}(\tilde{t})\,]\!]_t(u) \;\mid\; NewVar(v_1) \;\mid \cdots \mid\; NewVar(v_n)$$

NewVar (see figure 3) initializes a new logical variable. In $NewVar(v_i)$, the label v_i denotes the variable channel used to access the logical variable, and will initially communicate ξ, which denotes its being uninitialized. The restriction operator ν directly denotes variable scope, thereby defining a local environment. If this restriction is removed, then these variables can communicate outside of this expression.

Multiple environments are easily represented. Two terms $\mathtt{t}(u)$ resident in separate environments are denoted:

$$((\nu u)\, [\![\, \underline{\mathtt{t}(u)}\,]\!]_t \mid \mathcal{E}_1) \mid ((\nu u)\, [\![\, \underline{\mathtt{t}(u)}\,]\!]_t \mid \mathcal{E}_2)$$

Restriction delimits the scope as expected, and recursive definition of such terms within agent expressions are handled by label renaming in the π–calculus.

Environments are normally defined by the structure of a program. When defining the semantics of utilities such as unification, however, it is useful to be able to control the creation of environments to some extent. In particular, the ability to rename variables is useful, as it permits tentative operations to be performed on terms without destroying the originals. This is akin to creating temporary local memory, and copying desired terms into it for manipulation and testing. A term copying operator $\leadsto$ is defined for this purpose. Letting $\tilde{t}$ range over term expressions, then

$$\textbf{Copy} \quad \overline{u \leadsto v \mid [\![\, \tilde{t}\,]\!]_t(u) \mid \mathcal{E} \;\stackrel{\tau}{\rightarrow}\; [\![\, \tilde{t}\,]\!]_t(u) \mid [\![\, \tilde{t}\,]\!]_t(v) \mid \mathcal{E}}$$

Here, u is a channel which communicates some term to be copied, and v is to be a channel on which the copied term is to be communicated. After the transition, the environment is supplemented by the definition of a copy of this term on channel v, which uses fresh variables as part of the term retranslation. The net result is that a new term is output on channel v, which is identical to the original term on channel u, except that it refers to a structure with new variables.

The intuitive definition of $\leadsto$ in **Copy** above is an abstract account of its behavior. The $\leadsto$ operator is in fact modellable in the π–calculus, and one possible definition for it is in appendix A. (This definition of $\leadsto$ is not particularly pleasing, however, and a more aesthetic definition is desirable.) The idea behind any implemenation of term copying is to recursively traverse the term as translated in section 3.1, and while doing so, communicate a copy which reflects the original term's structure, while generating refreshed variables.

3.3 Unsafe sequential unification

The theoretical elegance of unification belies the complex procedural considerations encountered when deriving an operational semantics for it. An $=$ agent performs sequential term unification without occurs check. It takes two terms as arguments, and attempts to unify them in three steps:

1. The argument structures are copied into "safe" private memory.
2. Unification is attempted on the private copies.
3. If step 2 is successful, the original structures are unified. Otherwise, the whole unification step fails.

Given that logic variables are modelled as non-invertable channels, bindings to them are permanent. We do not want to bind logic variables unless it is certain that doing so is indeed desired. Therefore, in step 2, copies of the terms are first unified, which is a conditional pre–unification performed before the actual one.

The intended behaviour of $=$ is defined by these operator–level transitions:

$$\textbf{Unify}: \quad \frac{}{\underline{\tilde{t}_1} = \underline{\tilde{t}_2} \mid \mathcal{E} \xrightarrow{\tau.\bar{t}} \mathcal{E}\theta} \qquad\qquad \textbf{Fail}: \quad \frac{}{\underline{\tilde{t}_1} = \underline{\tilde{t}_2} \mid \mathcal{E} \xrightarrow{\tau.\bar{f}} \mathcal{E}}$$

$\mathcal{E}$ is the environment containing all the other logical variables executing concurrently with the unification operation. *Unify* models the case when the terms unify, in which case $\bar{t}$ denoting success is first transmitted, followed by the incorporation of the unifying substitution θ into the environment. Failure does not change $\mathcal{E}$, and is denoted by $\bar{f}$.

The $=$ definition is in figure 5. The terms referenced by u and v are first copied with $\leadsto$, and these copies are unified using $\stackrel{2}{=}$, which does the actual term traversing and binding of logical variables. If this conditional unification succeeds, then the original terms are unified. Otherwise, the terms are not unifiable, and the unification step ends in failure. In $\stackrel{2}{=}$, α ranges over $\mathcal{C}$. The $\stackrel{2}{=}$ operator performs a unification check on one level of the term structures of the arguments in u and v. Firstly, if x and y (the memory references) are the same, then unification holds trivially. Otherwise, the terms are read, and unification proceeds on the structures. For example, the first $\xi \Rightarrow \overline{u}y.\bar{t}$ expression means

$$u = v \stackrel{\text{def}}{=} (\nu u'v't')\, u \rightsquigarrow u' \mid v \rightsquigarrow v' \mid u' \stackrel{2}{=} v'\, [t'/t] \mid t'.\bar{t}.(u \stackrel{2}{=} v)$$

$$\begin{aligned} u \stackrel{2}{=} v \stackrel{\text{def}}{=} & u(x).v(y).x : [\ y \Rightarrow \bar{t}, \\ & \quad else \Rightarrow x(a).y(b).a : [\ \xi \Rightarrow \overline{u}y.\bar{t}, \\ & \qquad \alpha \Rightarrow b : [\ \xi \Rightarrow \overline{v}x.\bar{t}, \\ & \qquad\quad \alpha \Rightarrow (x \stackrel{\tau}{=} y), \\ & \qquad\quad else \Rightarrow \overline{f}\ ], \\ & \qquad \phi \Rightarrow b : [\ \phi \Rightarrow \bar{t},\ else \Rightarrow \overline{f}\ ]\ ] \end{aligned}$$

$$\begin{aligned} a \stackrel{\tau}{=} b \stackrel{\text{def}}{=} & a(x).b(y).x : [\ \phi \Rightarrow y : [\phi \Rightarrow \bar{t},\ else \Rightarrow \overline{f}], \\ & \quad else \Rightarrow y : [\ \phi \Rightarrow \overline{f},\ else \Rightarrow (x \stackrel{2}{=} y) \bullet (a \stackrel{\tau}{=} b)\ ]\ ] \end{aligned}$$

Figure 4: Unification

that the term at u is unbound, and therefore unifies with the other term; the other term is written to the channel u, thus unifying them. The rest of $\stackrel{2}{=}$ unifies the terms in a case-by-case fashion, and is fairly self-explanatory. The $\stackrel{\tau}{=}$ operator recursively applies $\stackrel{2}{=}$ to the non–empty tuples of arguments for each term. This is only done if both u and v have the same function name α, which ranges over program constants $\mathcal{C}$. Arity discrepancies result in failure. The correctness of the definitions in figure 5 can be verified by structural induction over terms to be unified.

Note that the recursive expression $(x \stackrel{2}{=} y) \bullet (a \stackrel{\tau}{=} b)$ in $\stackrel{\tau}{=}$ is sequential. This could be made parallel if more sophisticated environment contention schemes are modelled.

3.4 Atomic (safe) unification

$$u \stackrel{s}{=} v \stackrel{\text{def}}{=} (\nu t'f')\, \overline{get}.(\ u = v\ [t'/t, f'/f] \mid f'.\overline{f}.\overline{free}\ +\ t'.\bar{t}.\overline{free})$$

Figure 5: Atomic Unification

The definition of $=$ in section 3.3 is not safe. Problems arise when more than one mechanism tries to access and alter the environment simultaneously. This is because the accessing of terms requires traversal of the term structures and accessing of logical variable channels, which are not instrinsically atomic operations.

A safe atomic unification operator $\stackrel{s}{=}$ is defined in figure 6. This operator uses two semaphore signals, *get* and *free*, which permit the locking of the environment during the unification of terms by $=$. The following semaphore can be used in concert with agents which are performing concurrent unifications on the same environment:

$$Semaphore \stackrel{\text{def}}{=} get.free.Semaphore$$

An example of semaphore usage is in section 4.2.

4 Examples

4.1 Single unification call

Given two terms to be unified, if either term contains a variable reference, then the variable channel is defined in the environment $\mathcal{E}$. During unification, when a variable is bound to a new term, its variable channel is set to communicate this bound object. $\mathcal{E}$ will contain channel agents for all the logic variables which, after a successful unification step, will be adjusted to communicate their newly bound objects. To see the results of unification, one inspects the states of these channels by reading from them.

The following is an example of how unification affects the environment. Consider the unification of $t(X, b(Y), a(c))$ and $t(Z, W, a(Z))$. This is denoted:

$$\underline{\mathtt{t}(x, \mathtt{b}(y), \mathtt{a}(\mathtt{c}))} = \underline{\mathtt{t}(z, w, \mathtt{a}(z))} \mid NewVar(x) \mid NewVar(y) \mid NewVar(w) \mid NewVar(z) \mid \mathcal{E} \qquad (\dagger)$$

where $\mathcal{E}$ are other various agents in the environment. Before unification, querying any of the channels $\{x, y, w, z\}$ results in the broadcast of the variable's current state. For example, querying x with the expression

$$x(a).a(b).\overline{out}b \mid NewVar(x)$$

results in ξ being broadcast on out. During unification, $=$ binds the variable agents to their unifying terms. Given that the binding substitution for the above is $\theta = \{X \leftarrow Z,\ W \leftarrow b(Y),\ Z \leftarrow c\}$, then the new environment after unification is:

$$(\dagger) \stackrel{\tau.\bar{t}}{\rightarrow} Set(x, z) \mid NewVar(y) \mid Set(w, \underline{\mathtt{b}(y)}) \mid Set(z, \underline{\mathtt{c}}) \mid \mathcal{E}$$

Conceptually, this is equivalent to a new environment $\mathcal{E}'$, that has incorporated within it $\mathcal{E}\theta$.

4.2 Concurrently competing unifications

The $\stackrel{s}{=}$ agent incorporates a simple semaphore mechanism for synchronizing unifications which might compete to unify shared variables. Two semaphore signals, $\overline{get}$ and $\overline{free}$, are defined within $\stackrel{s}{=}$, and the following semaphore can be used in conjunction with a call to it:

$$Semaphore \stackrel{\text{def}}{=} get.free.Semaphore$$

Now $\stackrel{s}{=}$ will only proceed to unify two terms when access is permitted using get. Consider the expression:

$$(\nu\, get\ \ free)\ \underline{t(X)} \stackrel{s}{=} \underline{t(c)} \mid \underline{s(X)} \stackrel{s}{=} \underline{s(d)} \mid Semaphore \mid \mathcal{E} \qquad (\ddagger)$$

Both these unifications refer to the same X in the environment $\mathcal{E}$. If a contention scheme is not implemented, then it is possible for X to be simultaneously bound to conflicting terms. With a semaphore, only one of the two unifications will be allowed to access $\mathcal{E}$ at one instant in time. Assuming that X is uninitialized, then expanding ($\ddagger$):

$$(\ddagger) \ = \ (\nu\, get\ free)\ (\mathcal{E}\theta_1 \mid Semaphore) \ + \ (\mathcal{E}\theta_2 \mid Semaphore)$$

where $\theta_1 = X \leftarrow c$ and $\theta_2 = X \leftarrow d$.

When using multiple environments, separate unifications can be denoted in a single expression:

$$(\nu\ get\ free\ V_1)(\tilde{s}_1 = \tilde{s}_2 \mid \tilde{s}_3 = \tilde{s}_4 \mid Semaphore \mid \mathcal{E}_1) \mid$$
$$(\nu\ get\ free\ V_2)(\tilde{t}_1 = \tilde{t}_2 \mid \tilde{t}_3 = \tilde{t}_4 \mid Semaphore \mid \mathcal{E}_2)$$

where $V_i = Vars(\mathcal{E}_i)$. Having separate restrictions on the V_i means that they are mutually exclusive sets of variables ($V_1 \cap V_2 = \emptyset$).

4.3 Merging streams

Parallel logic programs often use logical variables to implement streams (eg. [Sha87], volume 1, part III). This example shows how the π–calculus model of logical variables represent streams, and how two streams can be merged.

Consider the following logic program clause:

$$p([1|W]) :- p(W).$$

Without going into details of the operational semantics of logic programs, when clause p is queried with an unbound logic variable, an infinite stream of 1's is generated on W. For example, the query "? – $p(X)$." results in the infinite list $[1,1,1,1,...]$ being bound onto X. This occurs because the expression $p(W)$ is a recursive call to p which binds W to the term $[1|W']$ in the recursive call, and this carries on *ad infinitum*[1].

A π–calculus translation of p is

$$P(u) \stackrel{\text{def}}{=} (\nu w)\ s = \underline{[1,w]} \mid P(w) \mid NewVar(w)$$

which is invoked by a query expression,

$$P(s) \mid NewVar(s)$$

Each invocation of agent P results in a new environment containing a fresh variable w. The query expression evolves as follows:

$$\begin{array}{lrl}
 & (\nu tf) & P(s) \mid NewVar(s) \\
= & (\nu tfw) & s = \underline{[1,w]} \mid P(w) \mid NewVar(w) \mid NewVar(s) \\
\xrightarrow{\tau} & (\nu tfw) & P(w) \mid NewVar(w) \mid Set(s, \underline{[1,w]}) \\
\xrightarrow{\tau} & (\nu tfww') & P(w') \mid NewVar(w') \mid Set(w, \underline{[1,w']}) \mid Set(s, \underline{[1,w]}) \\
\xrightarrow{\tau} & (\nu tfww'w'') & P(w'') \mid NewVar(w'') \mid Set(w', \underline{[1,w'']}) \mid \\
 & & \qquad Set(w, \underline{[1,w']}) \mid Set(s, \underline{[1,w]}) \\
\xrightarrow{\tau} & & \ldots
\end{array}$$

The environment being built reflects the binding $\{S \leftarrow [1,1,1,...]\}$.

Consider two streams as above, where P generates an infinite number of 1's, and Q an infinite number of 2's. The task is to nondeterministically merge these streams into a stream w,

$$(\nu tfuv)\ P(u) \mid Q(v) \mid Merge(u,v,w) \mid \mathcal{E}_0$$

where $\mathcal{E}_0$ defines u, v, and w. One model for $Merge$ is:

[1]The notation $[X|Y]$ is a standard logic program abbreviation for $'.'(X,Y)$, where '.' is a list constructor functor, X is a list element, Y is the tail, and $\emptyset$ is a null list. Thus, $[1,2,3] \equiv '.'(1,'.'(2,'.'(3,\emptyset)))$.

$$Merge(u,v,w) \stackrel{\text{def}}{=} (\nu t_1 t_2 f x a z)!Nonvar(u)[t_1/t] \mid !Nonvar(v)[t_2/t] \mid \\ (t_1.(u = \underline{[x|a]} \mid w = \underline{[x|z]} \mid Merge(a,v,z)) \\ + t_2.(v = \underline{[x|a]} \mid w = \underline{[x|z]} \mid Merge(u,a,z))) \mid \mathcal{E}_1$$

where $\mathcal{E}_1$ defines x, a, and z, and

$$Nonvar(u) \stackrel{\text{def}}{=} u(x).x(y).y : [\xi \Rightarrow \overline{f},\ else \Rightarrow \overline{t}]$$

The complication in merging these two streams arises from the fact that the stream variables from P and Q might remain unbound when $Merge$ is invoked. This is because P, Q, and $Merge$ execute in parallel, with no ordering implicit in their communication to one another. $Merge$ is therefore designed so that it only merges bound streams, which is the case when u or v are bound. This is determined by $Nonvar$. The merged stream on w is built iteratively, by non-deterministically choosing bound stream elements from P and Q, and inserting them to the front of w.

5 Discussion

This π–calculus semantics is a central component of a fuller semantics of concurrent logic program languages currently being developed [Ros92a]. The motivation for using the π–calculus is to investigate its appropriateness a kernel language for modelling a wide variety of concurrent logic program phenomena. We plan to use this semantics to refine efficient implementations of concurrent logic languages. So far, the active environment model of this paper is useful for modelling the communication which occurs between processes and logic variables. In addition, the π–calculus can model the control constructs of these languages, a flavour of which can be seen in the merge example of section 4.3. Another avenue being investigated is the modelling of various parallel unification algorithms, rather than the sequential unification done here. To do this, a more detailed contention scheme is necessary. One possibility is to put semaphores around individual logical variables, instead of entire local environments.

The domain modelled here is a simple one, and actual implementations of concurrent logic languages are considerably more complex due to efficiency considerations. For example, many committed logic languages use various variable protection schemes, such as in GHC [Ued86], which does not permit variables to be instantiated in the guard. This could be modelled in the π–calculus by redefining the semantics of logical variable channels, and requires an appropriately redefined unification algorithm. The semantics given here *could* be refined into more sophisticated models which are more amenable to efficient implementation on particular hardware. Given that the π–calculus has a well–founded semantics, encoding abstract machines in it can permit formal analyses of language design. Abstract machines such as that in [Lin84] shares similarities with ours, and could be encoded in the π–calculus if desired.

Two other process algebra models of concurrent logic programs are by Belmesk and Habbas [BH92] and de Boer and Palamidessi [dBP92], and a related approach is by Saraswat and Rinard [SR90]. All these papers take the Herbrand domain and unification to be abstract concepts which are used directly in semantic axioms. This paper differs by modelling terms and unification at the lower level of terms using a kernel of basic π–calculus operators, while still permitting their level of abstraction if desired. Such a level of description is needed when modelling operational characteristics of concurrent languages such as memory contention – concepts which are transparent when too abstract a view is taken.

Beckman uses CCS to model concurrent logic languages [Bec87], and Ross applies CCS to sequential Prolog [Ros92b]. In both these papers, CCS's value passing is not rich enough for directly modelling logical variables. As a result, Beckman redefines the composition "|" operator so that binding substitutions are automatically distributed to the entire environment. This paper shows how such a data domain can be implemented in a concurrent environment. One other related work is by Walker [Wal90], who uses the π–calculus for modelling imperative object–oriented programming languages.

Acknowledgements: Thanks to Robin Milner for hints on how to model logical variables, and to Robert Scott for his helpful advice on concurrent logic programming. This research was done at the University of Victoria (Canada).

References

[Bec87] L. Beckman. *Towards an Operational Semantics for Concurrent Logic Programming Languages.* PhD thesis, Uppsala University, 1987.

[BH92] M. Belmesk and Z. Habbas. A Process Calculus with Shared Variables. *Journal of Computers and Artificial Intelligence*, 1992. (to appear).

[CM81] W.F. Clocksin and C.S. Mellish. *Programming in Prolog.* Springer-Verlag, 1981.

[dBP92] F.S. de Boer and C. Palamidessi. A process algebra of concurrent constraint programming. In *Joint International Conference and Symposium on Logic Programming*, Washington, D.C., 1992. MIT Press.

[Lin84] G. Lindstrom. OR-parallelism on applicative architectures. In *2nd International Logic Programming Conference*, Uppsala, 1984.

[Llo84] J.W. Lloyd. *Foundations of Logic Programming.* Springer-Verlag, 1984.

[Mil89] R. Milner. *Communication and Concurrency.* Prentice Hall, 1989.

[Mil91] R. Milner. The Polyadic π-Calculus: A Tutorial. Technical Report ECS-LFCS-91-180, LFCS, U. of Edinburgh, 1991.

[MPW89a] R. Milner, J. Parrow, and D. Walker. A Calculus of Mobile Processes, Part I. Technical Report ECS-LFCS-89-85, LFCS, U. of Edinburgh, 1989.

[MPW89b] R. Milner, J. Parrow, and D. Walker. A Calculus of Mobile Processes, Part II. Technical Report ECS-LFCS-89-86, LFCS, U. of Edinburgh, 1989.

[Ros92a] B.J. Ross. A π–calculus Semantics of Committed Logic Program Control, 1992. In preparation.

[Ros92b] B.J. Ross. *An Algebraic Semantics of Prolog Control.* PhD thesis, University of Edinburgh, Scotland, 1992.

[Sha87] E.Y. Shapiro. *Concurrent Prolog vol. 1 and 2.* MIT Press, 1987.

[SR90] V. Saraswat and M. Rinard. Concurrent Constraint Programming. In *POPL*, pages 232–245, San Francisco, 1990.

[Ued86] K. Ueda. *Guarded Horn Clauses.* PhD thesis, University of Tokyo, 1986.

[Wal90] D. Walker. π-calculus Semantics of Object-Oriented Programming Languages. Technical Report ECS-LFCS-90-122, LFCS, U. of Edinburgh, 1990.

A Implementation of $\rightsquigarrow$

Figure 7 contains one possible π–calculus implementation of $\rightsquigarrow$, which uses some list processing utilities in figure 8. In $\rightsquigarrow$, the $\overset{2}{\rightsquigarrow}$ operator traverses the term to be copied, and echoes new instances of term components. Fresh channels and logic variables are generated during the traversal. However, a record must be kept of any new logical variable channels created, since multiple instances of a variable in a term should be denoted by the same fresh copy. Therefore, a list is checked via *mem* whether a variable has already been renamed. If not, then a new variable is created, and the old and new variable labels are saved in the list (via $\overline{add}$). Otherwise, it has been renamed already, and the new variable created previously is used instead. The $\overset{r}{\rightsquigarrow}$ operator recurs on the term arguments.

$$
\begin{aligned}
u \rightsquigarrow v &\overset{\text{def}}{=} (\nu\, done\; add\; mem\; m)\; u \overset{2}{\rightsquigarrow} v \mid List\\
u \overset{2}{\rightsquigarrow} v &\overset{\text{def}}{=} u(x).x(y).y : [\xi \Rightarrow \overline{mem}x \mid m(a).a : [\phi \Rightarrow (\nu w)\; Var(v,w) \mid !\overline{w}\xi \mid \overline{add}xw,\\
&\qquad\qquad\qquad\qquad\qquad\qquad\qquad\qquad else \Rightarrow \overline{v}a\;],\\
&\qquad\qquad\qquad else \Rightarrow (\nu w)\; Set(v,w) \mid !(\overline{w}y.\overline{done}\; Before\; x \overset{r}{\rightsquigarrow} w)\;]\\
x \overset{r}{\rightsquigarrow} w &\overset{\text{def}}{=} x(z).z : [\phi \Rightarrow \overline{w}\phi,\; else \Rightarrow (\nu a)\; z \overset{2}{\rightsquigarrow} a \mid \overline{w}a.\overline{done}\; Before\; x \overset{r}{\rightsquigarrow} w\;]
\end{aligned}
$$

Figure 6: Term copying

$$
\begin{aligned}
List &\overset{\text{def}}{=} (\nu x)\; \overline{x}\phi \mid L(x)\\
L(x) &\overset{\text{def}}{=} (\nu z)\; add(y_1, y_2).(L(z) \mid \overline{z}y_1y_2x)\\
&\qquad + mem(y).(L(x) \mid Member(y,x))\\
Member(v,x) &\overset{\text{def}}{=} x(a,b).a : [\phi \Rightarrow \overline{m}\phi,\; v \Rightarrow \overline{m}b,\\
&\qquad else \Rightarrow x(z).Member(v,z)\;]
\end{aligned}
$$

Figure 7: List utilities

The Total Order Assumption[1]

J.C.M. Baeten
Department of Computing Science, Eindhoven University of Technology, P.O.Box 513, 5600 MB Eindhoven, The Netherlands

The total order assumption (TOA) is the assumption that all execution sequences of observable actions or events are totally ordered by precedence. As long as in some cases TOA is to the point, total order and partial order semantics are both legitimate but lead to different theories. I argue that TOA is a simplifying assumption, and present an example of a total order theory without interleaving, and a partial order theory with interleaving.

1980 Mathematics Subject Classification (1985 revision): 68Q55, 68Q45, 68Q10.
1987 CR Categories: F.1.2, F.3.1, D.1.3, D.3.1.
Key words & Phrases: partial order, total order, true concurrency, interleaving, process algebra.
Note: This research was partially supported by ESPRIT Basic Research Action 7166, CONCUR2.

1. INTRODUCTION.

Semantics of concurrency like Petri nets [REI85] or event structures [WIN87] are often called *True Concurrency* semantics, in order to reflect the viewpoint that such semantics embody a truer treatment of concurrency than interleaving semantics like the standard semantics of CCS [MIL89], CSP [HOA85] or ACP [BAW90]. Another name for True Concurrency semantics is *partial order* semantics. In partial order semantics, execution paths of observable actions or events are partially ordered by precedence.

Semantics that do not classify as partial order semantics (like CCS, CSP, ACP semantics) are sometimes jokingly called False Concurrency semantics, but are usually referred to as *interleaving* semantics. An interleaving semantics is a semantics in which parallel composition of finite processes can be expressed in terms of some form of alternative composition and some form of sequencing. An interleaving semantics will contain some form of the *Expansion Theorem*, that can be used to remove parallel composition from a finite process expression, and can be used to unfold the parallel composition of recursively defined processes.

In this article, I will point out that the division of semantics into partial order semantics and interleaving semantics does not give a partition of the field. In order to get a clearer division, I propose to use the name *total order* semantics. A total order semantics is a semantics that satisfies the *Total Order Assumption (TOA)*. This

---

[1] This paper has also appeared in the Proceedings of the Workshop "What good is interleaving?", Sheffield, June 1992, University of Hildesheim Technical Report 1992.

assumption states that all execution sequences of observable actions or events are totally ordered by precedence. As long as in some cases TOA is to the point, total order and partial order semantics are both legitimate (but lead to different theories). I will argue that TOA is a simplifying assumption, and present an example of a total order theory without interleaving, and a partial order theory with interleaving.

ACKNOWLEDGEMENTS.
I thank J.A. Bergstra (University of Amsterdam) for extensively discussing the contents of this paper with me. At the workshop "What good is partial order" on 22 June 1992 at Sheffield, I discussed this paper with E. Best (University of Hildesheim), R. Milner (University of Edinburgh) and others. Further input came from S. Mauw (Eindhoven University of Technology).

2. THE TOTAL ORDER ASSUMPTION.

The division of semantics into partial order semantics like Petri nets [REI85] or event structures [WIN87], and interleaving semantics like the standard semantics of CCS [MIL80], CSP [HOA85] or ACP [BAW90] is not an exhaustive division. Moreover, the two notions overlap. In order to get a clearer division, I propose to use the name *total order* semantics. A total order semantics is a semantics that satisfies the *Total Order Assumption (TOA).*

THE TOTAL ORDER ASSUMPTION (TOA): All execution sequences of observable actions or events are totally ordered by precedence.

In contrast, in partial order semantics the execution sequences or execution paths need not be totally ordered, but are only partially ordered (by causality). We see that total order semantics form a special case of partial order semantics, just like, in mathematics, the theory of ordinary differential equations is a special case of the theory of partial differential equations, or linear mathematics is a special case of general mathematics. This subclass relation gives a better picture of the relation between the two types of semantics than opposites with emotional meaning like True vs. False, or Right vs. Wrong.

I want to emphasise the difference between interleaving semantics and total order semantics. I will do this by presenting, in section 4, a total order semantics without an expansion theorem, and presenting, in section 5, a partial order semantics with an expansion theorem.

There is sometimes a debate going on between advocates of partial order semantics and total order semantics as to which kind is better. I think this kind of debate is pointless. One does not ask if the theory of partial differential equations is better than the theory of ordinary differential equations. As long as both theories can be usefully applied in some cases, they are both legitimate (but have their own characteristics). The same holds for semantic theories, so we have the following thesis.

As long as in *some* cases TOA is to the point, total order and partial order semantics are both legitimate.

Thus, we have to avoid all claims of superiority when we talk about our favorite theory. I will never say that ACP preaches the TOA, or that ACP justifies the TOA, or even that ACP prefers the TOA. The only relevant question that we can ask about a semantics in one of the two categories is, whether or not it exploits the basic tenets of that class in a best possible way, whether or not it takes full advantage of the characteristics of that class. Thus, a relevant question is: does ACP exploit TOA in a best possible way? I think it does, and will give some motivation in the following.

A somewhat controversial thesis is the following.

TOA is a simplifying assumption.

Some people may argue that this thesis is not correct, as some methods of analysis are perhaps easier to apply in a more general framework. Specifically, methods for model checking in interleaving semantics usually consider the whole state space of a process, and the size of the state space can be very large compared to the size of the process expression. This blow-up in size is known as *combinatorial explosion*. In partial order semantics, we cannot calculate the whole state space, and easier methods may be found (see [ESP92]). Of course, we have to realise that since total order semantics is a special case of partial order semantics, any method found in the wider framework can be transfered to the more restricted framework.

The main argument in favor of the simplifying nature of the TOA is the *calculational* aspect: using the TOA, it is easier to calculate with process expressions, and it is easier to give correctness proofs in a linear form, amenable to automation. Such a calculus is often based on a set of axioms or laws. We give an example of an algebraic correctness proof in the following section 3. I claim that a proof like this calculation cannot be given in partial order semantics (at least, I have never seen one, nor have I seen a usable complete axiomatisation of a partial order semantics).

Another argument in favor of the simplifying nature of the TOA is extensibility: it is easier to extend a theory based on total order semantics to incorporate new features. As an example, we consider real time extensions in section 4. It is more difficult to achieve such extensions in the full generality of partial order semantics.

3. EXAMPLE: THE ALTERNATING BIT PROTOCOL.

The standard example in concurrency theory is the Alternating Bit Protocol. In this section, we give a purely calculational (in fact algebraic) proof of the correctness of this protocol. Note that no pictures or other visual aids are used in the proof. Essentially, we give the proof of [BEK86]. The specification is as follows. B is the set of Booleans, D is a finite data set, and we use the standard send/receive communication function of [BAW90] that only allows communications of the form $s_i(x) \mid r_i(x) = c_i(x)$ (i a port name).

Sender:

$S = S0 \cdot S1 \cdot S$

$Sb = \sum_{d \in D} r_1(d) \cdot Sb_d$ $\qquad b \in B$

$Sb_d = s_2(db) \cdot Tb_d$ $\qquad b \in B, d \in D$

$Tb_d = (r_6(1\text{-}b) + r_6(\text{error})) \cdot Sb_d + r_6(b)$ $\qquad b \in B, d \in D.$

Channels:

$K = \sum_{f \in D \times B} r_2(f) \cdot (i \cdot s_3(f) + i \cdot s_3(\text{error})) \cdot K$.

$L = \sum_{b \in B} r_5(b) \cdot (i \cdot s_6(b) + i \cdot s_6(\text{error})) \cdot L.$

Receiver:

$R = R1 \cdot R0 \cdot R$

$Rb = (r_3(\text{error}) + \sum_{d \in D} r_3(db)) \cdot s5(b) \cdot Rb + \sum_{d \in D} r_3(d(1\text{-}b)) \cdot s_4(d) \cdot s_5(1\text{-}b)$

$b \in B.$

Encapsulation:

$H = \{s_p(x), r_p(x) : p \in \{2,3,5,6\}, x \in D \times B \cup B \cup \{\text{error}\}\}.$

Abstraction:

$I = \{c_p(x) : p \in \{2,3,5,6\}, x \in D \times B \cup B \cup \{\text{error}\}\} \cup \{i\}.$

Specification of the protocol:

$ABP = \tau_I \circ \partial_H(S \parallel K \parallel L \parallel R).$

We want to prove: $ABP = \sum_{d \in D} r_1(d) \cdot s_4(d) \cdot ABP$.

We give the complete calculation that proves this fact. First, we derive a guarded recursive specification for $i_I \circ \partial_H(S \parallel K \parallel L \parallel R)$. The operator i_I is pre-abstraction (i.e. all internal actions are renamed into i, but no τ-laws can be applied). We will use the following abbreviations:

$X = i_I \circ \partial_H(S \parallel K \parallel L \parallel R)$

$X1_d = i_I \circ \partial_H(S0_d \cdot S1 \cdot S \parallel K \parallel L \parallel R)$

$X2_d = i_I \circ \partial_H(T0_d \cdot S1 \cdot S \parallel K \parallel L \parallel s_5(0) \cdot R0 \cdot R)$

$Y = i_I \circ \partial_H(S1 \cdot S \parallel K \parallel L \parallel R0 \cdot R)$

$Y1_d = i_I \circ \partial_H(S1_d \cdot S \parallel K \parallel L \parallel R0 \cdot R)$

$Y2_d = i_I \circ \partial_H(T1_d \cdot S \parallel K \parallel L \parallel s_5(1) \cdot R)$

$K'_f = (i \cdot s_3(f) + i \cdot s_3(\text{error})) \cdot K$

$L'_b = (i \cdot s_6(b) + i \cdot s_6(\text{error})) \cdot L.$

Each step in the following calculation is an application of the expansion theorem.

$$X = i_I \circ \partial_H(S \parallel K \parallel L \parallel R) = \sum_{d \in D} r_1(d) \cdot i_I \circ \partial_H(S0_d \cdot S1 \cdot S \parallel K \parallel L \parallel R) = \sum_{d \in D} r_1(d) \cdot X1_d$$

$$X1_d = i_I \circ \partial_H(S0_d \cdot S1 \cdot S \parallel K \parallel L \parallel R) = i \cdot i_I \circ \partial_H(T0_d \cdot S1 \cdot S \parallel K'_{d0} \parallel L \parallel R) =$$
$$= i \cdot (i \cdot i_I \circ \partial_H(T0_d \cdot S1 \cdot S \parallel s_3(d0) \cdot K \parallel L \parallel R) +$$
$$+ i \cdot i_I \circ \partial_H(T0_d \cdot S1 \cdot S \parallel s_3(error) \cdot K \parallel L \parallel R)) =$$
$$= i \cdot (i^2 \cdot i_I \circ \partial_H(T0_d \cdot S1 \cdot S \parallel K \parallel L \parallel s_4(d) \cdot s_5(0) \cdot R0 \cdot R) +$$
$$+ i^2 \cdot i_I \circ \partial_H(T0_d \cdot S1 \cdot S \parallel K \parallel L \parallel s_5(1) \cdot R1 \cdot R0 \cdot R)) =$$
$$= i \cdot (i^2 \cdot s_4(d) \cdot i_I \circ \partial_H(T0_d \cdot S1 \cdot S \parallel K \parallel L \parallel s_5(0) \cdot R0 \cdot R) +$$
$$+ i^3 \cdot i_I \circ \partial_H(T0_d \cdot S1 \cdot S \parallel K \parallel L'_1 \parallel R)) =$$
$$= i \cdot (i^2 \cdot s_4(d) \cdot X2_d + i^3 \cdot (i \cdot i_I \circ \partial_H(T0_d \cdot S1 \cdot S \parallel K \parallel s_6(1) \cdot L \parallel R) +$$
$$+ i \cdot i_I \circ \partial_H(T0_d \cdot S1 \cdot S \parallel K \parallel s_6(error) \cdot L \parallel R))) =$$
$$= i \cdot (i^2 \cdot s_4(d) \cdot X2_d + i^3 \cdot (i^2 \cdot i_I \circ \partial_H(S0_d \cdot S1 \cdot S \parallel K \parallel L \parallel R) +$$
$$+ i^2 \cdot i_I \circ \partial_H(S0_d \cdot S1 \cdot S \parallel K \parallel L \parallel R))) =$$
$$= i \cdot (i^2 \cdot s_4(d) \cdot X2_d + i^5 \cdot X1_d)$$

$$X2_d = i_I \circ \partial_H(T0_d \cdot S1 \cdot S \parallel K \parallel L \parallel s_5(0) \cdot R0 \cdot R) =$$
$$= i \cdot i_I \circ \partial_H(T0_d \cdot S1 \cdot S \parallel K \parallel L'_0 \parallel R0 \cdot R) =$$
$$= i \cdot (i \cdot i_I \circ \partial_H(T0_d \cdot S1 \cdot S \parallel K \parallel s_6(0) \cdot L \parallel R) +$$
$$+ i \cdot i_I \circ \partial_H(T0_d \cdot S1 \cdot S \parallel K \parallel s_6(error) \cdot L \parallel R)) =$$
$$= i \cdot (i^2 \cdot i_I \circ \partial_H(S1 \cdot S \parallel K \parallel L \parallel R0 \cdot R) + i^2 \cdot i_I \circ \partial_H(S0_d \cdot S1 \cdot S \parallel K \parallel L \parallel R0 \cdot R)) =$$
$$= i \cdot (i^2 \cdot Y + i^3 \cdot i_I \circ \partial_H(T0_d \cdot S1 \cdot S \parallel K'_{do} \parallel L \parallel R0 \cdot R)) =$$
$$= i \cdot (i^2 \cdot Y + i^3 \cdot (i \cdot i_I \circ \partial_H(T0_d \cdot S1 \cdot S \parallel s_3(d0) \cdot K \parallel L \parallel R0 \cdot R) +$$
$$+ i \cdot i_I \circ \partial_H(T0_d \cdot S1 \cdot S \parallel s_3(error) \cdot K \parallel L \parallel R0 \cdot R))) =$$
$$= i \cdot (i^2 \cdot Y + i^3 \cdot (i^2 \cdot i_I \circ \partial_H(T0_d \cdot S1 \cdot S \parallel K \parallel L \parallel s_5(0) \cdot R0 \cdot R) +$$
$$+ i^2 \cdot i_I \circ \partial_H(T0_d \cdot S1 \cdot S \parallel K \parallel L \parallel s_5(0) \cdot R0 \cdot R))) =$$
$$= i \cdot (i^2 \cdot Y + i^5 \cdot X2_d).$$

Likewise, we derive

$$Y = \sum_{d \in D} r_1(d) \cdot Y1_d$$

$$Y1_d = i \cdot (i^2 \cdot s_4(d) \cdot Y2_d + i^5 \cdot Y1_d)$$

$$Y2_d = i \cdot (i^2 \cdot X + i^5 \cdot Y2_d).$$

The equations derived for $X1_d$ and $X2_d$ show an internal loop of six steps. Application of Koomen's Fair Abstraction Rule ($KFAR_6$, see [BEK86] or [BAW90]) yields:

$$\tau \cdot \tau_I(X1_d) = \tau \cdot s_4(d) \cdot \tau_I(X2_d) \qquad \tau \cdot \tau_I(X2_d) = \tau \cdot \tau_I(Y)$$

Using this,

$$\tau_I(X) = \sum_{d \in D} r_1(d) \cdot \tau_I(X1_d) = \sum_{d \in D} r_1(d) \cdot \tau \cdot \tau_I(X1_d) = \sum_{d \in D} r_1(d) \cdot \tau \cdot s_4(d) \cdot \tau_I(X2_d) =$$

$$= \sum_{d\in D} r_1(d)\cdot s_4(d)\cdot\tau\cdot\tau_I(X2_d) = \sum_{d\in D} r_1(d)\cdot s_4(d)\cdot\tau\cdot\tau_I(Y),$$

so $\quad \tau_I(X) = \sum_{d\in D} r_1(d)\cdot s_4(d)\cdot\tau_I(Y).$

Similarly $\quad \tau_I(Y) = \sum_{d\in D} r_1(d)\cdot s_4(d)\cdot\tau_I(X).$

Application of the Recursive Specification Principle (RSP, see [BEK86] or [BAW90]) yields $\tau_I(X) = \tau_I(Y)$, and so we have obtained

$$ABP = \sum_{d\in D} r_1(d)\cdot s_4(d)\cdot ABP\,.$$

I claim that a proof of this nature cannot be given in partial order semantics. Further, I claim that proofs of this kind are very amenable to support by software tools, that can handle most of the calculations, and can incorporate many proof heuristics.

4. REAL TIME PROCESS ALGEBRA.

First, we take a look at ACPρ, the extension of ACP with real time aspects described in [BAB91a]. In total order semantics, an action or event is supposed to be observed at a certain moment in time. That is why such actions are called *atomic* and have no duration. If we want to talk about time explicitly, it makes sense therefore to attach the moment of observation to the atomic action. Here, we use absolute, global time.

So, if a is an atomic action and t a moment in time ($t \in \mathbb{R}_{\geq 0}$), then a(t) means that a takes place (is observed) at time t. ACPρ starts from these timed actions, and has the same operators as ACP. To give an example, we have

$$(a(2)\cdot c(4)) \parallel b(3) = a(2)\cdot b(3)\cdot c(4),$$

and we see that the execution sequence is totally ordered by time. If we use a totally ordered set like the non-negative reals for our time domain, it is natural to use a total order semantics.

ACPρ has interleaving axioms and an expansion theorem, as might be expected of a total order theory. We give some typical axioms:

$x \parallel y = x ⌊\!⌊ y + y ⌊\!⌊ x + x \mid y$

$a(t) ⌊\!⌊ x = a(t)\cdot x$ if $t < U(x)$ (i.e., if x can wait until t)

$a(t) ⌊\!⌊ x = \delta(U(x))$ (i.e., $a(t) ⌊\!⌊ x$ deadlocks at time $U(x)$) if $t \geq U(x)$ (i.e., if x cannot wait until t)

$a(t)\cdot x ⌊\!⌊ y = a(t)\cdot(x \parallel y)$ if $t < U(y)$

$a(t)\cdot x ⌊\!⌊ y = \delta(U(y))$ if $t \geq U(y)$.

An interesting feature of ACPρ (essential to describe many examples) is the presence of *integration*. The term $\int_{t\in(1,2)} a(t)$ represents the process that can execute action a somewhere between time 1 and time 2.

The theory with integration also has an expansion theorem. If we limit ourselves to *prefixed integration* (i.e. only integration over intervals, and the operand starts with an action containing the time variable), then we can still obtain a complete axiomatisation. As a result, we obtain that equality of finite processes is decidable (see [KLU91], [FOK92]).

Other real time process algebras do not have a concept of integration with explicit time variables like ACPρ. An example of this is the theory TCCS introduced by [MOT89]. Variants of TCCS are given in [MOT90] and [WAN90]. An interesting result about the dense time variant of TCCS was recently given in [GOL92]. Consider the TCCS term $\varepsilon(1)\cdot a \mid b$. In our notation, this would be the process expression

$$\int_{t\geq 1} a(t) \parallel \int_{v\geq 0} b(v).$$

By the expansion theorem of ACPρ, this expression can be reduced to

$$\int_{t\geq 1} a(t)\cdot \int_{v\geq 0} b(v) + \int_{v\geq 0} b(v)\cdot \int_{t\geq \max(1,v)} a(t).$$

[GOL92] shows that the term in TCCS cannot be reduced to an expression without parallel composition, and they show in general that there are TCCS-terms with n parallel components that cannot be written with fewer parallel components. Thus, there is no expansion theorem for TCCS, and TCCS has a total order semantics (because its semantics is based on a form of bisimulation) that is not an interleaving semantics. This result probably holds in general for all dense time process algebras without explicit time variables such as versions of ATP [NSY91].

The presence of an expansion theorem is a key feature for a total order theory, that allows calculations used in correctness proofs like the proof of the ABP above. The absence of an expansion theorem means therefore that the TOA is not optimally exploited, and we can draw the following conclusion.

TCCS exploits the TOA to a lesser extent than ACPρ.

Another interesting extension of a process algebra is the extension to probabilistic choice. An axiomatic approach to ACP with probabilistic choice, featuring an expansion theorem, is given in [BABS92]. I claim that the introduction of new features like time or probabilities is so much more complex in partial order semantics, that the key issues and main difficulties do not stand out so easily.

5. FROM TOA TO POA.

The theory ACPρ presented in the previous section can be further extended to take place coordinates into account [BAB91b]. Then, atomic actions are parametrized by four reals denoting a point in space and time.

First of all, we can consider this using classical, Newtonian space/time. We obtain a total order theory, with interleaving and an expansion theorem, of which all laws are invariant under Galilei transformations (classical change of coordinates). An example, with $x_0, x_1 \in \mathbb{R}^3$:

$$a(x_0, 1)\cdot b(x_0, 2)\cdot c(x_1, 3) \parallel d(x_0, 2)\cdot e(x_0, 3)\cdot f(x_1, 5) =$$
$$= a(x_0, 1)\cdot (b \mid d)(x_0, 2)\cdot (e(x_0) \,\&\, c(x_1))(3)\cdot f(x_1, 5).$$

Here, $b \mid d$ denotes *communication*, actions taking place at the same place and time, and $e(x_0)$ & $c(x_1)$ denotes *synchronisation*, actions taking place at the same time, but at a different place. The last construct is called a *multi-action*. This theory was used to describe communication between moving objects (e.g. satellite communication) in [BAB91c].

The situation changes if we consider relativistic, Einsteinian space/time. Then, we do not have a total ordering on events any longer, but only a partial ordering:

for $\alpha, \beta \in \mathbb{R}^4$: $\alpha < \beta$ iff for *all* observers the time of α is before the time of β. Consider fig. 1 (note that we start using pictures at the moment we turn to partial order semantics). The vertical axis is the time axis. The two other axes suggested by the drawing represent the three spatial dimensions.

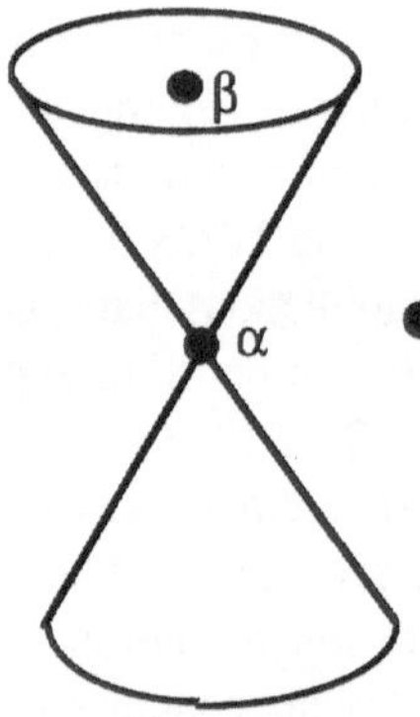

FIGURE 1.

In figure 1, we have $\alpha < \beta$ and we say β is inside the *positive light cone* of α (i.e. it is possible to travel from α to β with a speed less than the speed of light), whereas α and γ are incomparable, $\alpha \# \gamma$.

Now we can still use interleaving axioms as in the classical case, and obtain a total order theory with interleaving. To give an example, if $\alpha \# \gamma$ then we obtain $a(\alpha) \parallel c(\gamma) = a(\alpha)\cdot\delta + c(\gamma)\cdot\delta$, since, *after* executing a, c cannot be executed any more. This theory interleaves all actions on a single processor, we can call this *temporal interleaving* (see [BAB91b]). If we want a and c to execute independently in the example above, we cannot use the interleaving axioms as above, and we go to a partial order theory. This is the *multiple processor space/time process algebra* of [BAB91b]. The surprising observation of this paper is, that we can still formulate a formal expansion theorem, thus obtaining a partial order theory with an expansion theorem.

To this end, we add a new sequencing operator $\circ$, and $a(\alpha)\circ x$ now only excludes that part of x that precedes α, that is in the negative light cone of α (whereas $a(\alpha)\cdot x$ excludes that part of x that is not in the positive light cone of α, only allows points after α). If $\alpha \# \gamma$ then we obtain

$$a(\alpha) \parallel c(\gamma) = a(\alpha) \mathbin{⫗} c(\gamma) + c(\gamma) \mathbin{⫗} a(\alpha) = a(\alpha)\circ c(\gamma) + c(\gamma)\circ a(\alpha).$$

For any observer, this system will show only one behaviour, and not a choice of two behaviours. The sum here does not denote a choice between alternatives, but two different possibilities of observation. We have here a formal expansion with no clear operational intuition. Still, we have a partial order theory, with interleaving and an expansion theorem, of which all laws are Lorentz invariant (remain valid after a Lorentz transformation, a relativistic change of coordinates).

6. CONCLUDING REMARKS.

1. Total order semantics and partial order semantics are both legitimate (but lead to different theories).
2. In some cases, the Total Order Assumption is appropriate and useful.
3. The Total Order Assumption is a simplifying assumption.
4. Total order style mathematics (ACP, CCS, CSP, Floyd-Hoare logic, etc.) is an adequate exploitation of the Total Order Assumption.
5. Dense time process algebra requires explicit time variables for an optimal exploitation of the Total Order Assumption.
6. Partial order style mathematics (Petri nets, event structures, etc.) is very important in all cases where the TOA is not appropriate or to the point.

REFERENCES.

[BAB91a] J.C.M. BAETEN & J.A. BERGSTRA, *Real time process algebra,* Formal Aspects of Computing 3 (2), 1991, pp. 142-188.

[BAB91b] J.C.M. BAETEN & J.A. BERGSTRA, *Real space process algebra,* in: Proc. CONCUR'91, Amsterdam (J.C.M. Baeten & J.F. Groote, eds.), Springer LNCS 527, 1991, pp. 96-110.

[BAB91c] J.C.M. BAETEN & J.A. BERGSTRA, *Asynchronous communication in real space process algebra,* in: Proc. Formal Techniques in Real-Time and Fault-Tolerant Systems, Nijmegen 1992 (J. Vytopil, ed.), Springer LNCS 571, 1991, pp. 473-492.

[BABS92] J.C.M. BAETEN, J.A. BERGSTRA & S.A. SMOLKA, *Axiomatizing probabilistic processes: ACP with generative probabilities,* in: Proc. CONCUR'92, Stony Brook (W.R. Cleaveland, ed.), Springer LNCS 630, 1992, pp. 472-485.

[BAW90] J.C.M. BAETEN & W.P. WEIJLAND, *Process algebra,* Cambridge Tracts in Theor. Comp. Sci. 18, Cambridge University Press 1990.

[BEK86] J.A. BERGSTRA & J.W. KLOP, *Verification of an alternating bit protocol by means of process algebra,* in: Math. Methods of Spec. and Synthesis of Software Systems '85 (W. Bibel & K.P. Jantke, eds.), Springer LNCS 215, 1986, pp. 9-23.

[ESP92] J. ESPARZA, *Fast model checking using partial order semantics,* to appear in: Proc. Workshop "What good is partial order", Sheffield, Technical Report, University of Hildesheim 1992.

[FOK92] W.J. FOKKINK & A.S. KLUSENER, *Real time process algebra with prefixed integration,* report CS-R9219, CWI Amsterdam 1992.

[GOL92] J.C. GODSKESEN & K.G. LARSEN, *Real-time calculi and expansion theorems,* report, Aalborg University 1992.

[HOA85] C.A.R. HOARE, *Communicating sequential processes,* Prentice Hall 1985.

[KLU91] A.S. KLUSENER, *Completeness in real time process algebra,* in: Proc. CONCUR'91, Amsterdam (J.C.M. Baeten & J.F. Groote, eds.), Springer LNCS 527, 1991, pp. 376-392.

[MIL89] R. MILNER, *Communication and concurrency,* Prentice Hall 1989.

[MOT89] F. MOLLER & C. TOFTS, *A temporal calculus of communicating systems*, report LFCS-89-104, University of Edinburgh 1989.

[MOT90] F. MOLLER & C. TOFTS, *A temporal calculus of communicating systems*, in: Proc. CONCUR'90, Amsterdam (J.C.M. Baeten & J.W. Klop, eds.), Springer LNCS 458, 1990, pp. 401-415.

[NSY91] X. NICOLLIN, J. SIFAKIS & S. YOVINE, *From ATP to timed graphs and hybrid systems*, in: Proc. REX Workshop on Real Time: Theory in Practice, Mook 1991 (J.W. de Bakker, C. Huizing, W.P. de Roever & G. Rozenberg, eds.), Springer LNCS 600, 1992, pp. 549-572.

[REI85] W. REISIG, *Petri nets*, EATCS monograph on TCS, Springer Verlag 1985.

[WAN90] WANG YI, *Real-time behaviour of asynchronous agents*, in: Proc. CONCUR'90, Amsterdam (J.C.M. Baeten & J.W. Klop, eds.), Springer LNCS 458, 1990, pp. 502-520.

[WIN87] G. WINSKEL, *Event structures*, in: Petri nets: applications and relationships to other models of concurrency, Bad Honnef 1986 (W. Brauer, W. Reisig & G. Rozenberg, eds.), Springer LNCS 255, 1987, pp. 325-392.

Author Index

Published in 1990–91

AI and Cognitive Science '89, Dublin City University, Eire, 14–15 September 1989
A. F. Smeaton and G. McDermott (Eds.)

Specification and Verification of Concurrent Systems, University of Stirling, Scotland, 6–8 July 1988
C. Rattray (Ed.)

Semantics for Concurrency, Proceedings of the International BCS-FACS Workshop, Sponsored by Logic for IT (S.E.R.C.), University of Leicester, UK, 23–25 July 1990
M. Z. Kwiatkowska, M. W. Shields and R. M. Thomas (Eds.)

Functional Programming, Glasgow 1989
Proceedings of the 1989 Glasgow Workshop, Fraserburgh, Scotland, 21–23 August 1989
K. Davis and J. Hughes (Eds.)

Persistent Object Systems, Proceedings of the Third International Workshop, Newcastle, Australia, 10–13 January 1989
J. Rosenberg and D. Koch (Eds.)

Z User Workshop, Oxford 1989, Proceedings of the Fourth Annual Z User Meeting, Oxford, 15 December 1989
J. E. Nicholls (Ed.)

Formal Methods for Trustworthy Computer Systems (FM89), Halifax, Canada, 23–27 July 1989
Dan Craigen (Editor) and Karen Summerskill (Assistant Editor)

Security and Persistence, Proceedings of the International Workshop on Computer Architecture to Support Security and Persistence of Information, Bremen, West Germany, 8–11 May 1990
John Rosenberg and J. Leslie Keedy (Eds.)

Women into Computing: Selected Papers 1988–1990
Gillian Lovegrove and Barbara Segal (Eds.)

3rd Refinement Workshop (organised by BCS-FACS, and sponsored by IBM UK Laboratories, Hursley Park and the Programming Research Group, University of Oxford), Hursley Park, 9–11 January 1990
Carroll Morgan and J. C. P. Woodcock (Eds.)

Designing Correct Circuits, Workshop jointly organised by the Universities of Oxford and Glasgow, Oxford, 26–28 September 1990
Geraint Jones and Mary Sheeran (Eds.)

Functional Programming, Glasgow 1990
Proceedings of the 1990 Glasgow Workshop on Functional Programming, Ullapool, Scotland, 13–15 August 1990
Simon L. Peyton Jones, Graham Hutton and Carsten Kehler Holst (Eds.)

4th Refinement Workshop, Proceedings of the 4th Refinement Workshop, organised by BCS-FACS, Cambridge, 9–11 January 1991
Joseph M. Morris and Roger C. Shaw (Eds.)

AI and Cognitive Science '90, University of Ulster at Jordanstown, 20–21 September 1990
Michael F. McTear and Norman Creaney (Eds.)

Software Re-use, Utrecht 1989, Proceedings of the Software Re-use Workshop, Utrecht, The Netherlands, 23–24 November 1989
Liesbeth Dusink and Patrick Hall (Eds.)

Z User Workshop, 1990, Proceedings of the Fifth Annual Z User Meeting, Oxford, 17–18 December 1990
J.E. Nicholls (Ed.)

IV Higher Order Workshop, Banff 1990
Proceedings of the IV Higher Order Workshop, Banff, Alberta, Canada, 10–14 September 1990
Graham Birtwistle (Ed.)